DATE DUE

PROBLEMS IN STRUCTURAL INORGANIC CHEMISTRY

A Note from the Publisher

This volume was printed directly from a typescript prepared by the author, who takes full responsibility for its content and appearance. The Publisher has not performed his usual functions of reviewing, editing, typesetting, and proofreading the material prior to publication.

The Publisher fully endorses this informal and quick method of publishing lecture notes at a moderate price, and he wishes to thank the author for preparing the material for publication.

PROBLEMS IN STRUCTURAL INORGANIC CHEMISTRY

WILLIAM E. HATFIELD
University of North Carolina

RICHARD A. PALMER
Duke University

W. A. BENJAMIN, INC.
New York 1971

PROBLEMS IN STRUCTURAL INORGANIC CHEMISTRY

Standard Book Number 8053-3790-3 (Clothbound)
8053-3791-1 (Paperback)
Library of Congress Catalog Card Number
Manufactured in the United States of America
12345K43210

W. A. BENJAMIN, INC.
New York, New York 10016

PREFACE

Our purpose in writing this book has been to provide a balanced collection of problems in the modern, quantitative aspects of inorganic chemistry in a convenient and practical format. We have intended that these problems might serve, not only as supplemental material for established courses, but also as a guide for self-study. The level should be suitable for senior undergraduates and first-year graduate students. Perhaps the two most significant aspects of the book, which we believe set it apart from others presently available in the field of inorganic chemistry, are the following: First, the answers to about one-third of the problems are worked out in complete detail, while about one-third of the problems have tightly sketched answers, and the remaining problems have no answers given. Those using this book as a class text supplement may wish to assign problems from the latter two categories. Second, the bias of the problems is very strongly toward those with specific, unambiguous, and, in most cases, quantitative answers. Each chapter is designed to follow a logical progression and sequence with regard to difficulty and subject matter, with concepts developed in early problems used extensively in subsequent ones. Where possible, problems have been drawn from relevant current research. We have deliberately omitted any significant amount of text material, not only because the subjects are adequately covered in several readily available texts and other sources (which we have tried carefully to reference), but also because the inclusion of really adequate text would have required a book whose length and consequent cost were not consistent with our aims. It is primarily for this latter reason that we have elected this relatively informal publication format. Because of the large numbers of necessary drawings, symbols, and equations, publication by conventional techniques would have made the book prohibitively expensive.

We acknowledge the contributions and the useful comments and suggestions of our students and colleagues who have helped us during the past two years to check the solutions, and we hope that others may find these problems as useful as we have.

Notes to the Student

It is widely recognized that solving problems provides one of the most efficient means of learning the quantitative techniques used in chemistry and other sciences. The problems in this collection have been compiled with this in mind. We believe that they can profitably be used either as a self-study guide or a class text. They have been class tested on home-work sets, take-home examinations and in-class examinations. There are no "cute" or "trick" problems in the collection, and the range of difficulty proceeds from easy, review problems to fairly long, complicated computations. Each problem in the collection is important in that it presents an idea, illustrates a technique, or provides practice in recently learned material.

At the beginning of each chapter we have provided a short list of reference books and monographs which should prove useful. In addition, references are frequently given with the individual problems. These latter references may be used as a reading guide, but specific page numbers are not usually given. We feel that this is reasonable in that it encourages reading of the material in a more orderly manner so that the logical development of the subject can be appreciated. There is always the temptation to leaf through reference books and hunt for the specific formula or solution. Such a practice is time consuming and of little value in the long run.

We suggest the following procedure be adopted for self-study:

1. Read through the problems, work those that can be solved without additional preparation, and check your answers with those which we have provided.
2. For those problems requiring additional reading, we suggest that the problems be read carefully, and mental notes, at least, be made as a guide to the study of the listed references. After reading the reference material, return to the problems, again work as many as possible and check your results with the given solutions.
3. As a last resort *only*, use the solutions provided as an outline for your work.
4. Finally, we suggest that you try to devise additional problems of your own analogous to those given here but using different systems, different point groups, etc. Although *working* problems is very beneficial, *writing* significant, workable

problems is an even more demanding test of your understanding of the subject.

William E. Hatfield

Richard A. Palmer

November 15, 1970

ACKNOWLEDGMENTS

The following material has been reprinted with the permission of the publishers:

Appendix 4.1 from F. A. Cotton, *Chemical Applications of Group Theory*, John Wiley and Sons, Inc., New York, 1963.

Appendix 4.2 from B. N. Figgis, *Introduction to Ligand Fields*, John Wiley and Sons, Inc., New York, 1966.

Appendix 4.3 from E. B. Wilson, Jr., J. C. Decius, and P. C. Cross, *Molecular Vibrations*, McGraw-Hill Book Co., New York, 1955.

Tables 5.1, 5.2, 5.3, 5.6 from Linus Pauling, *The Nature of the Chemical Bond*, 3rd ed., copyright 1960 by Cornell University. Used by permission of Cornell University Press.

Table 5.8 from F. Basolo, and R. G. Pearson, *Mechanisms of Inorganic Reactions*, 2nd ed., John Wiley and Sons, Inc., New York, 1967.

Appendix 7.1 from J. Lewis, and R. G. Wilkins, *Modern Coordination Chemistry*, John Wiley and Sons, Inc., New York, 1960.

CONTENTS

CONTENTS

Chapter 1

NOMENCLATURE

The system of nomenclature currently used by inorganic chemists has evolved from less useful and less applicable systems devised by early practitioners of the science. It is interesting to trace this evolution through a selection of textbooks and research papers drawn from different periods of the scientific calendar. Although modern chemists must conform as closely as possible to the nomenclature system sanctioned by the International Union of Pure and Applied Chemistry, it is necessary to have some knowledge of obsolete names. Otherwise, reference to many of the classical papers required for systematic and complete literature searches would be of little value.

In this first chapter are collected a series of exercises which reviews most of the pertinent aspects of inorganic nomenclature. The most readily available, complete and authoritative reference to current usage is the American adaptation of the 1957 IUPAC report, "The Definitive Rules for Nomenclature of Inorganic Chemistry", *J. Am. Chem. Soc.*, 82, 5523 (1960). In addition most inorganic textbooks include useful abridged outlines of nomenclature rules.

PROBLEMS

1.1 *Names and symbols of elements.*

(a) Which of the following elements has the symbol

Tl? (i) tantalum, (ii) tellurium, (iii) thallium, (iv) thulium, (v) none of these.

(b) Which of the following elements has the symbol A? (i) actinium, (ii) americium, (iii) argon, (iv) astatine, (v) none of these.

(c) Which of the following is the symbol for tin? (i) Ti, (ii) W, (iii) Te, (iv) Sn, (v) Pb.

(d) Which of the following is the symbol for rhodium? (i) Rn, (ii) Rh, (iii) Rd, (iv) Rm, (v) none of these.

1.2 *Names of groups and series.* Identify the following: (a) the halogens, (b) the halides, (c) the chalcogens, (d) the lanthanides, (e) the actinides.

1.3 *Symbols of isotopes.* Give the complete correct symbols for the following: (a) the isotope of iron having 31 neutrons in its nucleus, (b) the three isotopes of hydrogen, (c) the element with Z = 12, (d) isotopically pure (oxygen-16) oxygenyl ion.

1.4 *Formulas.* Give the empirical formulas, molecular formulas, and structural formulas for the following: (a) white phosphorus, (b) phosphorus(III) oxide, (c) sulfur (in standard state), (d) diborane, (e) phosphorous acid.

1.5 *Sequence of atoms in formulas.* From the following series of formulas, write generalizations concerning the sequences of atoms in formulas: (a) B_2O_3, NH_3, PH_3, P_2O_3, Cl_2O, OF_2, ClF_3, ICl, B_2H_6, H_2O, P_3N_3, HCl, HI; (b) NaCl, $CaSO_4$, $FeCl_2$, $LiNO_3$; (c) NCS^-, BrCN, HOCl, $B(OH)_3$.

1.6 *The Stock notation.* In the Stock system the oxidation state of an element is represented by a Roman numeral placed in parentheses immediately following the name of the element. Use this system and name the following compounds: (a) $FeCl_2$, (b) $FeCl_3$, (c) MnO_2, (d) MoF_5, (e) CuO, (f) Fe_3O_4, (g) MoS_2, (h) $CrSO_4$.

1.7 *Trivial names.* Give formulas for the following compounds or ions: (a) diborane, (b) silane, (c) phosphine, (d) arsine, (e) stibine, (f) diphosphine, (g) nitrite, (h) sulfite.

1.8 *Names from formulas.* Give acceptable names for the following compounds: (a) U_3O_8, (b) N_2S_5, (c) $Na(SO_3F)$, (d) $K(PO_2F_2)$, (e) SiH_4, (f) PH_3, (g) AsH_3, (h) P_2H_4, (i) B_2H_6, (j) NH_2OH, (k) $PSCl_3$, (l) $ClSO_3H$, (m) HOCl, (n) H_2MnO_4, (o) $H_4P_2O_7$, (p) H_2CS_3, (q) NaH_2PO_4, (r) $KNaCO_3$, (s) LaOF, (t) $NaNbO_3$.

1.9 *Formulas from names.* Provide chemical formulas for the following compounds: (a) aluminum hexafluorosilicate, (b) aluminum orthophosphate, (c) ammonium bromate, (d) ammonium cyanate, (e) antimony(III) oxide, (f) barium arsenide, (g) barium iodate, (h) beryllium orthosilicate, (i) bismuth(III) oxide carbonate, (j) cadmium sulfite, (k) calcium phosphide, (l) calcium cyanamide, (m) calcium telluride, (n) chloric acid, (o) chromium(VI) oxide, (p) trichlorogermane, (q) permanganic acid, (r) nitrosyl chloride, (s) nitryl chloride, (t) phosphonium iodide.

1.10 *Formulas for ligands.* The following substances are used extensively as ligands in coordination chemistry. Draw structural formulas for each. (a) 1,10-phenanthroline, (b) 2,2'-bipyridine, (c) the acetylacetonate anion, (d) the ethylenediaminetetraacetate anion, (e) ethylenediamine, (f) propylenediamine, (g) diethylenetriamine, (h) thiourea, (i) oxine, (j) diarsine.

1.11 *Abbreviations for ligands.* Coordination chemists frequently use abbreviations such as *en* or *acac* in the formulas of compounds. Give names for the following abbreviations: (a) en, (b) pn, (c) py, (d) acac, (e) EDTA, (f) PDTA, (g) tu, (h) mnt, (i) bipy (j) tfac, (k) cptn, (l) cp, (m) tdt, (n) ox, (o) diars.

1.12 *Names from colors--an obsolete practice.* (Read: Bailar, *Chemistry of the Coordination Compounds*, Chapter 1) Early inorganic chemists used the color of complexes in a nomenclature system. Look up the color-names of the following complex ions: (a) $[Co(NH_3)_6]^{3+}$, (b) $[Co(NH_3)_5Cl]^{2+}$, (c) $[Co(NH_3)_5H_2O]^{3+}$, (d) *cis*-$[Co(NH_3)_4Cl_2]^+$, (e) *trans*-$[Co(NH_3)_4Cl_2]^+$.

1.13 *Names from discoverers.* Some compounds are still named after men who first prepared them. Look up the formulas for the following compounds: (a) Vaska's compound,

(b) Zeise's salt, (c) Magnus' green salt, (d) Mohr's salt.

1.14 *Trivial names.* Give chemical formulas for the following compounds: (a) ferrocene, (b) borazine, (c) phospham, (d) corundum, (e) rutile.

1.15 *Names from formulas.* Provide names for the following coordination compounds; (a) K_2CuCl_4, (b) $LiAlH_4$, (c) $[CoCl(NH_3)_5]SO_4$, (d) $[CoCl_2(NH_3)_4]Cl$, (e) $[Co(en)_3]Cl_3$, (f) $Cr(acac)_3$, (g) $K_3Cr(ox)_3$, (h) $[Co(en)_2Cl_2]Cl$, (i) $K[PtCl_3(C_2H_4)]$, (j) $[Pt(NH_3)_4][PtCl_4]$, (k) $[Cr(NH_3)_6]$-$[Cr(ox)_3]$, (l) $K_4Fe(CN)_6$, (m) *cis*-$Pt(NH_3)_2Cl_2$, (n) KCo(PDTA), (o) MnO_3Cl.

1.16 *Formulas from names.* Give formulas for the following coordination compounds: (a) chlorocarbonylbis(triphenylphosphine)iridium(I), (b) chloronitritotetraamminechromium(III) nitrate, (c) potassium *trans*-dichlorobis(oxalato)cobaltate(III), (d) hexaamminecobalt(III) pentachlorocuprate(II), (e) iron(III) acetylacetonate, (f) hexaaquovanadium(III) chloride, (g) mercury(II) tetrathiocyanatocobaltate(II), (h) potassium hexafluorochromate(II), (i) sodium tetracarbonylcobaltate(-I), (j) potassium oxopentachlorochromate(V), (k) potassium octacyanomolybdate(V), (l) acetylacetonatodichlorobis(triphenylphosphine)rhenium(III), (m) potassium enneahydridorhenate(VII), (n) sodium bis(thiosulfato)argentate(I).

1.17 *Nomenclature of complicated molecules.* Give a name or formula, as required, for the following compounds: (a) potassium cyanodicarbonylnitrosylcobaltate(0), (b) *asym*-di-μ-chloro-dichlorobis(triethylarsine)diplatinum(II),

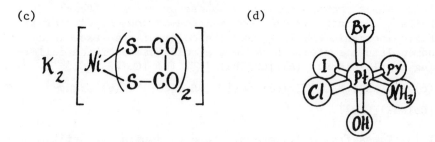

(c)

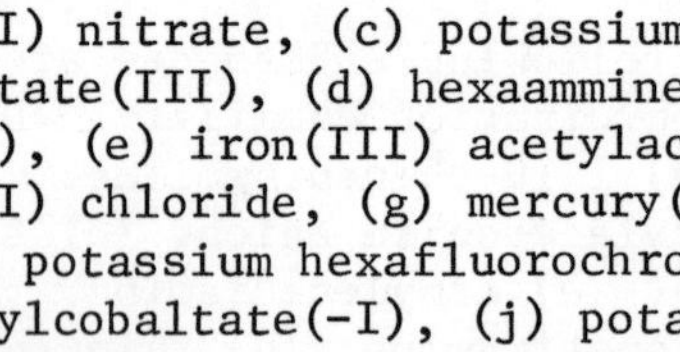

(d)

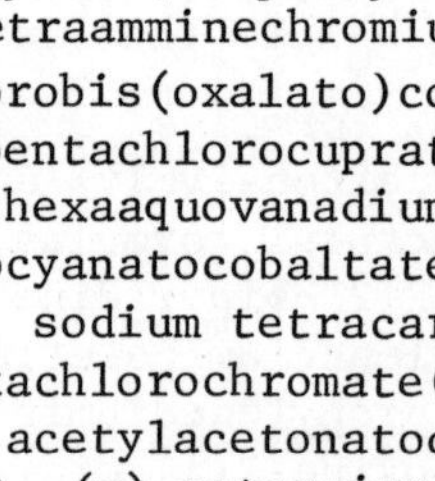

(e) $(NH_4)_3PW_{12}O_{40}$, (f) $(AsPh_4)_2Re_2Cl_8$.

SOLUTIONS

1.1 (a) Thallium has the symbol Tl.
(b) The symbol A is no longer used for argon; Ar is now the proper symbol.
(c) Sn is the symbol for tin.
(d) Rh is the symbol for rhodium.

1.2 (a) fluorine, chlorine, bromine, iodine, and astatine.
(b) fluoride, chloride, bromide, iodide, and astatide ions.
(c) oxygen, sulfur, selenium, tellurium, and polonium.
(d) elements 58 through 71 (Ce to Lu inclusive)
(e) elements 90 through 104 (Th to Lw inclusive)
Note the two different uses of the suffix "-ide" in (b), (d), and (e).

1.3 (a) $^{57}_{26}Fe$, (b) $^{1}_{1}H$, hydrogen; $^{2}_{1}H$, deuterium; $^{3}_{1}H$, tritium.

Note that deuterium is sometimes symbolized D and tritium, T. Hydrogen is the only element whose isotopes have such special names. (c) Mg, (d) $^{16}_{8}O_2^+$.

1.4

	empirical formula	molecular formula	structural formula
(a)	P	P_4	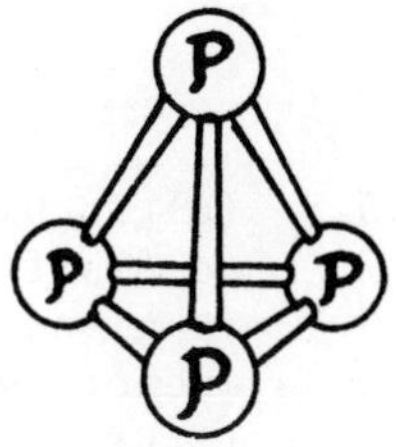
(b)	P_2O_3	P_4O_6	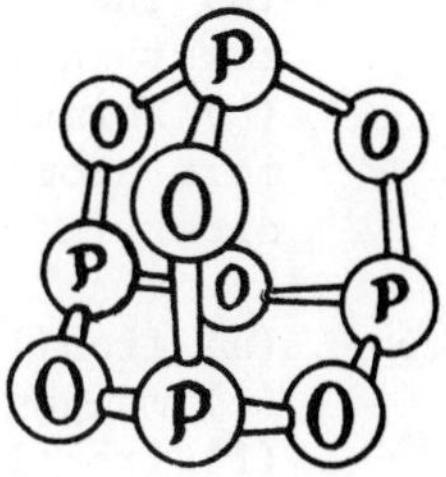

empirical formula	molecular formula	structural formula
(c) S	S_8	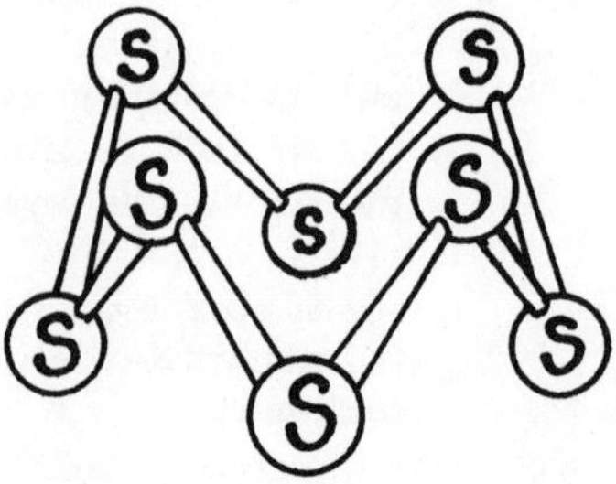
(d) BH_3	B_2H_6	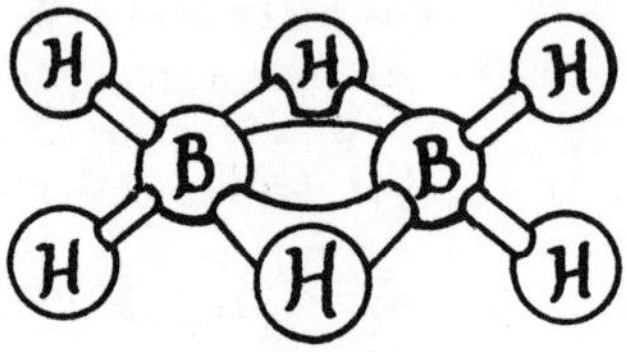
(e) H_3PO_3	H_3PO_3	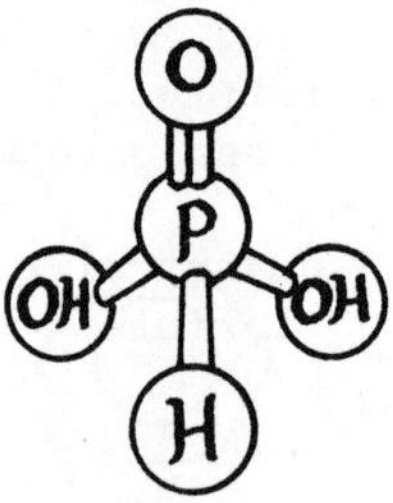

1.5 (a) In binary compounds of nonmetals the atom which occurs first in the sequence B, P, N, H, I, Cl, O, F is given first in the formula. (The complete sequence is B, Si, C, Sb, As, P, N, H, Te, Se, S, At, I, Br, Cl, O, F.)
(b) The electropositive constituent is always placed first in the formula.
(c) Formulas of compounds or ions containing three or more atoms are written with the atoms given in the order in which they are connected.

1.6 (a) iron(II) chloride, (b) iron(III) chloride, (c) manganese(IV) oxide, (d) molybdenum(V) fluoride, (f) iron(II) diiron(III) oxide, (h) chromium(II) sulfate. Note that the proportions of the constituents are indirectly, though clearly, indicated by this system.

1.7 (a) B_2H_6, (b) SiH_4, (c) PH_3, (d) AsH_3, (g) NO_2^-, (h) SO_3^{2-}

1.8 (a) Triuranium octaoxide, (b) dinitrogen pentasulfide, (c) sodium trioxofluorosulfate, (d) potassium dioxodifluorophosphate, (e) silane, (f) phosphine, (g) arsine, (h) diphosphine, (i) diborane, (m) hypochlorous acid, (n) manganic acid, (o) pyrophosphoric acid, (q) sodium dihydrogen phosphate, (r) potassium sodium carbonate, (s) lanthanum oxide fluoride.

1.9 (a) $Al_2(SiF_6)_3$, (b) $AlPO_4$, (c) NH_4BrO_3, (e) Sb_2O_3, (f) Ba_3As_2, (g) $Ba(IO_3)_2$, (i) $Bi_2O_2CO_3$, (j) $CdSO_3$, (k) Ca_3P_2, (l) $CaCN_2$, (n) $HClO_3$, (o) CrO_3, (q) $HMnO_4$, (s) NO_2Cl, (t) PH_4I.

1.10 (a)

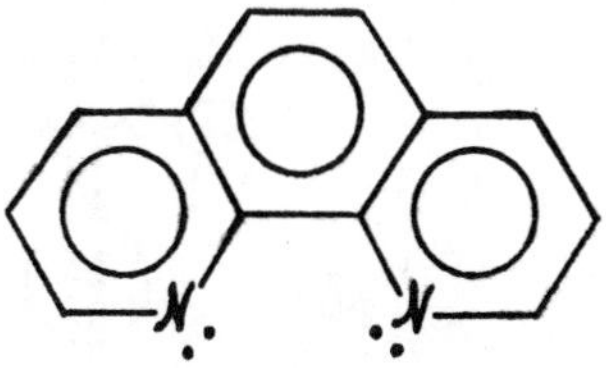

(b)

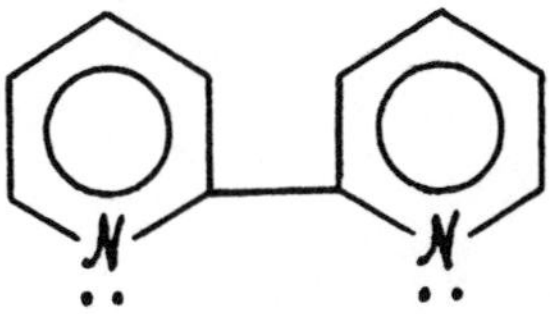

(c)

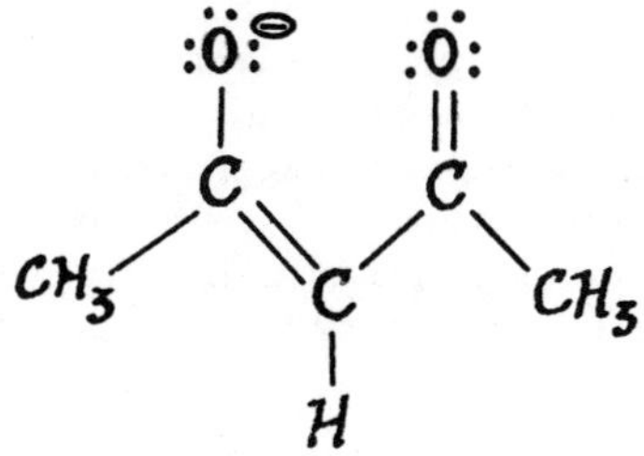

(e) $H_2\ddot{N}CH_2CH_2\ddot{N}H_2$ (f) $\ddot{N}H_2CH_2CH(CH_3)\ddot{N}H_2$

(j) 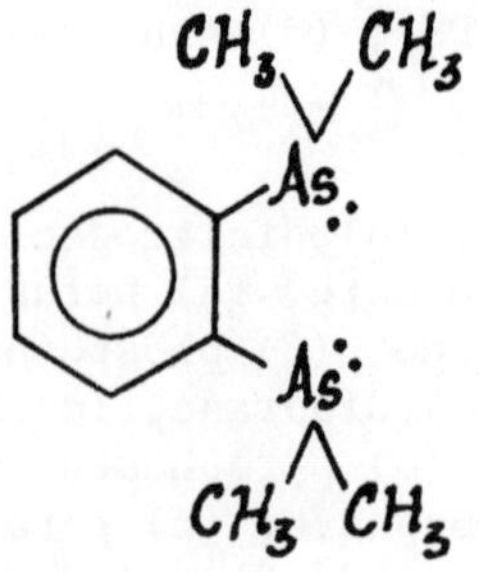

1.11 (a) ethylenediamine, (b) propylenediamine, (c) pyridine, (d) acetylacetonate anion, (e) ethylenediaminetetraacetate anion, (f) propylenediaminetetraacetate anion, (g) thiourea, (h) maleonitriledithiolate anion, (i) 2,2'-bipyridine, (o) *ortho*-phenylenebis(dimethylarsine).

1.12 (a) luteo ion, (b) purpureo ion, (c) roseo ion, (d) violeo ion.

1.13 (a) $Ir(CO)[(C_6H_5)_3P]_2Cl$, (b) $K[Pt(C_2H_4)Cl_3]$, (c) $[Pt(NH_3)_4][PtCl_4]$.

1.14 (a) $(C_5H_5)_2Fe$, (b) $B_3N_3H_6$, (c) PN_2H, (d) α-Al_2O_3.

1.15 (a) potassium tetrachlorocuprate(II), (b) lithium tetrahydridoaluminate, (c) chloropentaamminecobalt(III) sulfate, (d) dichlorotetraamminecobalt(III) chloride, (e) tris(ethylenediamine)cobalt(III) chloride, (f) tris(acetylacetonato)chromium(III), (h) dichlorobis(ethylenediamine)cobalt(III) chloride, (j) tetraammineplatinum(II) tetrachloroplatinate(II), (l) potassium hexacyanoferrate(II), (n) potassium propylenediaminetetraacetatocobaltate(III).

1.16 (a) $Ir(CO)[(C_6H_5)_3P]_2Cl$, (b) $[Cr(NH_3)_4(ONO)Cl]NO_3$, (c) *trans*-$K[Co(C_2O_4)_2Cl_2]$, (d) $[Co(NH_3)_6][CuCl_5]$, (e) $Fe(acac)_3$, (f) $[V(H_2O)_6]Cl_3$, (h) $K_4[CrF_6]$, (j) $K_2[CrOCl_5]$, (l) $Re(acac)[(C_6H_5)_3P]_2Cl_2$, (n) $Na_3[Ag(S_2O_3)_2]$.

1.17 (a) $K[Co(CO)_2(NO)(CN)]$,

(b) 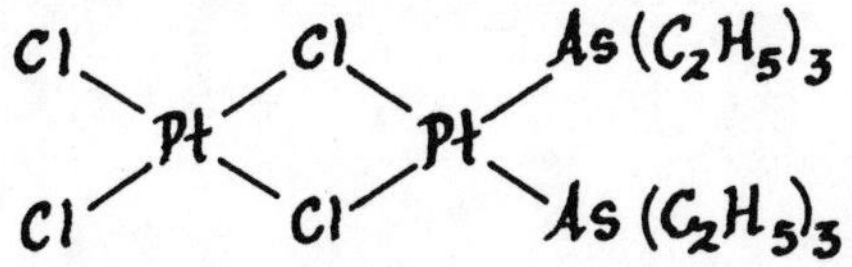

(c) potassium bis(dithiooxalato-S,S')nickelate(II),
(d) 6-hydroxo-1-bromo-4-chloro-5-iodo-3-ammine-2-pyridine platinum(IV).

Chapter 2

ELEMENTARY STRUCTURE AND STEREOCHEMISTRY

There are a number of correlative rules that are very useful for the prediction of molecular stoichiometry and geometry. These include the effective atomic number rule, full hybridization theory, and boron hydride *styx* rules. Although these empirical rules are not infallible and cannot be applied without discretion, they are, nevertheless, very useful. Most of the problems in this chapter are based on these correlative rules. In addition, there are some problems dealing with experimental methods that yield information about molecular geometry and stoichiometry.

In addition to general texts, valuable references include:

1. P. J. Wheatley, *The Determination of Molecular Structure*, 2nd ed., Oxford University Press, London, 1967.
2. J. C. D. Brand and J. C. Speakman, *Molecular Structure*, E. Arnold (Publishers) Ltd., London, 1960.

PROBLEMS

2.1 *Full hybridization theory*. (Read: Cotton and Wilkinson, Chapter 15; Douglas and McDaniel, Chapter 2). Predict the structures of the following molecules: (a) SF_4, (b) ClF_3, (c) IF_5, (d) SF_6, (e) $GeCl_4$, (f) $GeCl_3^-$, (g) PF_5.

2.2 *Effective atomic number rule*. (Read: Douglas and

McDaniel, Chapter 10; Cotton and Wilkinson, Chapter 27). Use the effective atomic number rule and evaluate n for the following compounds: (a) $(C_5H_5)V(CO)_n$; (b) $(C_5H_5)Mn(CO)_n$; (c) $(C_5H_5)Fe(CO)_n$; (d) $(C_5H_5)Ni(NO)_n$.

2.3 *Effective atomic number rule.* Among the compounds that iron forms with carbon monoxide are one with a CO to Fe ratio of 5 and another with a ratio of 4. On the basis of the *EAN* rule predict the molecular formulas for these two compounds.

2.4 *Effective atomic number rule.* Use the effective atomic number rule and decide how many carbonyls will be bonded to molybdenum in the product of the following reaction:

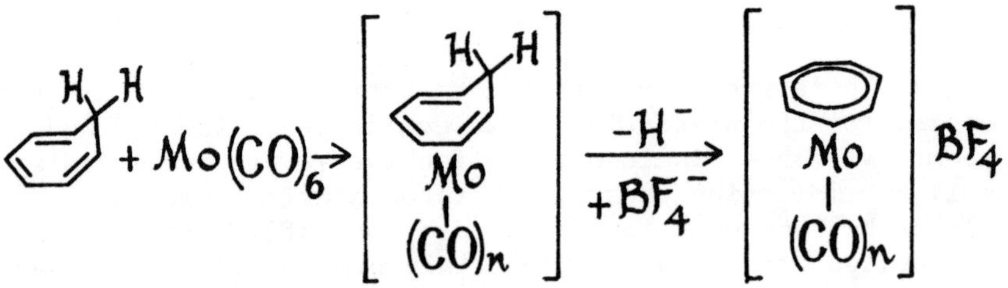

Hint: $C_7H_7^+$ donates six electrons

2.5 *Full hybridization theory and 3-center bond theory.* (Read: Cotton and Wilkinson, Chapter 15). Compare the predictions of the full hybridization and three-center bond theories for the following molecular species: (a) SF_4 (b) ClF_3, (c) XeF_6, (d) IF_7.

2.6 *Effective atomic number rule.* Use the *EAN* rule and predict the molecular formulas for the simple carbonyl compounds of Cr, Mn, Fe, Co, and Ni.

2.7 *STYX Rules.* (Read: Douglas and McDaniel, Chapter 8). Draw the structures for B_4H_{10} which satisfy the *styx* rules.

2.8 *STYX rules.* Describe the nature of the bonds in $B_6H_6^{2-}$ and predict a reasonable structure using the *styx* rules.

2.9 *Structures of boron hydrides.* The topological representation of the structure of B_5H_{11} is shown below. Explain how this structure is derived from a regular polyhedron.

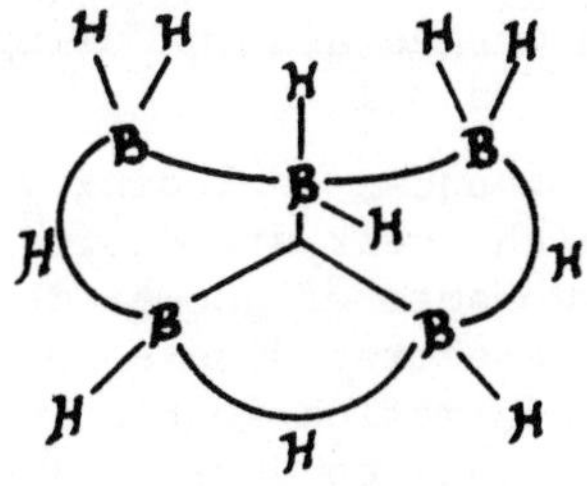

2.10 *Isomers of boron hydrides.* Draw structures of the isomers expected for

(a) $[B_{12}H_{10}X_2]^{2-}$ (Recall that $[B_{12}H_{12}]^{2-}$ has an icosahedral structure and that all twelve vertices are equivalent.)

(b) $[B_{10}H_9X]^{2-}$ (All ten boron atoms are not equivalent.)

2.11 *Geometrical isomers.* Draw structures for all the possible isomeric forms of the octahedral complex ML_3ABC.

2.12 *Polymerization isomers.* Phosphorus(V) chloride is monomeric in benzene, dimeric in carbon tetrachloride, and ionic in the solid state. Suggest structures for these three forms.

2.13 *Geometrical isomers.* Draw all geometrical isomers of $[Pt(NH_3)(NO_2)py(NH_2OH)]^+$. Which are optically active?

2.14 *Isomers of $M(AB)_3$.* Represent the unsymmetrical bidentate ligand trifluoroacetylacetonate ion, $tfac^-$, as A-B and sketch the structures of all geometrical and optical isomers of $M(tfac)_3$.

2.15 *Geometrical isomers.* When a mixture of 2,4-pentanedione (Hacac) and 1,1,1-trifluoro-2,4-pentanedione (Htfac) (See Chapter 1) is added to a slurry of $CrCl_3$ in acetone, the following reaction takes place:

$$n\text{Htfac} + (3-n)\text{Hacac} + \text{CrCl}_3\cdot 3\text{acetone} \xrightarrow{\text{benzene}} \text{Cr(tfac)}_n(\text{acac})_{3-n} + 3\text{HCl} + 3\text{acetone}$$

When all the HCl and acetone have been distilled away a solution containing all the possible compounds $Cr(tfac)_n(acac)_{3-n}$

(n = 0,1,2,3) is obtained. These can be separated by chromatography on acid washed alumina using a mixture of benzene and cyclohexane as solvent.

(a) How many compounds, including geometrical isomers, are there? Sketch their structures.
(b) Order the two sets of geometrical isomers based on magnitude of expected dipole moment.
(c) Using the information in Figure 2.1 calculate dipole moments for the geometrical isomers using simple vector addition of bond moments.

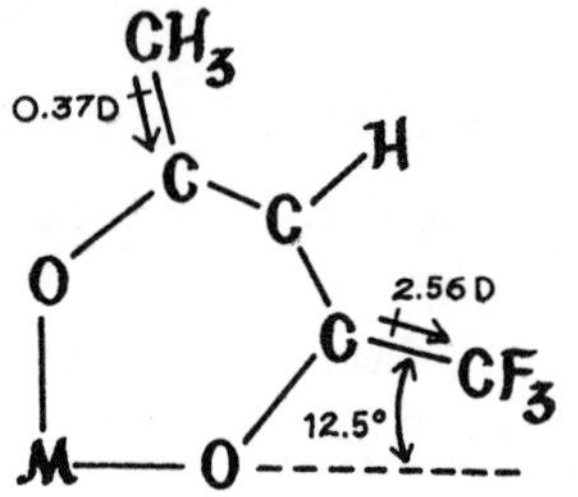

Figure 2.1 Bond dipole moments in the tfac chelate ring

2.16 *Optical isomers.* Which of the following molecules or molecule-ions may exist as optical isomers? (a) $[Co(en)_3]^{3+}$, (b) $[PtCl_3(NH_3)(H_2O)(Et_3P)]^+$, (c) $[Pt(en)_2]^{2+}$, (d) $[Co(en)_2Cl_2]^+$, (e) $[Co(NH_3)_4Cl_2]^+$, (f) ethylmethylsulfoxide, (g) *cis*-dichlorobis(triethylphosphine)platinum(II), (h)

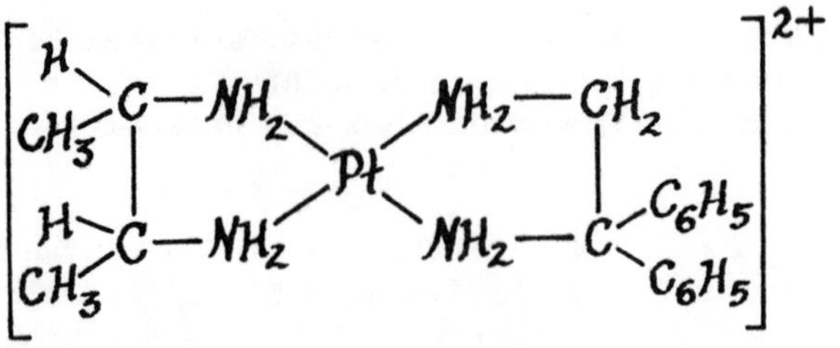

(square planar coordination)

2.17 *Isomers of M(AA)$_3$.* The separation of optical isomers for inert compounds of the type $M(AA)_3$ (where AA is a symmetrical bidentate ligand) has often been used as evidence for the octahedral arrangement for coordination number six. How should the inability to resolve the following inert complex be viewed?

$$Re \left(\begin{matrix} S-C-C_6H_5 \\ \parallel \\ S-C-C_6H_5 \end{matrix} \right)_3$$

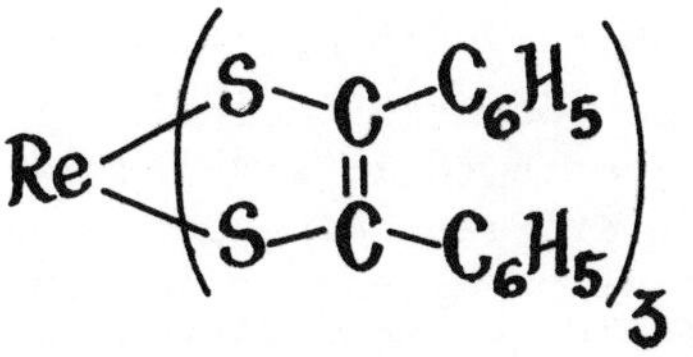

2.18 *Conformations of chelate cycles.* Sketch the various conformers of the bis(ethylenediamine)platinum(II) cation.

2.19 *Optical isomers of M(AB)$_2$.* Suggest reasons why it might be possible to partially resolve some copper(II) and nickel(II) β-ketoimines of the general formula

$$M(R-\overset{O}{\underset{|}{C}}=CH-\overset{NR'}{\underset{\parallel}{C}}-R'')_2$$

by adsorption of the complex on D-lactose followed by elution with benzene-petroleum ether.

2.20 *Dipole moments.* (Read: Wheatley, Chapter 11). One expression used in the evaluation of dipole moments of gases or dilute solutions of polar molecules in nonpolar solvents is

$$(\varepsilon-1)M/(\varepsilon+2)d = (4\pi/3)(\alpha_D + \mu^2/3kT) = P_M$$

where P_M is the total molar polarization, ε is the dielectric constant, M is the molecular weight, d is the density, N is Avogadro's number, α_D is the distortion polarizability, and μ is the dipole moment. From the data given in Table 2.1 for BrF_5, evaluate the dipole moment and predict a structure for the molecule.

Table 2.1

T, °K	ε	P_M, cc
345.6	1.006320	59.6
362.6	1.005824	57.7
374.9	1.005525	56.6
388.9	1.005180	55.0
402.4	1.004910	54.0
417.2	1.004603	52.5
430.8	1.004378	51.5

2.21 *Geometrical isomers.* Assign structures to the two geometrical isomers of dichlorobis(triethylarsine)platinum (II). One form is white, with mp 142°C, has a dipole moment of 10 D, and is sparingly soluble in benzene. The other form is yellow, with mp 121°C, is very soluble in benzene, and has a dipole moment near zero.

2.22 *Geometrical isomers.* When dichlorobis(triethylphosphine)palladium(II) is treated with $(NH_4)_2[PdCl_4]$, a substance with the empirical formula $Pd(Et_3P)Cl_2$ is formed. This substance has a sharp melting point, is soluble in benzene, in which it exhibits a molecular weight corresponding to a dimeric formulation, and has a dipole moment of 2.3 D. Explain these observations in terms of the isomeric forms possible for the dimer.

2.23 *Dipole moments and structures.* The gaseous compounds $BeCl_2$, BH_3, CCl_4, PCl_5, and SF_6 have zero dipole moments. Assign structures.

2.24 *Dipole moment and structure.* Explain why the linear, symmetrical molecule $HgCl_2$ exhibits a finite moment in dioxane.

2.25 *Conductivities.* (Read: Bailar, Chapter 2; Cotton and Wilkinson, Chapter 5). The equation representing electrical conductivities as a function of equivalent concentration, c, is commonly written in the form $\Lambda_e = \Lambda_o - B\sqrt{c}$ where Λ_e = equivalent conductivity at equivalent concentration c, Λ_o = equivalent conductivity at infinite dilution, and B is a constant for a particular electrolyte.

Using the conductivity data given below, for aqueous solutions of some cobalt complexes at 25°C, show how the above equation may be used to distinguish between various electrolyte types (*i.e.*, uni-univalent, uni-bivalent, *etc.*)

c(moles/l)	Λ_e *(ohms^{-1}moles^{-1})*		
	$[CoA_4(NO_2)_2]Cl^*$	$[CoA_5(NO_2)]Cl_2^*$	$[CoA_6]Cl_3^*$
.03125	93.7	-	124.6
.01562	98.6	116.4	135.9
.00781	102.9	123.1	147.9
.00391	106.8	128.7	155.7
.00195	-	133.9	163.9
.00098	-	136.6	169.9

* A = NH_3

2.26 *Conductivities.* Arrange the following compounds in order of anticipated decrease of their molar conductivities in aqueous solution: $Pt(NH_3)_6Cl_4$, $Pt(NH_3)_3Cl_4$, K_2PtCl_6, $Co(NH_3)_5NO_2Cl_2$, $Co(NH_3)_3(NO_2)_3$.

SOLUTIONS

2.1 (a) The sulfur atom contributes 6 electrons and the four fluorine atoms contribute 4 electrons. Thus, SF_4 has 10 electrons in the sulfur valence shell. The number of lone pairs equals the total number of valence electron pairs minus the number of bonding pairs. That is

$$10/2 - 4 = 1$$

Counting the stereochemically-active lone pair, the sulfur atom is 5-coordinated in SF_4. Possible structures are

(I) 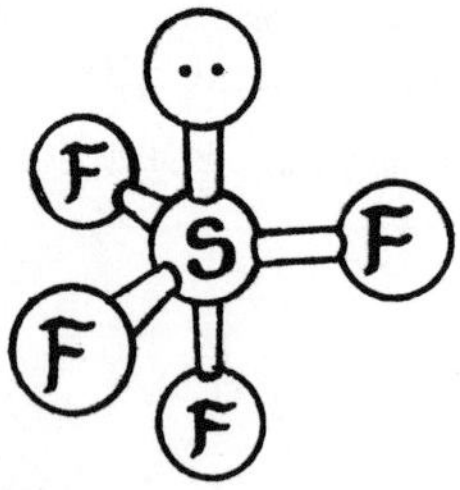

(II) 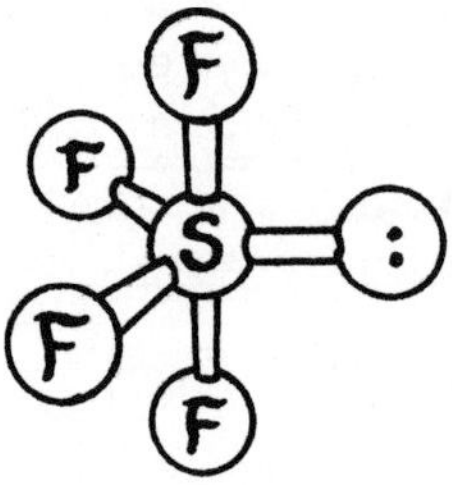

Now count the repulsions:

(I)	lone pair-lone pair, 0	(II)	ℓp-ℓp, 0
	lone pair-bonding pair, 3		ℓp-bp, 2
	bonding pair-bonding pair, 3		bp-bp, 4

The structure is predicted to be (II) since the interelectronic repulsions are lower. (Generally, ℓp-ℓp>ℓp-bp>bp-bp). Using the same reasoning, the following other structures are predicted:

(b) 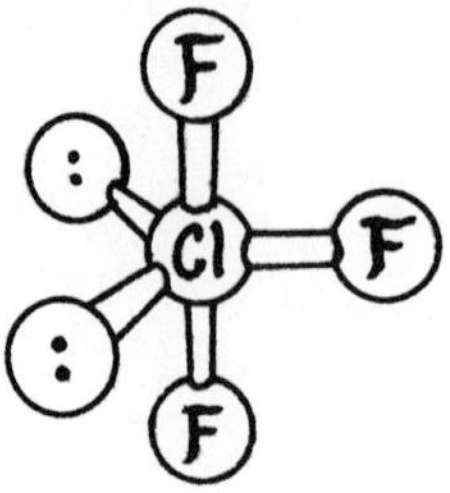

(c) 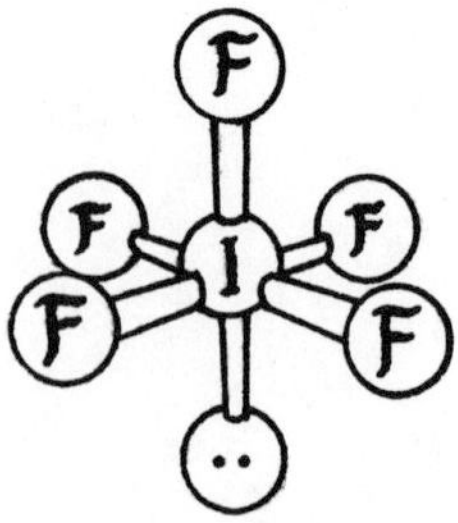

(d) 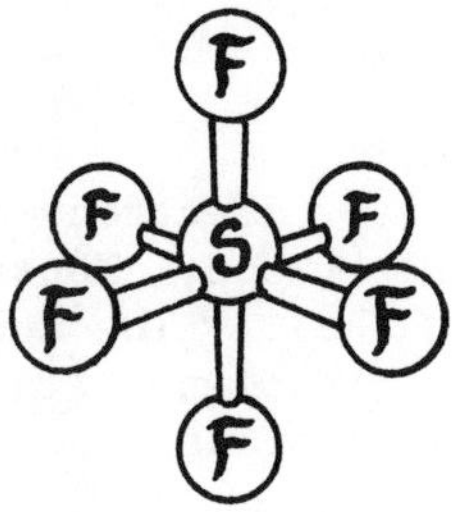

(e) 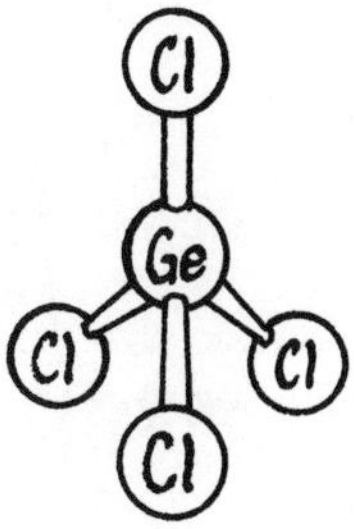

Note: These arguments are open to considerable question as regards the relative importance of the repulsions. (See Cotton and Wilkinson, p. 401 ff).

2.2 Note that the cyclopentadienyl ion is negatively charged.

(a) The $C_5H_5^-$ ion donates 6 electrons; V^+ has 4 valence electrons; and each CO donates 2 electrons. Thus $6 + 4 + 2n = 18$, $n = 4$, and the formula is $(C_5H_5)V(CO)_4$.

(b) $n = 3$

(c) 6 (from $C_5H_5^-$) + 7 (from Fe) + $2n$ (from CO) = 17, from which $n = 2$. The formula for the dimer is $(C_5H_5)_2Fe_2(CO)_4$.

2.3 For the compound with CO:Fe = 5, we have EAN = 8 (from Fe) + {2 x 5 (from CO) } = 18. Therefore, the compound is monomeric. For the compound with CO:Fe = 4, we have EAN = 8 (from Fe) + {2 x 4 (from CO) } = 16. This indicates that the compound should be trimeric.

2.5 (a) 3-center bond theory

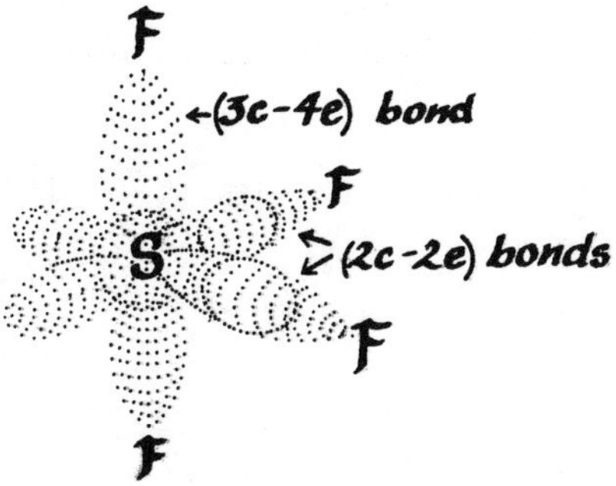

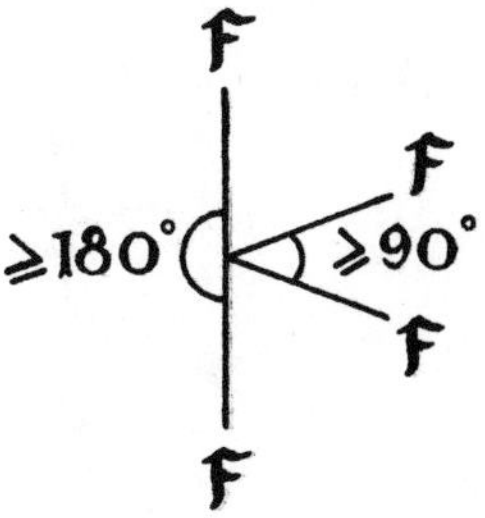

Full hybridization theory

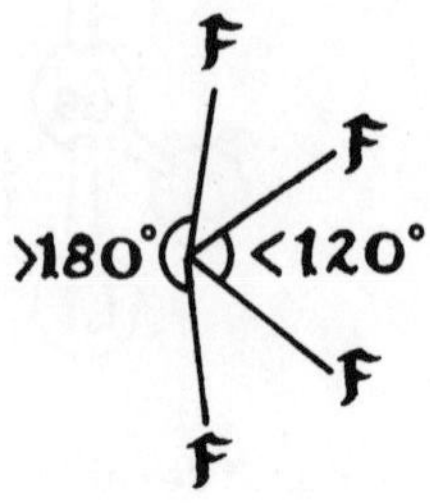

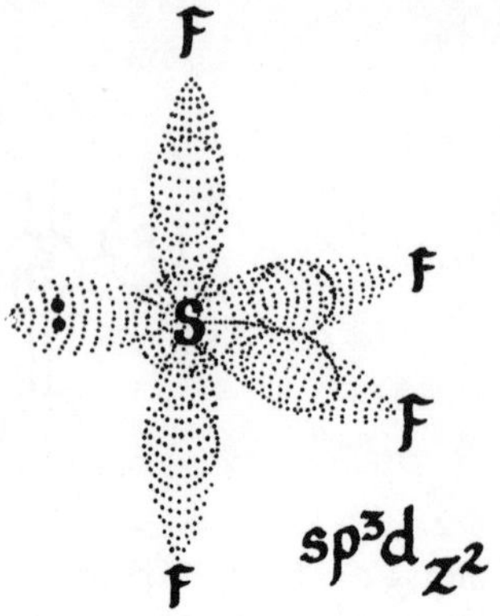

(b) Full hybridization theory

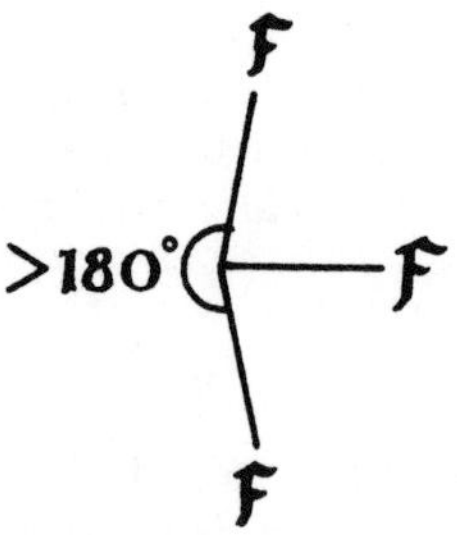

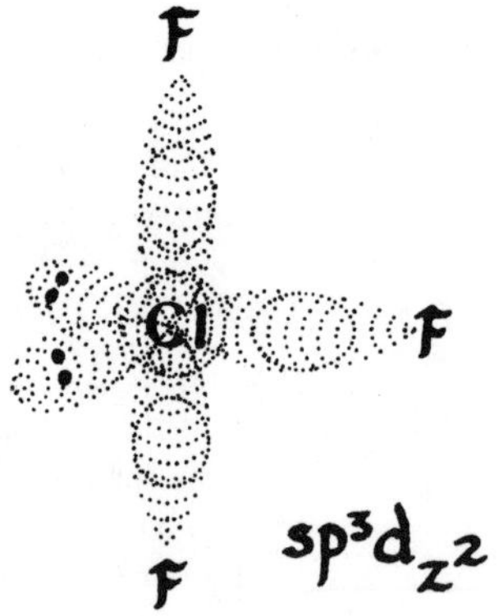

2.6 $Cr(CO)_6$, $Mn_2(CO)_{10}$, $Fe(CO)_5$.

2.7 According to the generalized formula B_pH_{p+q}, we have $p = 4$ and $q = 6$. From the equation $t + y = p - q/2$, we have $t + y = 4 - 3 = 1$. If $y = 0$, then $t = 1$, and from $s + t = p$ we have $s = 3$, and from $s + x = q$, we get $x = 3$. This gives for *styx* the sequence, 3103.

If $t = 0$ and $y = 1$, the sequence for *styx* is 4012. The structures consistent with these solutions are

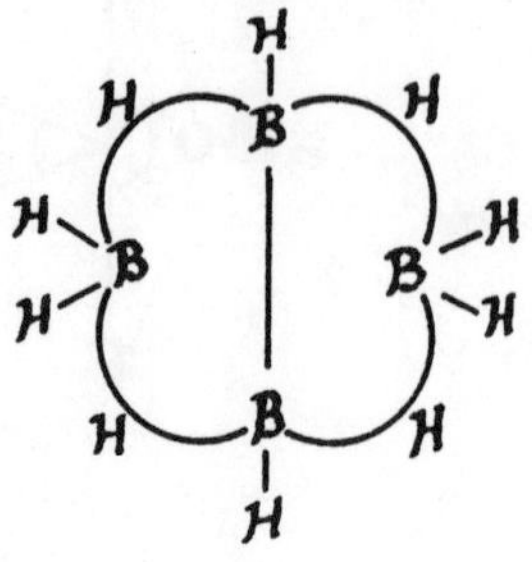

4012

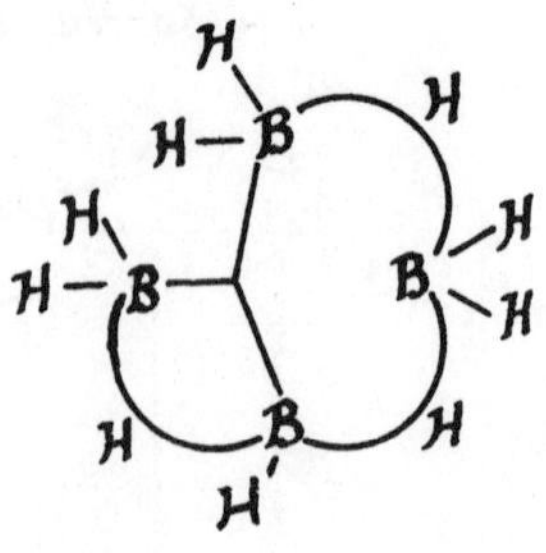

3103

The 4012 solution is more consistent with the experimentally determined structure of B_4H_{10}.

2.8 For $B_6H_6^{2-}$ we have from the general formula $[B_pH_{p+q+c}]^c$, $p = 6$ and $q = 2$. Using the equations of balance,

$$s + x = q + c = 0, \quad s + t = p + c = 4,$$

$$t + y = p - c - q/2 = 7$$

From the assumption that each boron has at least one hydrogen bonded to it, the conclusion must be that there are no "extra" hydrogens; that is, $x = 0$. Therefore, $s = 0$.

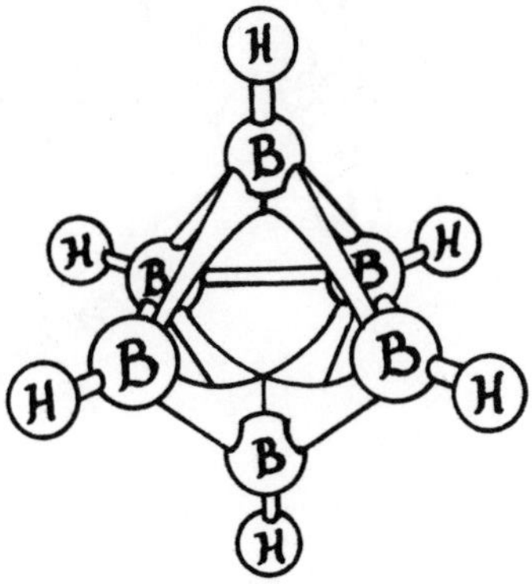

2.9 The structure is derived from an octahedron by removing the boron atom in the 6-position.

2.10 (a)

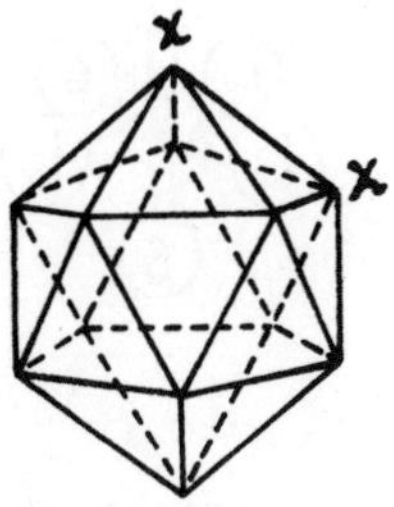

1,2-isomer

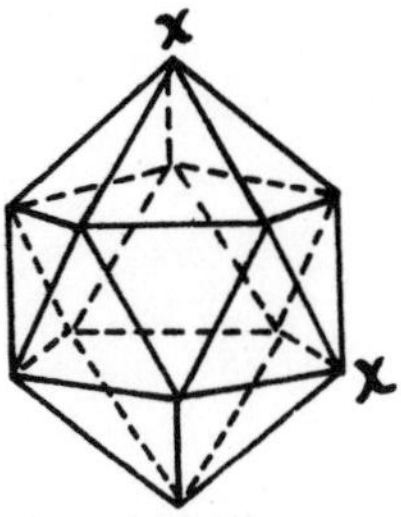

1,7-isomer

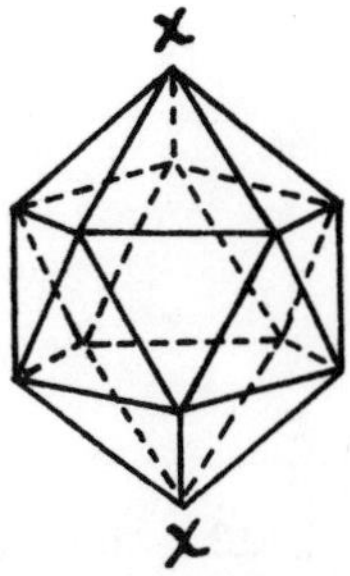

1,12-isomer

2.11 The most systematic way to arrive at the solution of this problem is to start with the ML_6 complex and to follow the scheme:

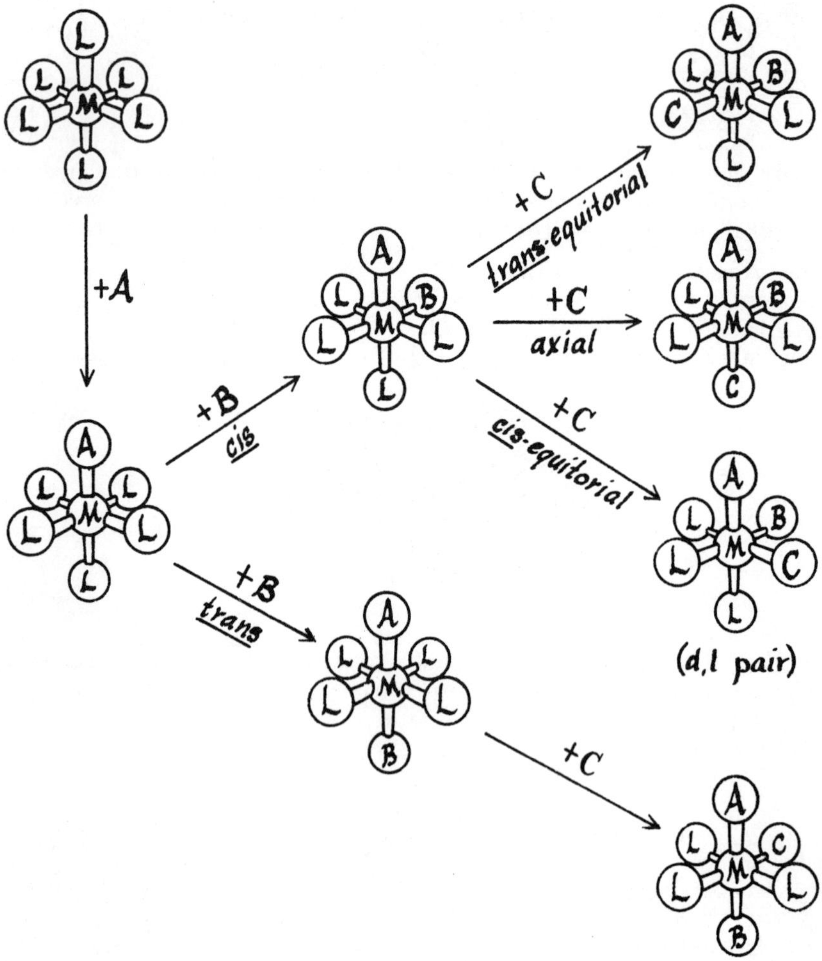

2.12 The structures are trigonal bipyramidal in benzene, two octahedra sharing an edge in carbon tetrachloride, and $[PCl_4]^+[PCl_6]^-$ in the solid state. (Although other ionic structures may be postulated, such as $PCl_4^+Cl^-$, the one above is found by experiment to be correct.)

2.14 There are two geometrical isomers and both are optically active.

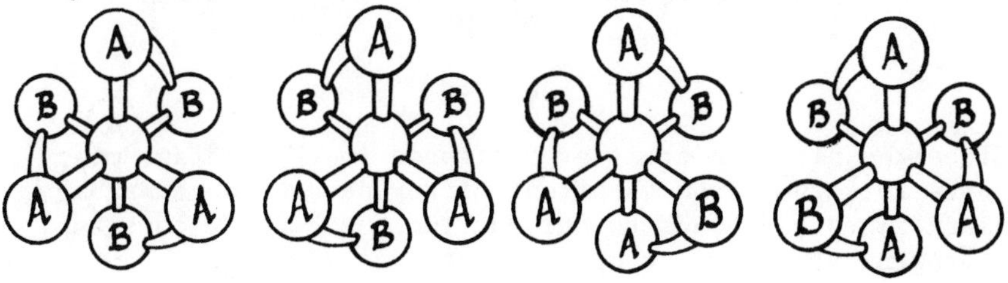

Δ-*cis* Λ-*cis* Δ-*trans* Λ-*trans*

Note that the *cis*-isomer has all three A ends on one side of the plane perpendicular to the C_3 axis, while the *trans*-form has only two A ends on one side of this plane. The Δ,Λ nomenclature refers to the nature of the screw generated by the vanes of the chelate cycles. Δ isomers are right-handed and Λ isomers are left-handed.

2.16 (a)

(b) the geometrical isomer with NH_3, H_2O, Et_3P in *cis*-positions

(c) one conformer only, but the isomers probably cannot be isolated

(d) *cis*-isomer only

(f) stereochemically-active lone pair

(h) If this were tetrahedral would it be optically active?

There are no forms of (e) or (g) which are enantiomorphic.

2.17 There are two important points here. First, the inability to resolve a potentially active compound cannot be used as structural evidence since the energy barrier to racemization is very low in some compounds, and consequently, racemization may be rapid. However, this compound does not have an octahedral configuration about rhenium; it is trigonal prismatic (D_{3h}) and thus does not have enantiomorphic forms.

2.18 The conformers can be seen best by profile diagrams.

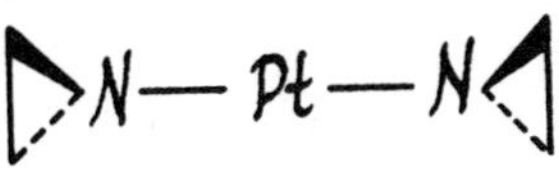

meso form

dl pair

Since the nitrogen atoms are eclipsed in these diagrams, bonds to carbon from the front nitrogens are shown as heavy black lines and bonds to carbon from the back or covered nitrogens are shown as broken lines.

2.19 The existence of isomers could arise from (1) a distortion from a planar configuration because of steric interaction of the terminal groups, or (2) the existence of a tetrahedral configuration, or (3) the existence of an octahedral geometry due to association either with another complex or with the solvent.

2.20 The expression

$$P_M = 4\pi N\mu^2/9kT + (4/3)(\pi N\alpha_D)$$

is of the general form $y = mx + b$, so if we plot P_M as a function of $1/T$ we get a straight line with a slope of $4\pi N\mu^2/9k$ and an intercept of $(4/3)(\pi N\alpha_D)$. The slope of the plot is 1.42×10^4, which gives

$$\mu^2 = \frac{(1.42 \times 10^4)\ (9)\ (1.38 \times 10^{-16})}{(4)\ (3.14)\ (6.02 \times 10^{23})} = 2.33 \times 10^{-36}$$

and

$$\mu = 1.52 \times 10^{-18} \text{ esu cm}$$

or

$$1.52\ D$$

Since the molecule has a finite moment it cannot have a planar or trigonal bipyramidal structure. A tetragonal pyramidal structure is consistent with the moment.

2.22 The possible isomers are

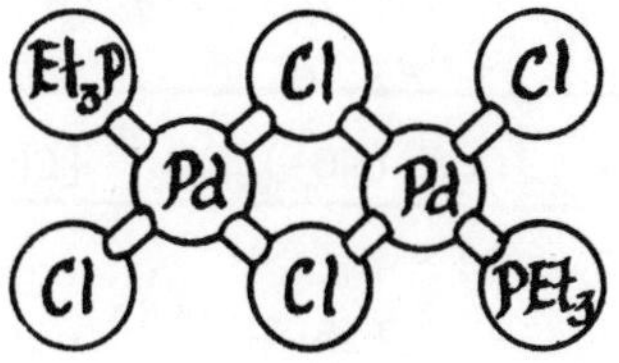

(I)

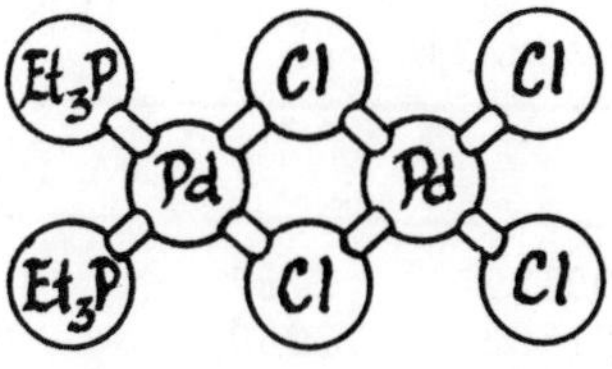

(II)

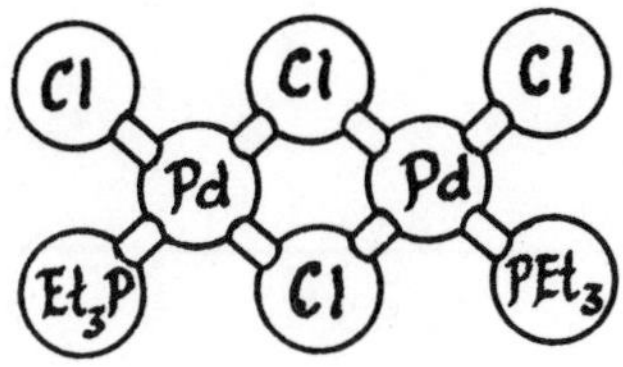

(III)

Since the solid material has a sharp melting point, it is likely that only one isomer occurs in the solid state. Since the *trans*-isomer(I) should have a dipole moment of zero, it can be concluded that the species in solution is not (I) alone. However, the solution may contain (II) only, (III) only, (II) and (III), or all three isomers.

2.25 In order to present the values for different electrolyte types on a single graph, it is necessary to determine Λ_o for each compound (from Λ_e vs $\sqrt{c}$ plot) and then to use the above equation in the form $\Lambda_o - \Lambda_e = B\sqrt{c}$.

From the resultant plots (Figures 2.2 and 2.3) it is clear that 1:1, 2:1, and 3:1 electrolytes may be immediately distinguished using this method. Since B is a function of Λ_o there may be some variation between the slopes of electrolytes of the same ion type but these are not sufficient to prevent positive identification.

At low concentrations it appears that the plot of Λ_e vs $\sqrt{c}$ deviates markedly from linearity. This effect is due to ion pair formation at these low concentrations and the points should be disregarded.

Thus the measurement of $\Lambda_o - \Lambda_e$ as a function of concentration will normally determine the ion type in addition to the molecular complexity.

Table 2.2

c (moles/l)	$\sqrt{c}$	Λ_e		
		$[CoA_4(NO_2)_2]Cl$*	$[CoA_5(NO_2)]Cl_2$*	$[CoA_6]Cl_3$*
.03125	.1767	93.7	-	124.6
.01562	.1250	98.6	116.4	135.9
.00781	.0884	102.9	123.1	147.9
.00391	.0625	106.8	128.7	155.7
.00195	.0442	-	133.9	163.9
.00098	.0312	-	136.6	169.9
Λ_o from graph of Λ_e vs $\sqrt{c}$		110.5	140.0	174.0
	$\sqrt{c}$	$\Lambda_o - \Lambda_e$	$\Lambda_o - \Lambda_e$	$\Lambda_o - \Lambda_e$
	.1767	16.8	-	49.4
	.1250	11.9	23.6	38.1
	.0884	7.6	16.9	26.1
	.0625	3.7	11.3	18.3
	.0442	-	6.1	10.1
	.0312	-	3.4	4.1
slope of $\Lambda_o - \Lambda_e$ vs $\sqrt{c}$ graph		92	189	298

*A = NH_3

2.26 $[Pt(NH_3)_6]Cl_4$ > $[Co(NH_3)_5NO_2]Cl_2$ ≃ K_2PtCl_6 > $[Pt(NH_3)Cl_3]Cl$ > $[Co(NH_3)_3(NO_2)_3]$.

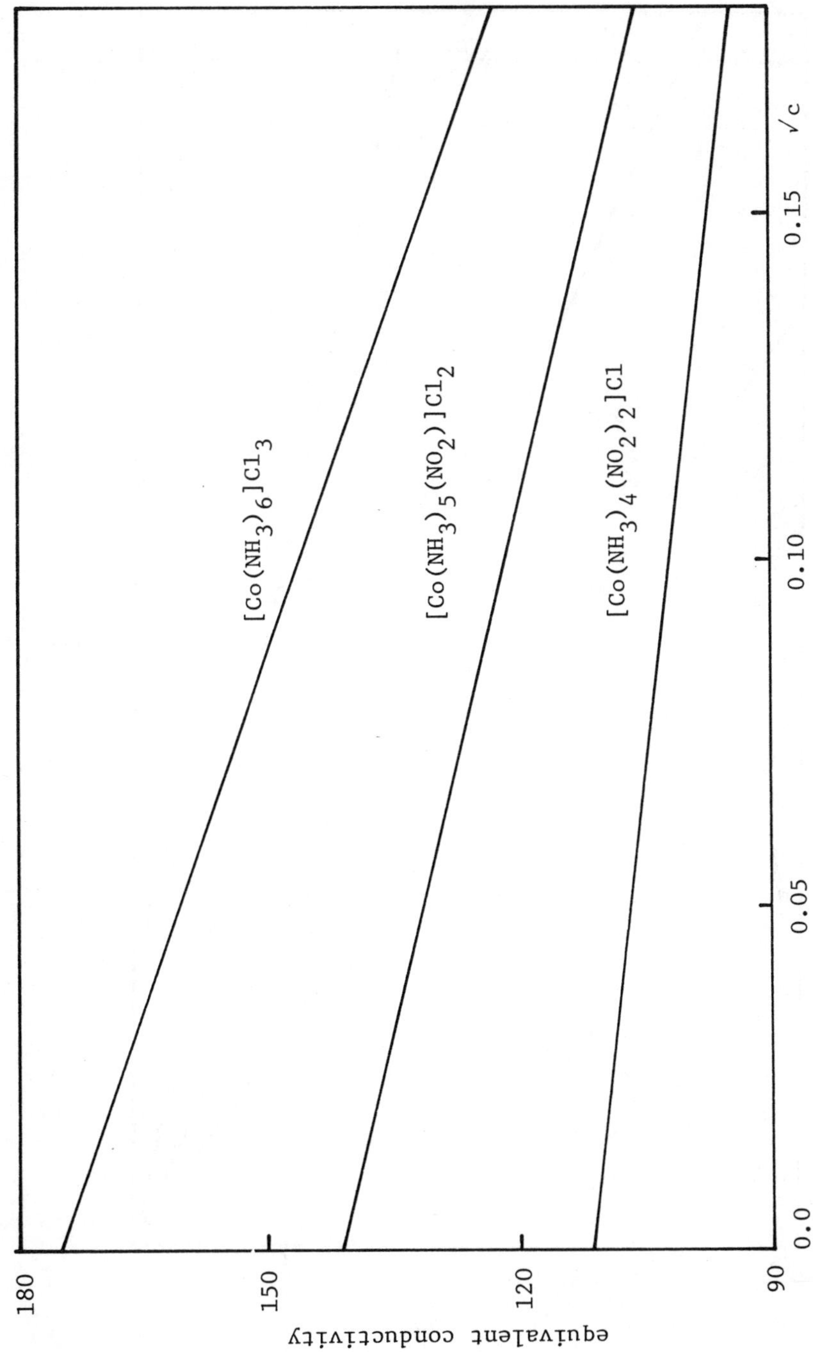
equivalent conductivity
180
150
120
90
0.0
0.05
0.10
0.15
√c
[Co(NH3)6]Cl3
[Co(NH3)5(NO2)]Cl2
[Co(NH3)4(NO2)2]Cl

Figure 2.2

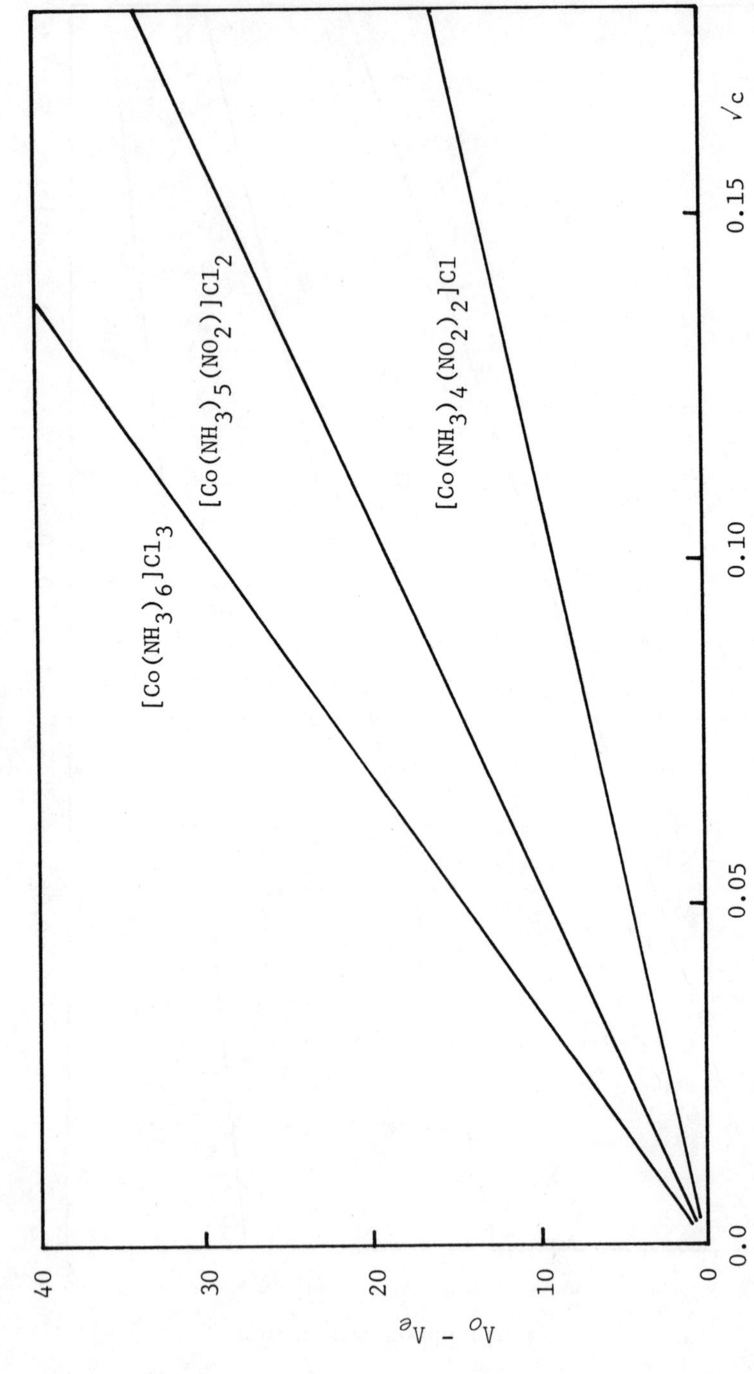
$[Co(NH_3)_6]Cl_3$
$[Co(NH_3)_5(NO_2)]Cl_2$
$[Co(NH_3)_4(NO_2)_2]Cl$
$\Lambda_o - \Lambda_e$
40
30
20
10
0
0.0
0.05
0.10
0.15
$\sqrt{c}$

Figure 2.3

Chapter 3

ATOMIC STRUCTURE

In this chapter we include some common problems arising in the quantum theory of atoms, particularly as the theory is applied to atomic spectra. We will develop here only those areas which we will find useful in later chapters. For more complete coverage of the theory of atomic structure the following references are recommended:

1. E. U. Condon and G. H. Shortley, *The Theory of Atomic Spectra*, 2nd ed., Cambridge University Press, New York, 1953.

2. H. Eyring, J. Walter, and G. E. Kimble, *Quantum Chemistry*, John Wiley and Sons, Inc., New York, 1944.

3. L. Pauling and E. B. Wilson, *Introduction to Quantum Mechanics*, McGraw-Hill Book Co., Inc., New York, 1935.

4. G. Herzberg, *Atomic Spectra and Atomic Structure*, Dover Publications, New York, 1944.

Brief discussions of atomic structure theory are given in many textbooks such as the following, which should be available to the student:

5. H. B. Gray, *Electrons and Chemical Bonding*, W. A. Benjamin, Inc., New York, 1964.

6. R. S. Drago, *Physical Methods in Inorganic Chemistry*, Reinhold Publishing Corp., New York, 1965.

7. B. E. Douglas and D. H. McDaniel, *Concepts and Models of Inorganic Chemistry*, Blaisdell Publishing Co., Waltham, Mass., 1965.

PROBLEMS

3.1 *Electronic configurations.* (Read: Cotton and Wilkinson, Chapter 1; Gray, Chapter 1; Herzberg, Chapter 3). Write the electronic configurations of the following atoms and ions: (a) Na, (b) Ca, (c) Mn, (d) Cu^+, (e) Sc, (f) Sc^{2+}, (g) Cr, (h) Pd, (i) Cu, (j) Ag.

3.2 *Russell-Saunders term symbols.* (Read: Cotton and Wilkinson, Chapter 1; Gray, Chapter 1; Douglas and McDaniel, Chapter 1). Write Russell-Saunders term symbols for the following states:

	orbital angular momentum, L	total spin, S
(a)	0	$5/2$
(b)	3	$3/2$
(c)	2	$1/2$
(d)	7	$1/2$
(e)	1	1

3.3 *Hund's rules.* By means of Hund's rules, select the ground states from the following possibilities:

(a) ^3F, ^3P, ^1P, ^1G; (b) ^5D, ^3H, ^3P, ^1I, ^1G;

(c) ^6S, ^4G, ^4P, ^2I.

3.4 *Russell-Saunders terms.* Write the Russell-Saunders term for the ground state of each of the following electronic configurations: (a) $1s^2 2s$, (b) $1s^2 2s^2$, (c) $1s^2 2s^2 2p$,

(d) $(Ne)3s$, (e) $(Ne)3s^2$, (f) $(Ne)3s^2 3p$, (g) $(Ar)4s$,

(h) $(Ar)3d^1 4s^2$, (i) $(Ar)3d^{10} 4s^2$.

3.5 *Russell-Saunders terms.* (Read: Gray, Chapter 1; see Herzberg, Chapter 3, Tables 10 and 11).

(a) Determine the Russell-Saunders terms which arise from the $3d^2$ electronic configuration.

(b) What terms arise from coulombic repulsion between electrons in the following configurations:

p^2, p^3, dp, sp^3

(c) What are the ground terms?

3.6 *Russell-Saunders terms.* Determine the ground state spectroscopic terms for transition metal ions with the electronic configurations d^1 through d^9.

3.7 *Hydrogen-like wave functions.* (Read: Pauling and Wilson, Chapter 5). Show that the hydrogen-like wave functions $\Psi(2p_{+1})$ and $\Psi(2p_{-1})$ are mutually orthogonal. The wave functions are:

$$\Psi(2p_{+1}) = (1/4\pi)(Z/a_0)^{3/2}(Zr/a_0)exp(-Zr/2a_0)(sin\ \theta) \times exp(i\phi)$$

$$\Psi(2p_{-1}) = (1/4\pi)(Z/a_0)^{3/2}(Zr/a_0)exp(-Zr/2a_0)(sin\ \theta) \times exp(-i\phi)$$

3.8 *Properties of the hydrogen atom.* (Read: Eyring, Walter, and Kimble, Chapter 6; Pauling and Wilson, Chapter 5). The normalized stationary state wave function for the ground state (1s) of the hydrogen-like atom or ion is

$$\Psi = (Z^3/\pi a_0{}^3)^{1/2} exp(-Zr/a_0)$$

where $a_0 = \hbar^2/e^2\mu$

μ = *reduced mass*, r = *radius*

The Hamiltonian for the hydrogen-like atom or ion is

$$H = -\frac{\hbar^2}{2\mu}\nabla^2 - \frac{Ze^2}{r}$$

Since there is no angular dependence, in polar coordinates

$$\nabla^2 = \frac{1}{r^2}\frac{\partial}{\partial r}\left(r^2 \frac{\partial}{\partial r}\right)$$

(a) Find the expected energy for the ground state of the hydrogen atom.
(b) Find the expectation value for r of the hydrogen atom.
(c) What is the standard deviation of r?
(Standard deviation is $[\langle r^2\rangle - \langle r\rangle^2]^{1/2}$.)

3.9 *Properties of hydrogen-like ions.* (Read: Gray, Chapter 1). Use the results of the Bohr theory and calculate the ionization potential for the single electron in the most stable orbital of the Be^{3+} ion, the radius of the ground state, and the radius of the first excited state.

3.10 *Properties of hydrogen-like ions.* Use a hydrogen-like wave function and calculate the average value of r for the ground state of Be^{3+}.

3.11 *Electron probability and distribution.* (Read: Pauling and Wilson, Chapter 5). According to a basic postulate of quantum mechanics $\Psi^*\Psi$ is proportional to the probability of finding the electron at a given position. What is the most probable position of the electron in the hydrogen atom with respect to the nucleus? Explain.

3.12 *Real and imaginary atomic orbitals.* (Read: Ballhausen, Chapter 4 and/or Figgis, Chapter 1). Transform the imaginary form of the p and d orbitals to the more familiar real orbitals in the x,y,z coordinate system.

3.13 *Spin-orbit coupling.* (Read: Ballhausen, Chapter 2). The operator for the one-electron spin-orbit interaction is

$$H' = \frac{\hbar^2}{2m^2c^2}\,\frac{1}{r}\,\frac{dU(r)}{dr}\,(\vec{\ell}\cdot\vec{s})$$

where

$$(\vec{\ell}\cdot\vec{s})_{op} = \frac{1}{2}(j^2 - \ell^2 - s^2)$$

Use perturbation theory and calculate the spin-orbit splitting between the $^2P_{1/2}$ and $^2P_{3/2}$ levels in the hydrogen atom. (Note this is a central field problem and $dU/dr = -e^2/r^2$).

3.14 *Raising and lowering operators.* (Read: Ballhausen, Chapter 2). The orbital angular momentum operators, $L_+ = (L_x + iL_y)$ and $L_- = (L_x - iL_y)$, when operating on the atomic function $|L,M_L,S,M_S\rangle$, yield a constant times $|L,M_L + 1,S,M_S\rangle$ and $|L,M_L - 1,S,M_S\rangle$, respectively. The analogous raising and lowering operators for spin angular momentum are $S_+ = (S_x + iS_y)$ and $S_- = (S_x - iS_y)$. The formulae are

$$(L_x + iL_y)|L,M_L,S,M_S\rangle = \hbar[(L + M_L + 1)(L - M_L)]^{1/2} \times |L,M_L + 1,S,M_S\rangle$$

$$(L_x - iL_y)|L,M_L,S,M_S\rangle = \hbar[L - M_L + 1)(L + M_L)]^{1/2} \times |L,M_L - 1,S,M_S\rangle$$

An important use of these operators is the generation of a complete set of wave functions for a Russell-Saunders term. For example, from the solution of problem 3.5, it was seen that the wave function for $|3,3,1,1>$ was $(2^+,1^+)$, while $|3,1,1,1>$ must contain $(2^+,-1^+)$ and $(1^+,0^+)$. Start with the function for $|3,3,1,1>$ and generate the seven wave functions for the ^3F state. (The orbitals in the wave functions must be written in standard form, that is, maximum M_L written first. For functions with the same M_L, that orbital with maximum M_S is written first. When it is necessary to carry out one exchange of two places, the component is multiplied by -1.)

3.15 *Spin-orbit splitting*. (Read: Ballhausen, Chapter 2). The spin-orbit interaction levels of the ^4F ground state of the free Cr^{3+} ion are

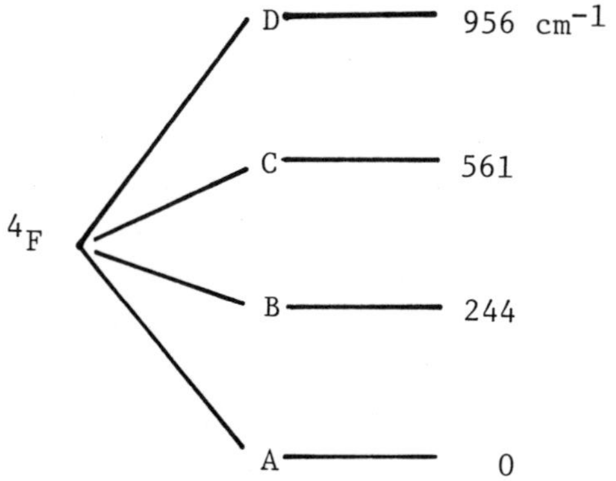

Label the states with the appropriate J quantum number, and describe any regularity observed between multiplet splittings and J quantum numbers.

3.16 *Interelectronic repulsion parameters*. (Read: Ballhausen, Chapter 2; and Figgis, Chapter 3). In the free gaseous ion spectrum of vanadium(III) (vanadium IV in spectroscopic notation) the transition energies which have been experimentally observed are

State	Energy	State	Energy
	cm^{-1}		cm^{-1}
$^3F_2$	0	$^3P_0$	13,121
$^3F_3$	318	$^3P_1$	13,238
$^3F_4$	730	$^3P_2$	13,453
^1D	10,960	^1G	18,389

In terms of F_0, F_2, and F_4 the energies are

$$E\ (^3F) = F_0 - 8F_2 - 9F_4$$
$$E\ (^3P) = F_0 + 7F_2 - 84F_4$$
$$E\ (^1D) = F_0 - 3F_2 + 36F_4$$
$$E\ (^1G) = F_0 + 4F_2 + F_4$$
$$E\ (^1S) = F_0 + 14F_2 + 126F_4$$

The F_n parameters are simply related to a second set of interelectronic repulsion parameters introduced by Racah. The relation of the F_n parameters to Racah's B and C is

$$B = F_2 - 5F_4$$
$$C = 35F_4$$

Calculate B and C for the free vanadium(III) ion using the information given. Note: The ratio C/B is fairly constant for transition metal ions and is 4 ± 0.5.

SOLUTIONS

3.1 (a) Na, $Z = 11$, $(1s)^2(2s)^2(2p)^6(3s)^1_;$ or more conveniently $[Ne]3s^1$.

(b) Ca, $Z = 20$, $(1s)^2(2s)^2(2p)^6(3s)^2(3p)^6(4s)^2_;$ or $[Ar](4s)^2$.

(c) Mn, $Z = 25$, $(1s)^2(2s)^2(2p)^6(3s)^2(3p)^6(3d)^5(4s)^2_;$ or $[Ar](3d)^5(4s)^2$.

(d) Cu^+, $Z = 29$, (Cu^+ has 28 electrons), $[Ar](3d)^{10}$.

(e) Sc, $Z = 21$, $[Ar](3d)^1(4s)^2$.

(f) Sc^{2+}, (19 electrons), $[Ar](3d)^1$.

(g) Cr, $Z = 24$, $[Ar](3d)^5(4s)^1$.

3.2 (a) ^6S, (b) ^4F, (c) ^2D, (d) ^2K.

3.3 (a) ^3F, (b) ^5D.

3.4 (a) ^2S, (b) ^1S, (c) ^2P, (d) ^2S, (e) ^1S, (f) ^2P.

3.5 (a) The problem is most efficiently handled by the use of a chart which lists all the micro-states which do not violate the Pauli Principle. For the d^2 configuration M_L can take on the values 4,3,2,1,0,-1,-2,-3, -4 and $M_S = 1,0,-1$. We shall write the micro-states as $(\overset{\pm}{m_{l_1}}, \overset{\pm}{m_{l_2}})$, and collect the permitted micro-states in Table 3.1.

Table 3.1

Micro-state Chart for *(3d)*^{2} Configuration

	M_S		
M_L	1	0	-1
4		$(\overset{+}{2},\overset{-}{2})$	
3	$(\overset{+}{2},\overset{+}{1})$	$(\overset{+}{2},\overset{-}{1})(\overset{-}{2},\overset{+}{1})$	$(\overset{-}{2},\overset{-}{1})$
2	$(\overset{+}{2},\overset{+}{0})$	$(\overset{+}{2},\overset{-}{0})(\overset{-}{2},\overset{+}{0})(\overset{+}{1},\overset{-}{1})$	$(\overset{-}{2},\overset{-}{0})$
1	$(\overset{+}{1},\overset{+}{0})(\overset{+}{2},-\overset{+}{1})$	$(\overset{+}{1},\overset{-}{0})(\overset{-}{1},\overset{+}{0})$ $(\overset{+}{2},-\overset{-}{1})(\overset{-}{2},-\overset{+}{1})$	$(\overset{-}{1},\overset{-}{0})(\overset{-}{2},-\overset{-}{1})$
0	$(\overset{+}{2},-\overset{+}{2})(\overset{+}{1},-\overset{+}{1})$	$(\overset{+}{2},-\overset{-}{2})(\overset{-}{2},-\overset{+}{2})$ $(\overset{+}{1},-\overset{-}{1})(\overset{-}{1},-\overset{+}{1})$ $(\overset{+}{0},\overset{-}{0})$	$(\overset{-}{2},-\overset{-}{2})(\overset{-}{1},-\overset{-}{1})$
-1	$(-\overset{+}{1},\overset{+}{0})(\overset{+}{1},-\overset{+}{2})$	$(-\overset{+}{1},\overset{-}{0})(-\overset{-}{1},\overset{+}{0})$ $(-\overset{+}{2},\overset{-}{1})(-\overset{-}{2},\overset{+}{1})$	$(-\overset{-}{1},\overset{-}{0})(\overset{-}{1},-\overset{-}{2})$
-2	$(-\overset{+}{2},\overset{+}{0})$	$(-\overset{+}{2},\overset{-}{0})(-\overset{-}{2},\overset{+}{0})$ $(-\overset{+}{1},-\overset{-}{1})$	$(-\overset{-}{2},\overset{-}{0})$
-3	$(-\overset{+}{2},-\overset{+}{1})$	$(-\overset{+}{2},-\overset{-}{1})(-\overset{-}{2},-\overset{+}{1})$	$(-\overset{-}{2},-\overset{-}{1})$
-4		$(-\overset{+}{2},-\overset{-}{2})$	

Now in a Russell-Saunders term there must be a complete set of micro-states, that is, a micro-state for every allowed combination of M_L and M_S. We note that the $(\overset{+}{2},\overset{+}{1})$ micro-state must arise from a ^3F term, so we proceed to eliminate the other twenty micro-states of the ^3F from the chart. Again for the sake of bookkeeping we proceed as follows:

^3F

$M_L = 3$	$M_S = 1\ (\overset{+}{2},\overset{+}{1})$	$M_L = -1$	$M_S = 1\ (\overset{+}{1},-\overset{+}{2})$
$= 3$	$= 0\ (\overset{+}{2},\overset{-}{1})$	$= -1$	$= 0\ (\overset{-}{1},-\overset{+}{2})$
$= 3$	$= -1\ (\overset{-}{2},\overset{-}{1})$	$= -1$	$= -1\ (\overset{-}{1},-\overset{-}{2})$
$M_L = 2$	$M_S = 1\ (\overset{+}{2},\overset{+}{0})$	$M_L = -2$	$M_S = 1\ (-\overset{+}{2},\overset{+}{0})$
$= 2$	$= 0\ (\overset{+}{2},\overset{-}{0})$	$= -2$	$= 0\ (-\overset{+}{2},\overset{-}{0})$
$= 2$	$= -1\ (\overset{-}{2},\overset{-}{0})$	$= -2$	$= -1\ (-\overset{-}{2},\overset{-}{0})$
$M_L = 1$	$M_S = 1\ (\overset{+}{2},-\overset{+}{1})$	$M_L = -3$	$M_S = 1\ (-\overset{+}{2},-\overset{+}{1})$
$= 1$	$= 0\ (\overset{+}{2},-\overset{-}{1})$	$= -3$	$= 0\ (-\overset{+}{2},-\overset{-}{1})$
$= 1$	$= -1\ (\overset{-}{2},-\overset{-}{1})$	$= -3$	$= -1\ (-\overset{-}{2},-\overset{-}{1})$
$M_L = 0$	$M_S = 1\ (\overset{+}{2},-\overset{+}{2})$		
$= 0$	$= 0\ (\overset{+}{2},-\overset{-}{2})$		
$= 0$	$= -1\ (\overset{-}{2},-\overset{-}{2})$		

Following the same procedure we find nine micro-states arising from a ^3P, nine from a ^1G, five from a ^1D, and one from a ^1S. Thus with the terms ^3F, ^3P, ^1G, ^1D, and ^1S we have accounted for all 45 micro-states in the chart. Note that the number of micro-states arising from a term is always $(2L + 1)(2S + 1)$.

(b) p^2; ^3P, ^1D, ^1S

sp^3; ^3D, ^1D, ^3P, ^1P, ^5S, ^3S

(c) The ground terms are p^2, ^3P; sp^3, ^5S.

3.6 For the d^1 case, $L = 2$ and $S = 1/2$; the spectroscopic term is ^2D. For the d^2 case there are two possible ways in which the spins of the electrons may be oriented--with the spins parallel (giving rise to a triplet state) or with the spins anti-parallel (giving rise to a singlet state). From Hund's first rule we know that the triplet state lies lowest. Next we must consider the orbital angular momentum. For this purpose we will use a set of five boxes with designated m_l values, and arrange the electrons in the boxes in the manner which will give rise to the greatest value of M_L (also greatest value of L).

	↿	↿			
m_l =	2	1	0	-1	-2

$$M_S = \sum_i m_s = 1/2 + 1/2 = 1$$

$$M_L = \sum_i m_l = 2 + 1 = 3$$

The maximum M_L value of 3 tells us that the ground state term is a ^3F. Using these principles we can now work out the ground state terms for the remainder of the series.

m_l	2	1	0	-1	-2	
d^3	↿	↿	↿			$2S + 1 = 4$, $L = 3$; ^4F
d^4	↿	↿	↿	↿		$2S + 1 = 5$, $L = 2$; ^5D
d^5	↿	↿	↿	↿	↿	$2S + 1 = 6$, $L = 0$; ^6S
d^6						$2S + 1 = 5$, $L = 2$; ^5D

3.7 We must show that

$$\int_0^\infty \int_0^\pi \int_0^{2\pi} \Psi^*(2p_{+1})\ \Psi(2p_{-1}) r^2 \sin\ \theta d\phi d\theta dr = 0$$

However, since

$$\int_0^{2\pi} exp(-2i\phi) d\phi = 0$$

then the total integrand is zero, and the functions are shown to be mutually orthogonal.

3.8 (a) The expected value of the energy is found from one of the postulates of quantum mechanics:

$$E = \frac{\int \Psi^* H \Psi d\tau}{\int \Psi^* \Psi d\tau}$$

Since our wave function is real and normalized

$$E = \int \Psi^* H \Psi d\tau$$

$$= \frac{1}{\pi a_0{}^3} \int exp(-r/a_0)(-\frac{\hbar^2}{2\mu}\nabla^2 - \frac{e^2}{r}) exp(-r/a_0) d\tau$$

First, let us evaluate $\nabla^2\Psi$.

$$\nabla^2\Psi = \frac{1}{r^2}\frac{\partial}{\partial r}(r^2 \frac{\partial\Psi}{\partial r})$$

$$= \frac{1}{r^2}\frac{\partial}{\partial r}[-\frac{r^2}{a_0} exp(-r/a_0)]$$

$$= (\frac{1}{a_0{}^2} - \frac{2}{a_0 r})\ exp(-r/a_0)$$

Recalling that $d\tau = 4\pi r^2 dr$ and substituting the expression for $\nabla^2\Psi$ into the energy equation we have

$$E = \frac{4}{a_0{}^3} \int_0^\infty exp(-r/a_0)(-\frac{\hbar^2}{2\mu a_0{}^2} + \frac{\hbar^2}{\mu a_0 r} - \frac{e^2}{r}) exp(-r/a_0)\ \text{x}$$

$$r^2 dr$$

To simplify matters let us make the following substitutions:

$$u = 2r/a_0; \; dr = (a_0/2)du; \; r^2 = a_0{}^2u^2/4$$

which yields

$$E = -\frac{\hbar^2}{a_0{}^2\mu}\int_0^\infty u^2 exp(-u)du + \frac{\hbar^2}{a_0{}^2\mu}\int_0^\infty u \; exp(-u)du - \frac{e^2}{a_0}\int_0^\infty u \; exp(-u)du$$

From a table of integrals we find:

$$\int_0^\infty u^n \; exp(-au)du = n!/a^{n+1}$$

Therefore,

$$E = -\frac{\hbar^2}{2a_0{}^2\mu} + \frac{\hbar^2}{a_0{}^2\mu} - \frac{e^2}{a_0} = \frac{\hbar^2}{2a_0{}^2\mu} - \frac{e^2}{a_0}$$

Substituting $a_0 = \dfrac{\hbar^2}{e^2\mu}$,

$$E = \frac{e^4\mu}{2\hbar^2} - \frac{e^4\mu}{\hbar^2} = -\frac{e^4\mu}{2\hbar^2}$$

(b) Now evaluate the expectation value for r.

$$<r> = \int_0^\infty \Psi^* r \Psi d\tau = \frac{4}{a_0{}^3}\int_0^\infty r^3 exp(-2r/a_0)dr$$

Again solving by substitution

$$u = 2r/a_0; \; dr = (a_0/2)du; \; r^3 = a_0{}^3u^3/8$$

$$<r> = (a_0/4)\int_0^\infty u^3 exp(-u)du = (a_0/4)(3\cdot 2\cdot 1) = 3a_0/2$$

(c) To find the standard deviation we must find $<r^2>$.

$$<r^2> = \int \Psi^* r^2 \Psi d\tau = (r/a_0{}^3)\int_0^\infty r^4 exp(-2r/a_0)dr$$

$$= (4/a_0{}^3)(a_0/2)^5(4\cdot 3\cdot 2\cdot 1)$$

$$= 3a_0{}^2$$

For the standard deviation, we have

$$\Delta r = (3a_0^2 - 9a_0^2/4)^{1/2}$$

$$= (\sqrt{3}/2)a_0$$

3.9 From the Bohr theory, we can derive the relationship

$$\Delta E = -(Z^2e^2/2a_0)(1/n_f^2 - 1/n_i^2)$$

where $n_f = \infty$, and $n_i = 1$. Thus,

$$\Delta E = 34.88 \times 10^{-11} \text{ ergs}$$

$$= 217.7 \text{ eV}$$

It can also be shown that

$$r = (n^2/Z)a_0$$

Therefore, for $n = 1$, $r = a_0/4 = .13A$;

and for $n = 2$, $r = 2^2a_0/4 = .529A$

3.10 Since there is no angular dependence, use the hydrogen-like radial wave function

$$R_{10}(r) = (Z/a_0)^{3/2}(2)\exp(-Zr/a_0)$$

It follows that $\langle r \rangle = 3a_0/8$

3.11 It is seen that $\Psi^*\Psi$ has its maximum value when $r = 0$. Therefore, the most probable position for the electron is at the nucleus. The probability distribution function $\Psi^*\Psi$ does not take into consideration the increase in the number of volume elements between r and $r + dr$ as r increases. The plot of the *radial* probability distribution gives a maximum at 0.53A, that is, at the Bohr radius.

3.12 The following relationships are necessary:

$$\exp(\pm im_l\phi) = \cos(m_l\phi) \pm i \sin(m_l\phi)$$

$$x = r \sin\theta \cos\theta, \; y = r \sin\theta \sin\phi, \; z = r \cos\theta$$

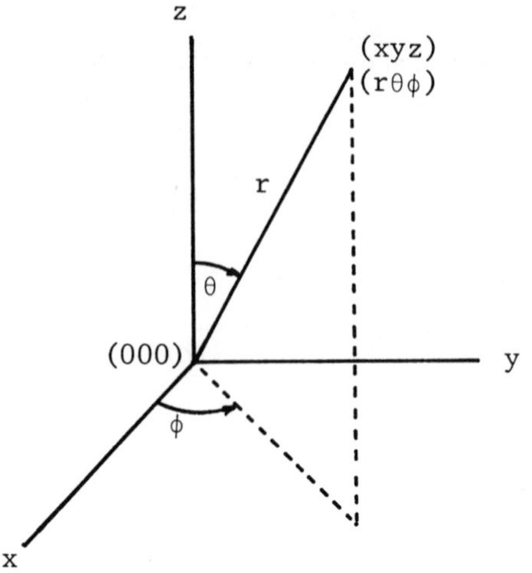

The technique is to take linear combinations of the imaginary orbitals such that the imaginary part drops out (or at least appears as only a phase factor--that is, multiplied into the whole function). Consider the combination:

$$[\Psi(n, l, +1) + \Psi(n, l, -1)] = p_+ + p_-$$

$$= R(n, l)Y_1{}^1 + R(n, l)Y_1{}^{-1}$$

$$= R(n, l)(3/4)^{1/2} \sin\theta(2\pi)^{-1/2}[\cos\phi + i \sin\phi]$$

$$+ R(n, l)(3/4)^{1/2} \sin\theta(2\pi)^{-1/2}[\cos\phi - i \sin\phi]$$

$$= 2R(n, l)(3/4)^{1/2}(2\pi)^{-1/2} \sin\theta \cos\phi$$

$$= 2R(n, l)(3/4)^{1/2}(2\pi)^{-1/2}(x/r)$$

$$= [(2R(n, l)/r)(3/4)^{1/2}(2\pi)^{-1/2}]\, x \equiv p_x$$

A normalizing coefficient of $(1/2)^{1/2}$ is necessary. Likewise $(1/i\sqrt{2})[\Psi(n,l,+1) - \Psi(n,l,-1)] = p_y$ (i appears in the normalizing coefficient). p_0 is already a real function and becomes p_z. For the d orbitals, letting $(m_l) = \Psi(n,2,m_l)$, some of the linear combinations are:

$$(0) = (1/2)(3z^2 - r^2)$$

$$(1/i\sqrt{2})[(1) - (-1)] = \sqrt{3}yz$$

$$(1/\sqrt{2})[(2) + (-2)] = (\sqrt{3}/2)(x^2 - y^2)$$

Common factors have been omitted in the above.

3.13 For hydrogen $^2P_{1/2},\ _{3/2}$, use $R_{21}(r)$. Therefore,

$$\langle H'\rangle = (h^2e^2/m^2c^2)(1/48)(1/a_0)^5 \int_0^\infty \exp(-r/a_0)rdr\ x$$

$$\{j(j+1) - l(l+1) - s(s+1)\}(1/2)$$

$$= 3.02\ x\ 10^{-5}[j(j+1) - l(l+1) - s(s+1)]\ x$$

$$(1/2)eV$$

Note: $\dfrac{erg^2sec^2esu^2}{g^2cm^2sec^{-2}cm^3}\ x\ 6.2419\ x\ 10^{11}eV/erg = eV$

For $^2P_{1/2}$; $j = 1/2,\ l = 1,\ s = 1/2$, and $(1/2)[j(j+1) - l(l+1) - s(s+1)] = -1$

$^2P_{3/2}$; $j = 3/2,\ l = 1,\ s = 1/2$, and $(1/2)[j(j+1) - l(l+1) - s(s+1)] = 1/2$

The spin-orbit splitting is

$$\Delta E = (3/2)(3.02\ x\ 10^{-5}) = 4.53\ x\ 10^{-5}eV$$

$$= (4.53\ x\ 10^{-5}eV)(8067\ cm^{-1}/eV)$$

$$= 0.37\ cm^{-1}$$

3.14 From the micro-state chart

$$\Psi(3,3,1,1) = (2^+,1^+)$$

where it is assumed that the one-electron functions are normalized. By application of L_ on both m_l values in the two-electron function we get

$$L_-\Psi(3,3,1,1) = \sqrt{6}(3,2,1,1)$$

$$= \sqrt{4}(1^+,1^+) + \sqrt{6}(2^+,0^+)$$

However, $(1^+,1^+)$ is not permitted because of the Pauli principle and $\Psi(3,2,1,1) = (2^+,0^+)$. Applying L_- again:

$$L_-\Psi(3,2,1,1) = \sqrt{10}\Psi(3,1,1,1)$$
$$= \sqrt{4}(1^+,0^+) + \sqrt{6}(2^+,-1^+)$$
$$\therefore \Psi(3,1,1,1) = \sqrt{2/5}(1^+,0^+) + \sqrt{3/5}(2^+,-1^+)$$

Applying L_- again:

$$L_-\Psi(3,1,1,1) = \sqrt{12}\Psi(3,0,1,1)$$
$$= \sqrt{2/5}\ \sqrt{6}(1^+,-1^+) + \sqrt{3/5}\{\sqrt{4}(1^+,-1^+) + \sqrt{4}(2^+,-2^+)\}$$
$$\therefore \Psi(3,0,1,1) = \sqrt{4/5}(1^+,-1^+) + \sqrt{1/5}(2^+,-2^+)$$

Finally

$$L_-\Psi(3,0,1,1) = \sqrt{12}\Psi(3,-1,1,1)$$
$$\therefore \Psi(3,-1,1,1) = \sqrt{2/5}(0^+,-1^+) + \sqrt{3/5}(1^+,-2^+)$$

The micro-state chart reveals that

$$\Psi(3,-2,1,1) = (0^+,-2^+)$$
$$\Psi(3,-3,1,1) = (-1^+,-2^+)$$

Therefore the seven functions are

$$\Psi(3,3,1,1) = (2^+,1^+)$$
$$\Psi(3,2,1,1) = (2^+,0^+)$$
$$\Psi(3,1,1,1) = \sqrt{2/5}(1^+,0^+) + \sqrt{3/5}(2^+,-1^+)$$
$$\Psi(3,0,1,1) = \sqrt{4/5}(1^+,-1^+) + \sqrt{1/5}(2^+,-2^+)$$
$$\Psi(3,-1,1,1) = \sqrt{2/5}(0^+,-1^+) + \sqrt{3/5}(1^+,-2^+)$$
$$\Psi(3,-2,1,1) = (0^+,-2^+)$$
$$\Psi(3,-3,1,1) = (-1^+,-2^+)$$

3.16 Using the Landé interval rule the centers of gravity of the ^3F and ^3P multiplets are at $4\lambda = 420\ cm^{-1}$ and at $13,238 + \lambda = 13,343\ cm^{-1}$ (taking $\lambda = \zeta/2S = 105\ cm^{-1}$). Then,

$$E(^3P) - E(^3F) = 12,923\ cm^{-1} = 15F_2 - 75F_4$$

$$E(^1D) - E(^3F) = 10,540\ cm^{-1} = 5F_2 + 45F_4$$

or,

$$F_2 - 5F_4 = 863\ cm^{-1}$$

$$\underline{-(F_2 + 9F_4 = 2110)}$$

$$-14F_4 = -1247\ cm^{-1};\quad F_4 = 89\ cm^{-1}$$

$$F_2 - 5(89\ cm^{-1}) = 863\ cm^{-1}$$

$$F_2 = 1308\ cm^{-1}$$

Therefore

$$B = F_2 - 5F_4 = 863\ cm^{-1}$$

$$C = 35F_4 = 3120\ cm^{-1}$$

$$C/B = 3.6$$

This result is characteristic; that is, the difference between terms of the same multiplicity is expressable as a function of only one Racah parameter, whereas two Racah parameters are required for differences between terms of different multiplicity. Two F_n parameters are generally required in either case.

The ratio of B for the free ion as found above to B found for the complexed ion, B *complex*/B *free ion*, is β, the nephelauxetic parameter. (See problem 7.20).

Chapter 4

MOLECULAR SYMMETRY AND GROUP THEORY

The use of symmetry notions in structural and dynamic inorganic chemistry is extremely important. In fact, the potential of such common research tools as vibrational and electronic spectroscopy can be realized only if the information is analyzed in terms of the symmetry properties of the molecules under study.

The problems in this chapter are designed to give an operational familiarity with aspects of group theory related to molecular symmetry. Chapter 10 deals with the elements of crystal lattice symmetry. In addition there are a number of problems based on these concepts and tools in other chapters.

Especially valuable reference books are the following:

1. F. A. Cotton, *Chemical Applications of Group Theory*, Interscience Publishers, New York, 1963.
2. R. Hochstrasser, *Molecular Aspects of Symmetry*, W. A. Benjamin, Inc., New York, 1966.
3. D. Schonland, *Molecular Symmetry*, D. van Nostrand, New York, 1965.
4. H. H. Jaffé and M. Orchin, *Symmetry in Chemistry*, John Wiley and Sons, New York, 1965.

PROBLEMS

4.1 *Symmetry elements.* (Read: Cotton, Chapter 2; Hochstrasser, Chapter 1; Schonland, Chapter 2). Identify all the symmetry elements in the following molecules: (a) H_2O, (b) CH_4, (c) NO_3^-.

4.2 *Point groups.* (Read: Cotton, Chapter 3; Hochstrasser, Chapter 3; Schonland, Chapter 2). Classify the following molecules into the appropriate point groups: (a) ferrocene, (b) $PtCl_4{}^{2-}$, (c) SF_4, (d) staggered configuration of $H_3B–PH_3$, (e) $CuCl_5{}^{3-}$ (trigonal bipyramid).

(f)

(g)

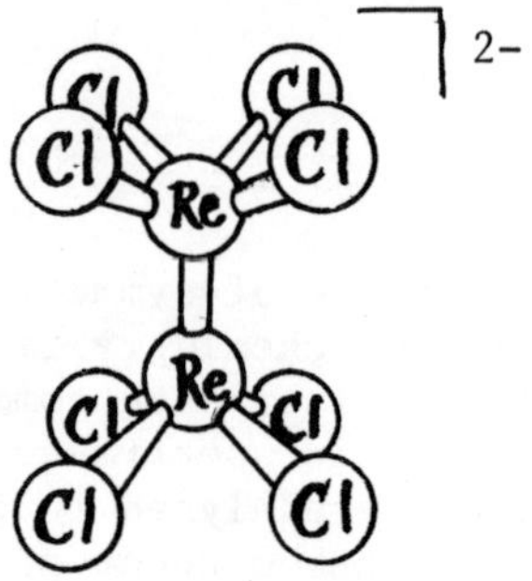

(h)

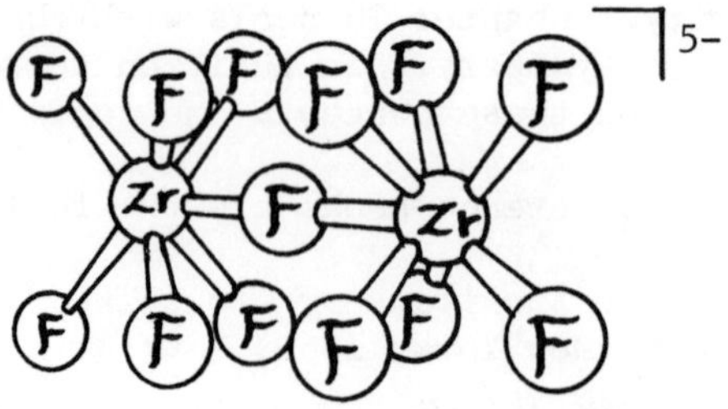

(i) *cis*-dichlorodiammineplatinum(II) (Ignore the hydrogen atoms.), (j) the *trans*-dichlorotetraamminecobalt(III) ion, (k) $Cr(en)_3{}^{3+}$, (l) $Co(PDTA)^-$ (Consider the cobalt and the donor atoms only.), (m) permanganyl chloride.

4.3 *Point groups.* Give examples of molecules which belong to the following point groups: (a) $C_{\infty v}$, (b) D_3, (c) C_s, (d) D_{2d}

4.4 *Point groups.* Consider the molecule MA_6 in which the atoms A are disposed about M at the vertices of a regular octahedron.

(a) Changing only one or more of the bond lengths, MA, illustrate how you might reduce the symmetry of MA_6 to C_{4v}, D_{4h}, C_{3v}, and C_{2v}.

(b) Changing only one or more of the A-M-A bond angles show how you could reduce the symmetry to D_{2h}, D_{3d}, and D_{3h}.

(c) Leaving all the donor atoms in this molecule (the A's) at their octahedral points, but exchanging one or more of the A's for either B, A——A, or A——B (where A——A and A——B are bidentate ligands) show how you could construct species with the following symmetries: D_{4h}, C_{4v}, D_3, C_{3v}, C_3, D_{2h}, C_{2v}, C_2, C_s, and C_1.

4.5 *Group properties.* (Read: Cotton, Chapter 2; Hochstrasser, Chapter 1; Schonland, Chapter 2). Use the following form and construct the group multiplication table for the operations generated by the symmetry elements of the group C_{3v}. In the construction of the table, first perform the operation listed in the row (across the top), then do the operation in the column.

C_{3v}	E	C_3	C_3^2	σ_a	σ_b	σ_c
E					σ_b	
C_3						
C_3^2						
σ_a		σ_a				
σ_b				C_3		
σ_c						

4.6 *Group properties.* Use the results of problem 4.5 and demonstrate that the symmetry operations generated by the symmetry elements of a triangular based pyramid obey the rules defining a group.

4.7 *Class structure.* Determine the class structure of the C_{3v} point group.

4.8 *Group properties.* The set of numbers 1, -1, i, -i can be shown to constitute a group.

(a) What is the group multiplication operation?

(b) What is the identity element?

(c) What is the reciprocal of each element?

(d) Is the group Abelian?

(e) What sub groups are contained in the group?

(f) What is the class structure?

4.9 *Group properties.* Consider the following group:

Δ *is the identity*

$\square 0 = \nabla = 0 \square$

$\square \square = 0 = \nabla\nabla$

$0^{-1} = 0$

(a) Make up the complete multiplication table.

(b) Verify that this is a proper group.

(c) Is the group Abelian?

(d) What are the sub groups?

(e) What is the class structure?

4.10 *Group properties.* Consider the following proposed groups. Which of them are proper groups and which are not? Why?

(a) The collection of all positive and negative integers and zero with arithmatic addition as the group multiplication operation.

(b) The collection of all positive integers and all positive integral fractions with arithmatic division as the operation.

(c) The collection of all positive and negative integers and zero with arithmatic subtraction as the operation.

(d) The collection of all positive integers and all positive integral fractions with arithmatic multiplication as the operation.

4.11 *Group properties.* Consider the trigonal prismatic molecule tris(cis-diphenylethene-1,2-dithiolato)rhenium (R. Eisenberg and J. A. Ibers, *J. Am. Chem. Soc.*, 87, 3776 (1965).)

(a) Draw a model of the molecule, list the symmetry operations, and define the point group to which the molecule belongs. Prove that these operations conform to the four group postulates.

(b) Determine the class structure of this group.

(c) Determine the subgroups of this group.

(d) Show which subgroups are Abelian.

4.12 *Group properties.* Do the numbers -1, 0, +1 constitute a group with arithmatic addition as the group multiplication operation?

4.13 *Matrix manipulations.* (Read: Cotton, Chapter 4 and Appendix I; Hochstrasser, Chapter 2; Schonland, Chapter 4). Perform the following matrix multiplications:

(a) $\begin{pmatrix} 1 & 2 & 3 & 4 \end{pmatrix} \begin{pmatrix} 1 \\ 2 \\ 3 \\ 4 \end{pmatrix}$ (b) $\begin{pmatrix} 1 \\ 2 \\ 3 \\ 4 \end{pmatrix} \begin{pmatrix} 1 & 2 & 3 & 4 \end{pmatrix}$

(c) $\begin{pmatrix} 1 & 0 & 2 \\ 2 & 1 & 0 \\ 3 & 2 & 1 \end{pmatrix} \begin{pmatrix} 2 & 2 & 1 \\ 1 & 0 & 3 \\ 3 & 1 & 3 \end{pmatrix}$

4.14 *Matrix manipulations.* For the matrix $\mathbb{A}$ given below, find $\mathbb{A}^{-1}$ and verify that $\mathbb{A} \cdot \mathbb{A}^{-1} = \mathbb{1}$.

$$A = \begin{pmatrix} 1 & 6 & 2 \\ 0 & 3 & 1 \\ 2 & 1 & 2 \end{pmatrix}$$

4.15 *Representation matrices.* (Read: Cotton, Chapter 4; Hochstrasser, Chapter 4; Schonland, Chapter 4). Consider a triangular based pyramid. Take the three Cartesian unit vectors x, y, and z as a basis set and construct the matrices representing the operations of the group. Show that the matrices obey the rules which define a group.

4.16 *Characters of representation matrices as group representations.*

(a) Construct the representation made up by the characters of the representation matrices of Problem 4.15.

(b) Is the representation reducible? If so, reduce the representation into its irreducible components.

4.17 *Reduction of representations.* Decompose the following reducible representations into their irreducible components:

(a)

D_{3h}	E	$2C_3$	$3C_2$	σ_h	$2S_3$	$3\sigma_v$
Γ_a	5	2	1	3	0	3
Γ_b	3	0	-1	-3	0	1
Γ_c	3	0	-1	3	0	-1

(b)

C_{2v}	E	C_2	$\sigma_v(xz)$	$\sigma_v'(yz)$
Γ_a	2	0	0	2
Γ_b	4	0	0	0
Γ_c	3	1	3	1

(c)

C_{2h}	E	C_2	i	σ_h
Γ_a	3	-1	1	-3
Γ_b	2	0	-2	0
Γ_c	5	1	1	1

4.18 *Transformation properties of atomic orbitals.* The atomic orbitals p_x, p_y, and p_z transform as do x, y, and z. Refer to the appropriate character tables and determine the transformation properties of the p-orbitals in the following point groups:

(a) C_{2v}, (b) O_h, (c) D_{4h}.

4.19 *Transformation properties of atomic orbitals.* The atomic d-orbitals transform as do the various squares and binary products of x, y, and z. How does the d_{z^2} orbital transform in the following point groups?

(a) O_h, (b) C_{3v}, (c) D_{3d}.

4.20 *Transformation properties of atomic orbitals.* Determine from the character tables the symmetry labels of the d-orbitals in molecules which belong to the following point groups:

(a) O_h, (b) T_d, (c) D_{4h}, (d) D_3.

4.21 *Direct products.* (Read: Cotton, Chapter 5; Hochstrasser, Chapter 5; Schonland, Chapter 6). Find the direct products indicated below. If the direct product representation is reducible, reduce it to its irreducible components.

(a) T_d, $T_2 \times T_1$; (b) C_{2v}, $A_1 \times B_1$; (c) D_{2h}, $B_{1g} \times B_{2g} \times B_{3u}$.

SOLUTIONS

4.1 (a)

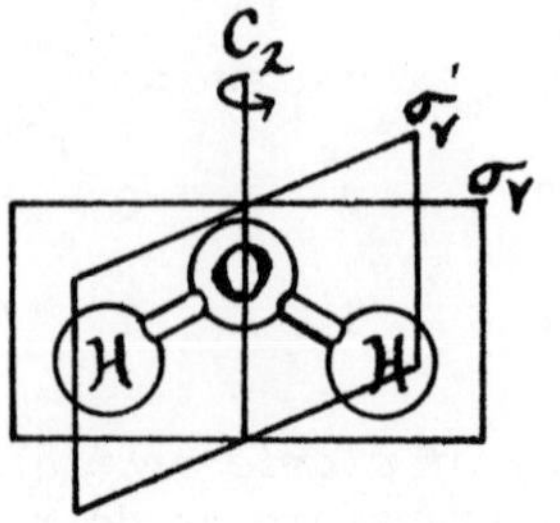

C_2 axis as shown (conventionally counterclockwise)

σ_v'- perpendicular to the plane of the page

(σ_v and σ_v' intersect along the C_2 axis)

(b) A C_3 axis colinear with each C-H bond; three C_2 axes, one bisecting each pair of opposite H-C-H angles; an S_4 axis colinear with each C_2; six σ_d planes described by the carbon and two of the hydrogen atoms (note that there are only six ways four equivalent atoms may be permuted in pairs).

4.2 (a)

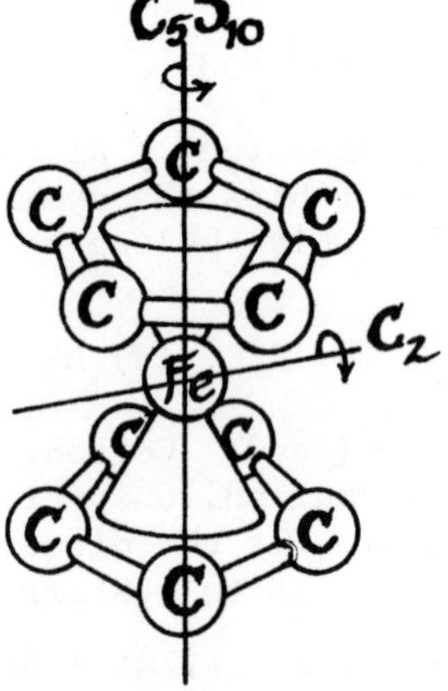

Proceeding systematically

(1) The molecule does not belong to one of the special groups.

(2) The highest order rotation axis is C_5.

(3) There is a colinear S_{10} axis, and there are also other symmetry elements.

(4) There are five C_2 axes in the plane perpendicular to the C_5 axis. Therefore the molecule belongs to one of the D groups

(5) There is no σ_h but there are five σ_d planes; therefore the point group is D_{5d}.

Proceeding in like fashion the other answers are found to be as follows:

(b) D_{4h}, (c) C_{2v}, (f) D_{5h}, (g) D_{4h}, (i) C_{2v}, (j) D_{4h}, (k) D_3, (m) C_{3v}.

4.3 (a) Any of a number of unsymmetrical linear molecules, for example, HCN.

(b) If the chelate cycles are considered, then $[M(en)_3]^{n+}$ or $M(acac)_3$.

(c) There are only E and σ in C_s, so HOD is an example.

4.4 (a) To reduce symmetry of MA_6 from O_h to C_{4v}, increase or decrease one bond length; from O_h to D_{4h}, increase or decrease, simultaneously, two *trans* bond lengths; from O_h to C_{3v}, increase or decrease, simultaneously, three *cis* M-A bond lengths.

(b) To reduce symmetry of MA_6 from O_h to D_{2h}, compress two angles in the following manner:

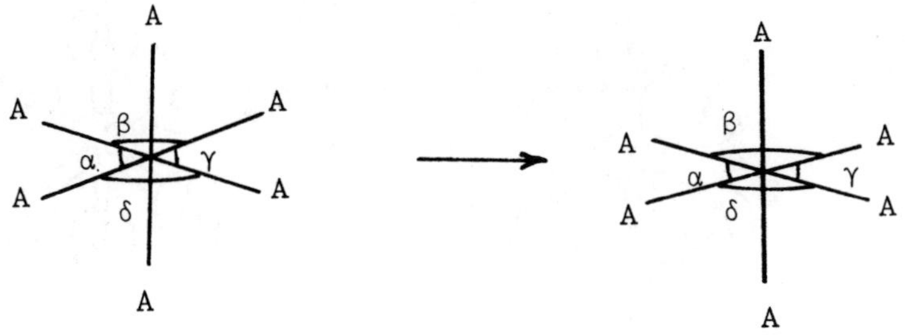

$\alpha = \beta = \gamma = \delta = 90°$ $\alpha = \gamma < 90°$, $\beta = \delta > 90°$

To reduce symmetry to D_{3d}, compress or elongate the molecule along the C_3 axis as shown.

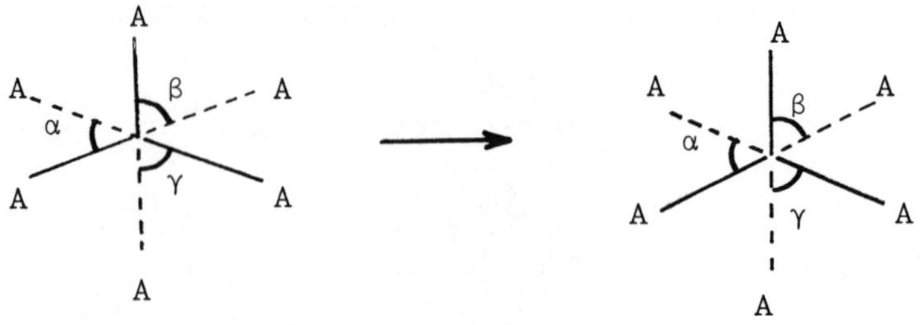

$\alpha = \beta = \gamma = 90^{\circ}$ $\qquad$ $\alpha = \beta = \gamma \neq 90^{\circ}$

(Note that dotted lines signify bonds to atoms which are below the plane and full lines represent bonds to atoms above the plane of the page.)

(c)

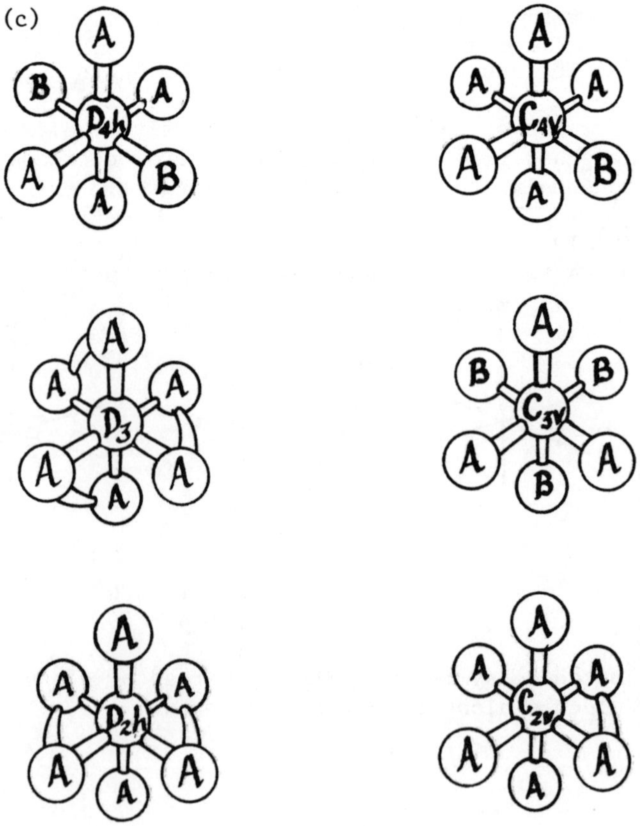

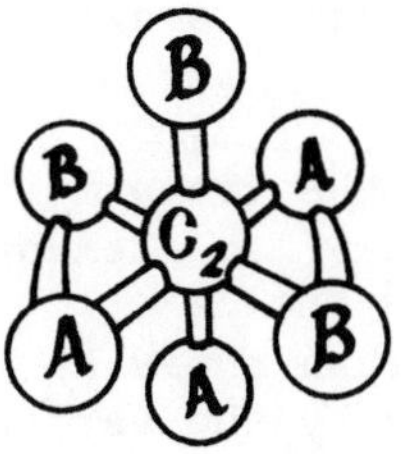

Several possibilities exist for C_{2v} and C_2.

4.5 The entries in the table are more readily ascertained by use of diagrams. For example, consider the product $C_3{}^2\sigma_c$. The operation on the right is always performed first, and the symmetry elements are not moved by the symmetry operation. Therefore,

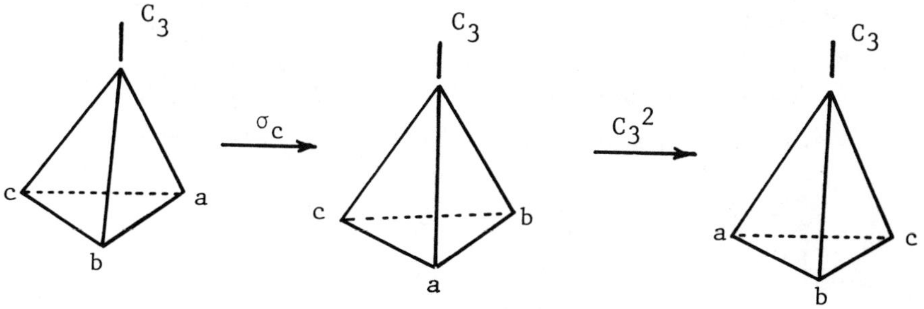

The completed table is:

C_{3v}	E	C_3	$C_3{}^2$	σ_a	σ_b	σ_c
E	E	C_3	$C_3{}^2$	σ_a	σ_b	σ_c
C_3	C_3	$C_3{}^2$	E	σ_b	σ_c	σ_a
$C_3{}^2$	$C_3{}^2$	E	C_3	σ_c	σ_a	σ_b
σ_a	σ_a	σ_c	σ_b	E	$C_3{}^2$	C_3
σ_b	σ_b	σ_a	σ_c	C_3	E	$C_3{}^2$
σ_c	σ_c	σ_b	σ_a	$C_3{}^2$	C_3	E

As a check on the table note that no symmetry operation occurs more than once in either a single row or column in the body of the table.

4.6 (a) The results in the group multiplication table verify that the square of an element or the product of any two elements is an element of the group. Thus, no elements other than those already present in the group are generated by the various multiplications.

(b) The second rule requires that group multiplication obey the associative law. We will examine the representative case $(\sigma_a C_3)\sigma_b \stackrel{?}{=} \sigma_a(C_3\sigma_b)$. Arbitrarily, we do the left side first. From the multiplication table we see that $\sigma_a C_3 = \sigma_c$. Now we have the product $\sigma_c\sigma_b$ which gives C_3. If the associative law holds, then $\sigma_a(C_3\sigma_b)$ must also yield C_3. From the table $C_3\sigma_b = \sigma_c$ and $\sigma_a\sigma_c = C_3$. These results confirm the equality.

(c) The identity operation is defined, since it leaves the body in its original orientation.

(d) Each operation has a reciprocal which is a member of the group. For example

$\sigma_a\sigma_a^{-1} = E$, $\sigma_b\sigma_b^{-1} = E$, $\sigma_c\sigma_c^{-1} = E$, and

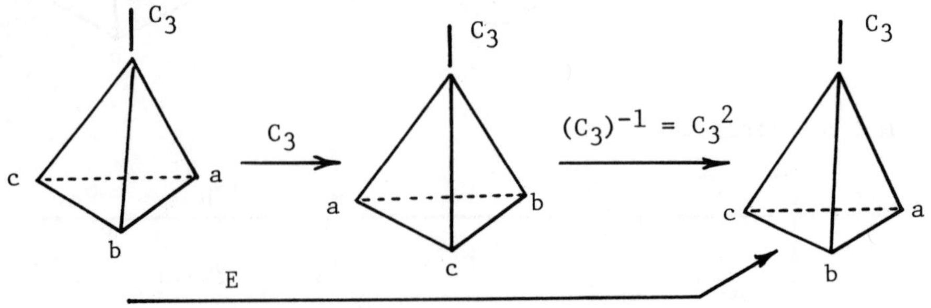

4.8 (a) Arithmatic multiplication.

(b) The number 1 is the identity element.

(c) The reciprocal of 1 is 1; of i is -i; of -1 is -1; and of -i is i.

(d) Yes, because multiplication is commutative.

(e) The subgroups are {1} and {1, -1}.

(f) Each element forms a class unto itself.

4.10 The only proper groups are (a) and (d), since addition and multiplication are associative, whereas subtraction and division are not associative. The identity elements in (a) and (d) are zero and one, respectively. Zero will also fit the definition of an identity element in the pseudo group in (c), but there is no identity element for (b). Show why.

4.11 (a) As shown by x-ray diffraction (*loc. cit.*) the molecule has a trigonal prismatic array of sulfur donors surrounding the central rhenium. The chelate rings are planar. Therefore, the point group is D_{3h}. In locating all the symmetry elements and operations, note that only two of the $S_3{}^n$ operations are actually unique. The best way to start this problem is to work out the complete 12 x 12 multiplication table. The table contains all the necessary proof that the operations conform to the four basic group postulates. Further, it is useful in sorting out the classes and subgroups.

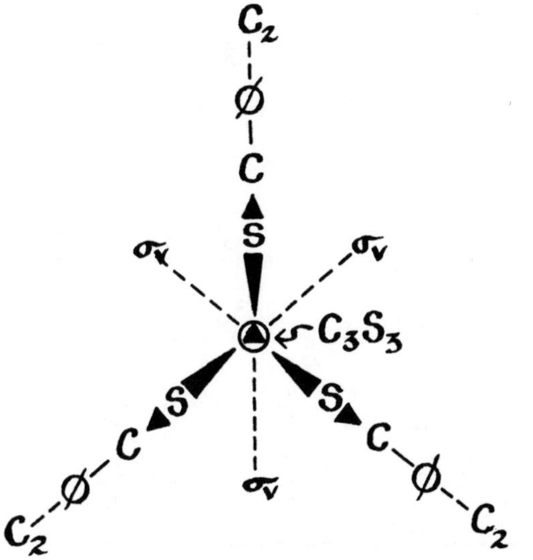

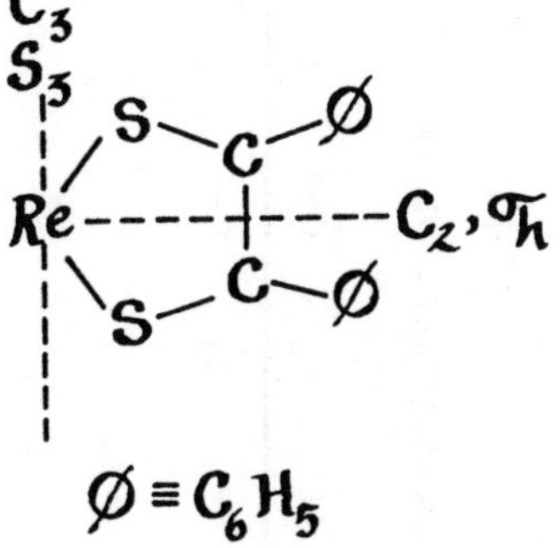

Group multiplication table

D_{3h}	E	C_3	C_3^2	σ_h	C_2	C_2'	C_2''	S_3	S_3^5	σ_v	σ_v'	σ_v''
E	E	C_3	C_3^2	σ_h	C_2	C_2'	C_2''	S_3	S_3^5	σ_v	σ_v'	σ_v''
C_3	C_3	C_3^2	E	S_3	C_2''	C_2	C_2'	S_3^5	σ_h	σ_v''	σ_v	σ_v'
C_3^2	C_3^2	E	C_3	S_3^5	C_2'	C_2''	C_2	σ_h	S_3	σ_v'	σ_v''	σ_v
σ_h	σ_h	S_3	S_3^5	E	σ_v	σ_v'	σ_v''	C_3	C_3^2	C_2	C_2'	C_2''
C_2	C_2	C_2'	C_2''	σ_v	E	C_3	C_3^2	σ_v'	σ_v''	σ_h	S_3	S_3^5
C_2'	C_2'	C_2''	C_2	σ_v'	C_3^2	E	C_3	σ_v''	σ_v	S_3^5	σ_h	S_3
C_2''	C_2''	C_2	C_2'	σ_v''	C_3	C_3^2	E	σ_v	σ_v'	S_3	S_3^5	σ_h
S_3	S_3	S_3^5	σ_h	C_3	σ_v''	σ_v	σ_v'	C_3^2	E	C_2''	C_2	C_2'
S_3^5	S_3^5	σ_h	S_3	C_3^2	σ_v'	σ_v''	σ_v	E	C_3	C_2'	C_2''	C_2
σ_v	σ_v	σ_v'	σ_v''	C_2	σ_h	S_3	S_3^5	C_2'	C_2''	E	C_3	C_3^2
σ_v'	σ_v'	σ_v''	σ_v	C_2'	S_3^5	σ_h	S_3	C_2''	C_2	C_3^2	E	C_3
σ_v''	σ_v''	σ_v	σ_v'	C_2''	S_3	S_3^5	σ_h	C_2	C_2'	C_3	C_3^2	E

(b) The class structure of D_{3h} is extensive. To delineate each class (except E which is always a class to itself) proceed as follows: For an element A, find all the triple products $B^{-1} \times A \times B$ where B is each of the other elements. For example,

$$C_3^{-1}C_3C_3 = C_3$$
$$(C_3^2)^{-1}C_3C_3^2 = C_3$$
$$\sigma_h^{-1}C_3\sigma_h = C_3$$

All elements appearing in this series of products belong to the same class. Thus, the first two classes are {E} and $\{C_3, C_3^2\}$. Proceeding in like fashion the rest of the classes can be found. The character table reveals that there are six irreducible representations in D_{3h}. Therefore, there must be six classes. Since the classes of a group exhaust the operations of the group, we can be satisfied when we have found the following classes: {E}; $\{C_3, C_3^2\}$; $\{\sigma_h\}$; $\{C_2, C_2', C_2''\}$; $\{S_3, S_3^5\}$; $\{\sigma_v, \sigma_v', \sigma_v''\}$.

(c) There are numerous subgroups of D_{3h}, that is, collections of two or more elements which themselves obey the group postulates; in fact, there are fourteen. You should be able to locate the following: D_3, C_{3h}, C_{3v}, C_3, three C_{2v}'s, three C_2's, and four C_s's. Note that the subgroups are not mutually exclusive, although the classes are.

(d) The Abelian subgroups are those in which group multiplication is commutative. What two sorts of groups does this include? All but five of the subgroups of D_{3h} are Abelian. Which are these?

4.12 No. Why?

4.13 (a) $(1 + 4 + 9 + 16) = (30)$

(b) $$\begin{pmatrix} 1 & 2 & 3 & 4 \\ 2 & 4 & 6 & 8 \\ 3 & 6 & 9 & 12 \\ 4 & 8 & 12 & 16 \end{pmatrix}$$

(c) $$\begin{pmatrix} 2+0+6 & 2+0+2 & 1+0+6 \\ 4+1+0 & 4+0+0 & 2+3+0 \\ 6+2+3 & 6+0+1 & 3+6+3 \end{pmatrix} = \begin{pmatrix} 8 & 4 & 7 \\ 5 & 4 & 5 \\ 11 & 7 & 12 \end{pmatrix}$$

4.14 We follow the steps below to construct $\mathbb{A}^{-1}$. The cofactor matrix is

$$\begin{pmatrix} \begin{vmatrix} 3 & 1 \\ 1 & 2 \end{vmatrix} & -\begin{vmatrix} 0 & 1 \\ 2 & 2 \end{vmatrix} & \begin{vmatrix} 0 & 3 \\ 2 & 1 \end{vmatrix} \\ -\begin{vmatrix} 6 & 2 \\ 1 & 2 \end{vmatrix} & \begin{vmatrix} 1 & 2 \\ 2 & 2 \end{vmatrix} & -\begin{vmatrix} 1 & 6 \\ 2 & 1 \end{vmatrix} \\ \begin{vmatrix} 6 & 2 \\ 3 & 1 \end{vmatrix} & -\begin{vmatrix} 1 & 2 \\ 0 & 1 \end{vmatrix} & \begin{vmatrix} 1 & 6 \\ 0 & 3 \end{vmatrix} \end{pmatrix} = \begin{pmatrix} 5 & 2 & -6 \\ -10 & -2 & 11 \\ 0 & -1 & 3 \end{pmatrix}$$

The determinant of $\mathbb{A}$ is

$$(1 \times 3 \times 2) + (6 \times 1 \times 2) + (2 \times 1 \times 0) - (2 \times 3 \times 2) - (1 \times 1 \times 1) - (2 \times 6 \times 0) = 6 + 12 + 0 - 12 - 1 - 0$$
$$= 5$$

Therefore, $\mathbb{A}^{\dagger} / |\mathbb{A}|$ is

$$\begin{pmatrix} 1 & 2/5 & -6/5 \\ -2 & -2/5 & 11/5 \\ 0 & -1/5 & 3/5 \end{pmatrix}$$

which is transposed to give $\mathbb{A}^{-1}$

$$\begin{pmatrix} 1 & -2 & 0 \\ 2/5 & -2/5 & -1/5 \\ -6/5 & 11/5 & 3/5 \end{pmatrix}$$

For the second part of the problem

$$\mathbb{A}\ \mathbb{A}^{-1} = \begin{pmatrix} 1 & 6 & 2 \\ 0 & 3 & 1 \\ 2 & 1 & 2 \end{pmatrix}\begin{pmatrix} 1 & -2 & 0 \\ 2/5 & -2/5 & -1/5 \\ -6/5 & 11/5 & 3/5 \end{pmatrix}$$

$$= \begin{pmatrix} 1 & 0 & 0 \\ 0 & 1 & 0 \\ 0 & 0 & 1 \end{pmatrix} = \mathbb{1}$$

4.15 Use the coordinate system shown below. Notice that the z vector, which is perpendicular to the plane of the page, is unaffected by the operations. The representation matrices are

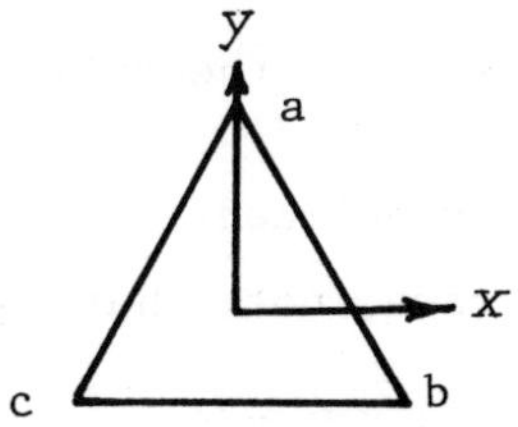

$$E\begin{pmatrix} x \\ y \\ z \end{pmatrix} = \begin{pmatrix} 1 & 0 & 0 \\ 0 & 1 & 0 \\ 0 & 0 & 1 \end{pmatrix}\begin{pmatrix} x \\ y \\ z \end{pmatrix}$$

$$\sigma_a\begin{pmatrix} x \\ y \\ z \end{pmatrix} = \begin{pmatrix} -1 & 0 & 0 \\ 0 & 1 & 0 \\ 0 & 0 & 1 \end{pmatrix}\begin{pmatrix} x \\ y \\ z \end{pmatrix}$$

$$\sigma_b\begin{pmatrix} x \\ y \\ z \end{pmatrix} = \begin{pmatrix} 1/2 & -\sqrt{3}/2 & 0 \\ -\sqrt{3}/2 & -1/2 & 0 \\ 0 & 0 & 1 \end{pmatrix}\begin{pmatrix} x \\ y \\ z \end{pmatrix}$$

$$\sigma_c\begin{pmatrix} x \\ y \\ z \end{pmatrix} = \begin{pmatrix} 1/2 & \sqrt{3}/2 & 0 \\ \sqrt{3}/2 & -1/2 & 0 \\ 0 & 0 & 1 \end{pmatrix}\begin{pmatrix} x \\ y \\ z \end{pmatrix}$$

$$C_3 \begin{pmatrix} x \\ y \\ z \end{pmatrix} = \begin{pmatrix} \cos 120^\circ & -\sin 120^\circ & 0 \\ \sin 120^\circ & \cos 120^\circ & 0 \\ 0 & 0 & 1 \end{pmatrix} \begin{pmatrix} x \\ y \\ z \end{pmatrix} =$$

$$\begin{pmatrix} -1/2 & -\sqrt{3}/2 & 0 \\ \sqrt{3}/2 & -1/2 & 0 \\ 0 & 0 & 1 \end{pmatrix} \begin{pmatrix} x \\ y \\ z \end{pmatrix}$$

$$C_3{}^2 \begin{pmatrix} x \\ y \\ z \end{pmatrix} = \begin{pmatrix} -1/2 & \sqrt{3}/2 & 0 \\ -\sqrt{3}/2 & -1/2 & 0 \\ 0 & 0 & 1 \end{pmatrix} \begin{pmatrix} x \\ y \\ z \end{pmatrix}$$

Now check to see if these matrices obey the rules which define a group:

(i) The identity element is defined.

(ii) As a representative case, note that $C_3C_3 = C_3{}^2$

$$\begin{pmatrix} -1/2 & -\sqrt{3}/2 & 0 \\ \sqrt{3}/2 & -1/2 & 0 \\ 0 & 0 & 1 \end{pmatrix} \begin{pmatrix} -1/2 & -\sqrt{3}/2 & 0 \\ \sqrt{3}/2 & -1/2 & 0 \\ 0 & 0 & 1 \end{pmatrix} =$$

$$\begin{pmatrix} -1/2 & \sqrt{3}/2 & 0 \\ -\sqrt{3}/2 & -1/2 & 0 \\ 0 & 0 & 1 \end{pmatrix}$$

That is, the square or product of two elements yields a member of the group.

(iii) Since σ_a is the reciprocal of σ_a, see if $\sigma_a\sigma_a = E$.

$$\begin{pmatrix} -1 & 0 & 0 \\ 0 & 1 & 0 \\ 0 & 0 & 1 \end{pmatrix}\begin{pmatrix} -1 & 0 & 0 \\ 0 & 1 & 0 \\ 0 & 0 & 1 \end{pmatrix} = \begin{pmatrix} 1 & 0 & 0 \\ 0 & 1 & 0 \\ 0 & 0 & 1 \end{pmatrix}$$

(iv) Finally check the associative law:

$(\sigma_a C_3)\sigma_b = \sigma_a(C_3\sigma_b)$

$$\left[\begin{pmatrix} -1 & 0 & 0 \\ 0 & 1 & 0 \\ 0 & 0 & 1 \end{pmatrix}\begin{pmatrix} -1/2 & -\sqrt{3}/2 & 0 \\ \sqrt{3}/2 & -1/2 & 0 \\ 0 & 0 & 1 \end{pmatrix}\right]\begin{pmatrix} 1/2 & -\sqrt{3}/2 & 0 \\ -\sqrt{3}/2 & -1/2 & 0 \\ 0 & 0 & 1 \end{pmatrix} \stackrel{?}{=}$$

$$\begin{pmatrix} -1 & 0 & 0 \\ 0 & 1 & 0 \\ 0 & 0 & 1 \end{pmatrix}\left[\begin{pmatrix} -1/2 & -\sqrt{3}/2 & 0 \\ \sqrt{3}/2 & -1/2 & 0 \\ 0 & 0 & 1 \end{pmatrix}\begin{pmatrix} 1/2 & -\sqrt{3}/2 & 0 \\ -\sqrt{3}/2 & -1/2 & 0 \\ 0 & 0 & 1 \end{pmatrix}\right]$$

$$\begin{pmatrix} 1/2 & \sqrt{3}/2 & 0 \\ \sqrt{3}/2 & -1/2 & 0 \\ 0 & 0 & 1 \end{pmatrix}\begin{pmatrix} 1/2 & -\sqrt{3}/2 & 0 \\ -\sqrt{3}/2 & -1/2 & 0 \\ 0 & 0 & 1 \end{pmatrix} \stackrel{?}{=}$$

$$\begin{pmatrix} -1 & 0 & 0 \\ 0 & 1 & 0 \\ 0 & 0 & 1 \end{pmatrix}\begin{pmatrix} 1/2 & \sqrt{3}/2 & 0 \\ \sqrt{3}/2 & -1/2 & 0 \\ 0 & 0 & 1 \end{pmatrix}$$

$$\begin{pmatrix} -1/2 & -\sqrt{3}/2 & 0 \\ \sqrt{3}/2 & -1/2 & 0 \\ 0 & 0 & 1 \end{pmatrix} = \begin{pmatrix} -1/2 & -\sqrt{3}/2 & 0 \\ \sqrt{3}/2 & -1/2 & 0 \\ 0 & 0 & 1 \end{pmatrix}$$

Representative cases obey the group postulates. The student may work through the entire group to see that there are no exceptions.

4.16 (a) The character of a matrix is the sum of the elements on the principle diagonal. Therefore,

$\chi(E) = 3$, $\chi(C_3) = 0$, $\chi(C_3{}^2) = 0$, $\chi(\sigma_a) = 1$, $\chi(\sigma_b) = 1$, $\chi(\sigma_c) = 1$.

Note that the characters of representation matrices belonging to the same class are equal (See 4.7); this permits the characters to be grouped under operations belonging to the same class.

C_{3v}	E	$2C_3$	$3\sigma_v$
$\Gamma_{x,y,z}$	3	0	1

(b) The representation is reducible. By means of the decomposition formula and the character table for the group C_{3v} we find

The number of A_1 irreducible representations in $\Gamma_{x,y,z} = (1/6)[(1)(1)(3) + (2)(1)(0) + (3)(1)(1)] = 1.$

The number of A_2 irreducible representations in $\Gamma_{x,y,z} = (1/6)[(1)(1)(3) + (2)(1)(0) + (3)(-1)(1)] = 0.$

The number of E irreducible representations in $\Gamma_{x,y,z} = (1/6)[(1)(2)(3) + (2)(-1)(0) + (3)(0)(1)] = 1.$

So, $\Gamma_{x,y,z} = A_1 + E$, *i.e.*, the reducible representation spanned by the basis set x,y,z transforms as the A_1 and E irreducible representations.

4.17 (a) $\Gamma_a = 2A_1 + E' + A_2''$, $\Gamma_b = A_2'' + E''$, $\Gamma_c = A_2' + E'$.

(b) $\Gamma_a = A_1 + B_2$, $\Gamma_b = A_1 + A_2 + B_1 + B_2$, $\Gamma_c = 2A_1 + B_1$.

4.18 (a) The p_z-orbital transforms as A_1; p_x, as B_1; and p_y, as B_2 in C_{2v}.

(b) The set of three p-orbitals forms a basis for the representation T_{1u} in O_h.

4.19 (a) In O_h, d_{z^2} alone is not a symmetry orbital; as a set d_{z^2} and $d_{x^2-y^2}$ form a basis for E_g.

(b) A_1.

4.20 (a) O_h, (d_{xz}, d_{yz}, d_{xy}) transform as T_{2g} and $(d_{z^2}, d_{x^2-y^2})$, as E_g.

(b) T_d, (d_{xz}, d_{yz}, d_{xy}) transform as T_2 and $(d_{z^2}, d_{x^2-y^2})$, as E.

(c) D_{4h}, d_{z^2} transforms as A_{1g}; $d_{x^2-y^2}$, as B_{1g}; d_{xy}, as B_{2g}; and (d_{xz}, d_{yz}), as E_g.

4.21 (a)

T_d	E	$8C_3$	$3C_2$	$6S_4$	$6\sigma_d$
T_1	3	0	-1	1	-1
T_2	3	0	-1	-1	1
$T_1 \times T_2$	9	0	1	-1	-1

The number of A_1 irreducible representations in $T_1 \times T_2 = (1/24)[1(9)(1) + 8(0)(1) + 3(1)(1) + 6(-1)(1) + 6(-1)(1)] = 0$.

The number of A_2 representations $= (1/24)[1(9)(1) + 8(0)(1) + 3(1)(1) + 6(-1)(-1) + 6(-1)(-1)] = 1$.

Finally, $T_2 \times T_1 = A_2 + E + T_1 + T_2$.

(b)

C_{2v}	E	C_2	$\sigma_v(xz)$	$\sigma_v(yz)$
A_1	1	1	1	1
B_1	1	-1	1	-1
$A_1 \times B_1$	1	-1	1	-1

Therefore, $A_1 \times B_1 = B_1$.

APPENDIX 4.1

CHARACTER TABLES

1. The Nonaxial Groups

C_1	E
A	1

C_s	E	σ_h		
A'	1	1	x, y, R_z	x^2, y^2, z^2, xy
A''	1	-1	z, R_x, R_y	yz, xz

C_i	E	i		
A_g	1	1	R_x, R_y, R_z	$x^2, y^2, z^2, xy, xz, yz$
A_u	1	-1	x, y, z	

2. The C_n Groups

C_2	E	C_2		
A	1	1	z, R_z	x^2, y^2, z^2, xy
B	1	-1	x, y, R_x, R_y	yz, xz

$\varepsilon = \exp(2\pi i/3)$

C_3	E	C_3	C_3^2		
A	1	1	1	z, R_z	$x^2 + y^2, z^2$
E	$\begin{Bmatrix} 1 \\ 1 \end{Bmatrix}$	$\begin{matrix} \varepsilon \\ \varepsilon^* \end{matrix}$	$\begin{matrix} \varepsilon^* \\ \varepsilon \end{matrix}$	$(x, y)(R_x, R_y)$	$(x^2 - y^2, xy)(yz, xz)$

C_4	E	C_4	C_2	C_4^3		
A	1	1	1	1	z, R_z	$x^2 + y^2, z^2$
B	1	-1	1	-1		$x^2 - y^2, xy$
E	$\begin{Bmatrix} 1 \\ 1 \end{Bmatrix}$	$\begin{matrix} i \\ -i \end{matrix}$	$\begin{matrix} -1 \\ -1 \end{matrix}$	$\begin{matrix} -i \\ i \end{matrix}$	$(x, y)(R_x, R_y)$	(yz, xz)

The C_n Groups *(continued)*

C_5	E	C_5	C_5^2	C_5^3	C_5^4		$\varepsilon = \exp(2\pi i/5)$
A	1	1	1	1	1	z, R_z	$x^2 + y^2, z^2$
E_1	$\begin{Bmatrix} 1 \\ 1 \end{Bmatrix}$	$\begin{matrix} \varepsilon \\ \varepsilon^* \end{matrix}$	$\begin{matrix} \varepsilon^2 \\ \varepsilon^{2*} \end{matrix}$	$\begin{matrix} \varepsilon^{2*} \\ \varepsilon^2 \end{matrix}$	$\begin{matrix} \varepsilon^* \\ \varepsilon \end{matrix}$	$(x, y)(R_x, R_y)$	(yz, xz)
E_2	$\begin{Bmatrix} 1 \\ 1 \end{Bmatrix}$	$\begin{matrix} \varepsilon^2 \\ \varepsilon^{2*} \end{matrix}$	$\begin{matrix} \varepsilon^* \\ \varepsilon \end{matrix}$	$\begin{matrix} \varepsilon \\ \varepsilon^* \end{matrix}$	$\begin{matrix} \varepsilon^{2*} \\ \varepsilon^2 \end{matrix}$		$(x^2 - y^2, xy)$

C_6	E	C_6	C_3	C_2	C_3^2	C_6^5		$\varepsilon = \exp(2\pi i/6)$
A	1	1	1	1	1	1	z, R_z	$x^2 + y^2, z^2$
B	1	-1	1	-1	1	-1		
E_1	$\begin{Bmatrix} 1 \\ 1 \end{Bmatrix}$	$\begin{matrix} \varepsilon \\ \varepsilon^* \end{matrix}$	$\begin{matrix} -\varepsilon^* \\ -\varepsilon \end{matrix}$	$\begin{matrix} -1 \\ -1 \end{matrix}$	$\begin{matrix} -\varepsilon \\ -\varepsilon^* \end{matrix}$	$\begin{matrix} \varepsilon^* \\ \varepsilon \end{matrix}$	(x, y) (R_x, R_y)	(xz, yz)
E_2	$\begin{Bmatrix} 1 \\ 1 \end{Bmatrix}$	$\begin{matrix} -\varepsilon^* \\ -\varepsilon \end{matrix}$	$\begin{matrix} -\varepsilon \\ -\varepsilon^* \end{matrix}$	$\begin{matrix} 1 \\ 1 \end{matrix}$	$\begin{matrix} -\varepsilon^* \\ -\varepsilon \end{matrix}$	$\begin{matrix} -\varepsilon \\ -\varepsilon^* \end{matrix}$		$(x^2 - y^2, xy)$

C_7	E	C_7	C_7^2	C_7^3	C_7^4	C_7^5	C_7^6		$\varepsilon = \exp(2\pi i/7)$
A	1	1	1	1	1	1	1	z, R_z	$x^2 + y^2, z^2$
E_1	$\begin{Bmatrix} 1 \\ 1 \end{Bmatrix}$	$\begin{matrix} \varepsilon \\ \varepsilon^* \end{matrix}$	$\begin{matrix} \varepsilon^2 \\ \varepsilon^{2*} \end{matrix}$	$\begin{matrix} \varepsilon^3 \\ \varepsilon^{3*} \end{matrix}$	$\begin{matrix} \varepsilon^{3*} \\ \varepsilon^3 \end{matrix}$	$\begin{matrix} \varepsilon^{2*} \\ \varepsilon^2 \end{matrix}$	$\begin{matrix} \varepsilon^* \\ \varepsilon \end{matrix}$	(x, y) (R_x, R_y)	(xz, yz)
E_2	$\begin{Bmatrix} 1 \\ 1 \end{Bmatrix}$	$\begin{matrix} \varepsilon^2 \\ \varepsilon^{2*} \end{matrix}$	$\begin{matrix} \varepsilon^{3*} \\ \varepsilon^3 \end{matrix}$	$\begin{matrix} \varepsilon^* \\ \varepsilon \end{matrix}$	$\begin{matrix} \varepsilon \\ \varepsilon^* \end{matrix}$	$\begin{matrix} \varepsilon^3 \\ \varepsilon^{3*} \end{matrix}$	$\begin{matrix} \varepsilon^{2*} \\ \varepsilon^2 \end{matrix}$		$(x^2 - y^2, xy)$
E_3	$\begin{Bmatrix} 1 \\ 1 \end{Bmatrix}$	$\begin{matrix} \varepsilon^3 \\ \varepsilon^{3*} \end{matrix}$	$\begin{matrix} \varepsilon^* \\ \varepsilon \end{matrix}$	$\begin{matrix} \varepsilon^2 \\ \varepsilon^{2*} \end{matrix}$	$\begin{matrix} \varepsilon^{2*} \\ \varepsilon^2 \end{matrix}$	$\begin{matrix} \varepsilon \\ \varepsilon^* \end{matrix}$	$\begin{matrix} \varepsilon^{3*} \\ \varepsilon^3 \end{matrix}$		

C_8	E	C_8	C_4	C_2	C_4^3	C_8^3	C_8^5	C_8^7		$\varepsilon = \exp(2\pi i/8)$
A	1	1	1	1	1	1	1	1	z, R_z	$x^2 + y^2, z^2$
B	1	-1	1	1	1	-1	-1	-1		
E_1	$\begin{Bmatrix} 1 \\ 1 \end{Bmatrix}$	$\begin{matrix} \varepsilon \\ \varepsilon^* \end{matrix}$	$\begin{matrix} i \\ -i \end{matrix}$	$\begin{matrix} -1 \\ -1 \end{matrix}$	$\begin{matrix} -i \\ i \end{matrix}$	$\begin{matrix} -\varepsilon^* \\ -\varepsilon \end{matrix}$	$\begin{matrix} -\varepsilon \\ -\varepsilon^* \end{matrix}$	$\begin{matrix} \varepsilon^* \\ \varepsilon \end{matrix}$	(x, y) (R_x, R_y)	(xz, yz)
E_2	$\begin{Bmatrix} 1 \\ 1 \end{Bmatrix}$	$\begin{matrix} i \\ -i \end{matrix}$	$\begin{matrix} -1 \\ -1 \end{matrix}$	$\begin{matrix} 1 \\ 1 \end{matrix}$	$\begin{matrix} -1 \\ -1 \end{matrix}$	$\begin{matrix} -i \\ i \end{matrix}$	$\begin{matrix} i \\ -i \end{matrix}$	$\begin{matrix} -i \\ i \end{matrix}$		$(x^2 - y^2, xy)$
E_3	$\begin{Bmatrix} 1 \\ 1 \end{Bmatrix}$	$\begin{matrix} -\varepsilon \\ -\varepsilon^* \end{matrix}$	$\begin{matrix} i \\ -i \end{matrix}$	$\begin{matrix} -1 \\ -1 \end{matrix}$	$\begin{matrix} -i \\ i \end{matrix}$	$\begin{matrix} \varepsilon^* \\ \varepsilon \end{matrix}$	$\begin{matrix} \varepsilon \\ \varepsilon^* \end{matrix}$	$\begin{matrix} -\varepsilon^* \\ -\varepsilon \end{matrix}$		

3. The D_n Groups

D_2	E	$C_2(z)$	$C_2(y)$	$C_2(x)$		
A	1	1	1	1		x^2, y^2, z^2
B_1	1	1	-1	-1	z, R_z	xy
B_2	1	-1	1	-1	y, R_y	xz
B_3	1	-1	-1	1	x, R_x	yz

D_3	E	$2C_3$	$3C_2$		
A_1	1	1	1		$x^2 + y^2, z^2$
A_2	1	1	-1	z, R_z	
E	2	-1	0	$(x, y)(R_x, R_y)$	$(x^2 - y^2, xy)(xz, yz)$

D_4	E	$2C_4$	$C_2(=C_4^2)$	$2C_2'$	$2C_2''$		
A_1	1	1	1	1	1		$x^2 + y^2, z^2$
A_2	1	1	1	-1	-1	z, R_z	
B_1	1	-1	1	1	-1		$x^2 - y^2$
B_2	1	-1	1	-1	1		xy
E	2	0	-2	0	0	$(x, y)(R_x, R_y)$	(xz, yz)

D_5	E	$2C_5$	$2C_5^2$	$5C_2$		
A_1	1	1	1	1		$x^2 + y^2, z^2$
A_2	1	1	1	-1	z, R_z	
E_1	2	2 cos 72°	2 cos 144°	0	$(x, y)(R_x, R_y)$	(xz, yz)
E_2	2	2 cos 144°	2 cos 72°	0		$(x^2 - y^2, xy)$

D_6	E	$2C_6$	$2C_3$	C_2	$3C_2'$	$3C_2''$		
A_1	1	1	1	1	1	1		$x^2 + y^2, z^2$
A_2	1	1	1	1	-1	-1	z, R_z	
B_1	1	-1	1	-1	1	-1		
B_2	1	-1	1	-1	-1	1		
E_1	2	1	-1	-2	0	0	$(x, y)(R_x, R_y)$	(xz, yz)
E_2	2	-1	-1	2	0	0		$(x^2 - y^2, xy)$

4. The C_{nv} Groups

C_{2v}	E	C_2	$\sigma_v(xz)$	$\sigma_v'(yz)$		
A_1	1	1	1	1	z	x^2, y^2, z^2
A_2	1	1	-1	-1	R_z	xy
B_1	1	-1	1	-1	x, R_y	xz
B_2	1	-1	-1	1	y, R_x	yz

C_{3v}	E	$2C_3$	$3\sigma_v$		
A_1	1	1	1	z	$x^2 + y^2, z^2$
A_2	1	1	-1	R_z	
E	2	-1	0	$(x, y)(R_x, R_y)$	$(x^2 - y^2, xy)(xz, yz)$

C_{4v}	E	$2C_4$	C_2	$2\sigma_v$	$2\sigma_d$		
A_1	1	1	1	1	1	z	$x^2 + y^2, z^2$
A_2	1	1	1	-1	-1	R_z	
B_1	1	-1	1	1	-1		$x^2 - y^2$
B_2	1	-1	1	-1	1		xy
E	2	0	-2	0	0	$(x, y)(R_x, R_y)$	(xz, yz)

C_{5v}	E	$2C_5$	$2C_5^2$	$5\sigma_v$		
A_1	1	1	1	1	z	$x^2 + y^2, z^2$
A_2	1	1	1	-1	R_z	
E_1	2	2 cos 72°	2 cos 144°	0	$(x, y)(R_x, R_y)$	(xz, yz)
E_2	2	2 cos 144°	2 cos 72°	0		$(x^2 - y^2, xy)$

C_{6v}	E	$2C_6$	$2C_3$	C_2	$3\sigma_v$	$3\sigma_d$		
A_1	1	1	1	1	1	1	z	$x^2 + y^2, z^2$
A_2	1	1	1	1	-1	-1	R_z	
B_1	1	-1	1	-1	1	-1		
B_2	1	-1	1	-1	-1	1		
E_1	2	1	-1	-2	0	0	$(x, y)(R_x, R_y)$	(xz, yz)
E_2	2	-1	-1	2	0	0		$(x^2 - y^2, xy)$

5. The C_{nh} Groups

C_{2h}	E	C_2	i	σ_h		
A_g	1	1	1	1	R_z	x^2, y^2, z^2, xy
B_g	1	-1	1	-1	R_x, R_y	xz, yz
A_u	1	1	-1	-1	z	
B_u	1	-1	-1	1	x, y	

$\varepsilon = \exp(2\pi i/3)$

C_{3h}	E	C_3	C_3^2	σ_h	S_3	S_3^5		
A'	1	1	1	1	1	1	R_z	$x^2 + y^2, z^2$
E'	$\begin{Bmatrix}1 \\ 1\end{Bmatrix}$	$\begin{matrix}\varepsilon \\ \varepsilon^*\end{matrix}$	$\begin{matrix}\varepsilon^* \\ \varepsilon\end{matrix}$	$\begin{matrix}1 \\ 1\end{matrix}$	$\begin{matrix}\varepsilon \\ \varepsilon^*\end{matrix}$	$\begin{Bmatrix}\varepsilon^* \\ \varepsilon\end{Bmatrix}$	(x, y)	$(x^2 - y^2, xy)$
A''	1	1	1	-1	-1	-1	z	
E''	$\begin{Bmatrix}1 \\ 1\end{Bmatrix}$	$\begin{matrix}\varepsilon \\ \varepsilon^*\end{matrix}$	$\begin{matrix}\varepsilon^* \\ \varepsilon\end{matrix}$	$\begin{matrix}-1 \\ -1\end{matrix}$	$\begin{matrix}-\varepsilon \\ -\varepsilon^*\end{matrix}$	$\begin{Bmatrix}-\varepsilon^* \\ -\varepsilon\end{Bmatrix}$	(R_x, R_y)	(xz, yz)

C_{4h}	E	C_4	C_2	C_4^3	i	S_4^3	σ_h	S_4		
A_g	1	1	1	1	1	1	1	1	R_z	$x^2 + y^2, z^2$
B_g	1	-1	1	-1	1	-1	1	-1		$x^2 - y^2, xy$
E_g	$\begin{Bmatrix}1 \\ 1\end{Bmatrix}$	$\begin{matrix}i \\ -i\end{matrix}$	$\begin{matrix}-1 \\ -1\end{matrix}$	$\begin{matrix}-i \\ i\end{matrix}$	$\begin{matrix}1 \\ 1\end{matrix}$	$\begin{matrix}i \\ -i\end{matrix}$	$\begin{matrix}-1 \\ -1\end{matrix}$	$\begin{Bmatrix}-i \\ i\end{Bmatrix}$	(R_x, R_y)	(xz, yz)
A_u	1	1	1	1	-1	-1	-1	-1	z	
B_u	1	-1	1	-1	-1	1	-1	1		
E_u	$\begin{Bmatrix}1 \\ 1\end{Bmatrix}$	$\begin{matrix}i \\ -i\end{matrix}$	$\begin{matrix}-1 \\ -1\end{matrix}$	$\begin{matrix}-i \\ i\end{matrix}$	$\begin{matrix}-1 \\ -1\end{matrix}$	$\begin{matrix}-i \\ i\end{matrix}$	$\begin{matrix}1 \\ 1\end{matrix}$	$\begin{Bmatrix}i \\ -i\end{Bmatrix}$	(x, y)	

C_{5h}	E	C_5	C_5^2	C_5^3	C_5^4	σ_h	S_5	S_5^7	S_5^3	S_5^9		$\varepsilon = \exp(2\pi i/5)$
A'	1	1	1	1	1	1	1	1	1	1	R_z	$x^2 + y^2, z^2$
E_1'	1	ε	ε^2	ε^{2*}	ε^*	1	ε	ε^2	ε^{2*}	ε^*	(x, y)	
	1	ε^*	ε^{2*}	ε^2	ε	1	ε^*	ε^{2*}	ε^2	ε		
E_2'	1	ε^2	ε^*	ε	ε^{2*}	1	ε^2	ε^*	ε	ε^{2*}		$(x^2 - y^2, xy)$
	1	ε^{2*}	ε	ε^*	ε^2	1	ε^{2*}	ε	ε^*	ε^2		
A''	1	1	1	1	1	-1	-1	-1	-1	-1	z	
E_1''	1	ε	ε^2	ε^{2*}	ε^*	-1	$-\varepsilon$	$-\varepsilon^2$	$-\varepsilon^{2*}$	$-\varepsilon^*$	(R_x, R_y)	(xz, yz)
	1	ε^*	ε^{2*}	ε^2	ε	-1	$-\varepsilon^*$	$-\varepsilon^{2*}$	$-\varepsilon^2$	$-\varepsilon$		
E_2''	1	ε^2	ε^*	ε	ε^{2*}	-1	$-\varepsilon^2$	$-\varepsilon^*$	$-\varepsilon$	$-\varepsilon^{2*}$		
	1	ε^{2*}	ε	ε^*	ε^2	-1	$-\varepsilon^{2*}$	$-\varepsilon$	$-\varepsilon^*$	$-\varepsilon^2$		

C_{6h}	E	C_6	C_3	C_2	C_3^2	C_6^5	i	S_3^5	S_6^5	σ_h	S_6	S_3		$\varepsilon = \exp(2\pi i/6)$
A_g	1	1	1	1	1	1	1	1	1	1	1	1	R_z	$x^2 + y^2, z^2$
B_g	1	-1	1	-1	1	-1	1	-1	1	-1	1	-1		
E_{1g}	1	ε	$-\varepsilon^*$	-1	$-\varepsilon$	ε^*	1	ε	$-\varepsilon^*$	-1	$-\varepsilon$	ε^*	(R_x, R_y)	(xz, yz)
	1	ε^*	$-\varepsilon$	-1	$-\varepsilon^*$	ε	1	ε^*	$-\varepsilon$	-1	$-\varepsilon^*$	ε		
E_{2g}	1	$-\varepsilon^*$	$-\varepsilon$	1	$-\varepsilon^*$	$-\varepsilon$	1	$-\varepsilon^*$	$-\varepsilon$	1	$-\varepsilon^*$	$-\varepsilon$		$(x^2 - y^2, xy)$
	1	$-\varepsilon$	$-\varepsilon^*$	1	$-\varepsilon$	$-\varepsilon^*$	1	$-\varepsilon$	$-\varepsilon^*$	1	$-\varepsilon$	$-\varepsilon^*$		
A_u	1	1	1	1	1	1	-1	-1	-1	-1	-1	-1	z	
B_u	1	-1	1	-1	1	-1	-1	1	-1	1	-1	1		
E_{1u}	1	ε	$-\varepsilon^*$	-1	$-\varepsilon$	ε^*	-1	$-\varepsilon$	ε^*	1	ε	$-\varepsilon^*$	(x, y)	
	1	ε^*	$-\varepsilon$	-1	$-\varepsilon^*$	ε	-1	$-\varepsilon^*$	ε	1	ε^*	$-\varepsilon$		
E_{2u}	1	$-\varepsilon^*$	$-\varepsilon$	1	$-\varepsilon^*$	$-\varepsilon$	-1	ε^*	ε	-1	ε^*	ε		
	1	$-\varepsilon$	$-\varepsilon^*$	1	$-\varepsilon$	$-\varepsilon^*$	-1	ε	ε^*	-1	ε	ε^*		

6. The D_{nh} Groups

D_{2h}	E	$C_2(z)$	$C_2(y)$	$C_2(x)$	i	$\sigma(xy)$	$\sigma(xz)$	$\sigma(yz)$		
A_g	1	1	1	1	1	1	1	1		x^2, y^2, z^2
B_{1g}	1	1	−1	−1	1	1	−1	−1	R_z	xy
B_{2g}	1	−1	1	−1	1	−1	1	−1	R_y	xz
B_{3g}	1	−1	−1	1	1	−1	−1	1	R_x	yz
A_u	1	1	1	1	−1	−1	−1	−1		
B_{1u}	1	1	−1	−1	−1	−1	1	1	z	
B_{2u}	1	−1	1	−1	−1	1	−1	1	y	
B_{3u}	1	−1	−1	1	−1	1	1	−1	x	

D_{3h}	E	$2C_3$	$3C_2$	σ_h	$2S_3$	$3\sigma_v$		
A_1'	1	1	1	1	1	1		$x^2 + y^2, z^2$
A_2'	1	1	−1	1	1	−1	R_z	
E'	2	−1	0	2	−1	0	(x, y)	$(x^2 - y^2, xy)$
A_1''	1	1	1	−1	−1	−1		
A_2''	1	1	−1	−1	−1	1	z	
E''	2	−1	0	−2	1	0	(R_x, R_y)	(xz, yz)

D_{4h}	E	$2C_4$	C_2	$2C_2'$	$2C_2''$	i	$2S_4$	σ_h	$2\sigma_v$	$2\sigma_d$		
A_{1g}	1	1	1	1	1	1	1	1	1	1		$x^2 + y^2, z^2$
A_{2g}	1	1	1	−1	−1	1	1	1	−1	−1	R_z	
B_{1g}	1	−1	1	1	−1	1	−1	1	1	−1		$x^2 - y^2$
B_{2g}	1	−1	1	−1	1	1	−1	1	−1	1		xy
E_g	2	0	−2	0	0	2	0	−2	0	0	(R_x, R_y)	(xz, yz)
A_{1u}	1	1	1	1	1	−1	−1	−1	−1	−1		
A_{2u}	1	1	1	−1	−1	−1	−1	−1	1	1	z	
B_{1u}	1	−1	1	1	−1	−1	1	−1	−1	1		
B_{2u}	1	−1	1	−1	1	−1	1	−1	1	−1		
E_u	2	0	−2	0	0	−2	0	2	0	0	(x, y)	

D_{5h}	E	$2C_5$	$2C_5^2$	$5C_2$	σ_h	$2S_5$	$2S_5^3$	$5\sigma_v$		
A_1'	1	1	1	1	1	1	1	1		$x^2 + y^2, z^2$
A_2'	1	1	1	-1	1	1	1	-1	R_z	
E_1'	2	2 cos 72°	2 cos 144°	0	2	2 cos 72°	2 cos 144°	0	(x, y)	
E_2'	2	2 cos 144°	2 cos 72°	0	2	2 cos 144°	2 cos 72°	0		$(x^2 - y^2, xy)$
A_1''	1	1	1	1	-1	-1	-1	-1		
A_2''	1	1	1	-1	-1	-1	-1	1	z	
E_1''	2	2 cos 72°	2 cos 144°	0	-2	-2 cos 72°	-2 cos 144°	0	(R_x, R_y)	(xz, yz)
E_2''	2	2 cos 144°	2 cos 72°	0	-2	-2 cos 144°	-2 cos 72°	0		

D_{6h}	E	$2C_6$	$2C_3$	C_2	$3C_2'$	$3C_2''$	i	$2S_3$	$2S_6$	σ_h	$3\sigma_d$	$3\sigma_v$		
A_{1g}	1	1	1	1	1	1	1	1	1	1	1	1		$x^2 + y^2, z^2$
A_{2g}	1	1	1	1	-1	-1	1	1	1	1	-1	-1	R_z	
B_{1g}	1	-1	1	-1	1	-1	1	-1	1	-1	1	-1		
B_{2g}	1	-1	1	-1	-1	1	1	-1	1	-1	-1	1		
E_{1g}	2	1	-1	-2	0	0	2	1	-1	-2	0	0	(R_x, R_y)	(xz, yz)
E_{2g}	2	-1	-1	2	0	0	2	-1	-1	2	0	0		$(x^2 - y^2, xy)$
A_{1u}	1	1	1	1	1	1	-1	-1	-1	-1	-1	-1		
A_{2u}	1	1	1	1	-1	-1	-1	-1	-1	-1	1	1	z	
B_{1u}	1	-1	1	-1	1	-1	-1	1	-1	1	-1	1		
B_{2u}	1	-1	1	-1	-1	1	-1	1	-1	1	1	-1		
E_{1u}	2	1	-1	-2	0	0	-2	-1	1	2	0	0	(x, y)	
E_{2u}	2	-1	-1	2	0	0	-2	1	1	-2	0	0		

7. The D_{nd} Groups

D_{2d}	E	$2S_4$	C_2	$2C_2'$	$2\sigma_d$		
A_1	1	1	1	1	1		x^2+y^2, z^2
A_2	1	1	1	-1	-1	R_z	
B_1	1	-1	1	1	-1		x^2-y^2
B_2	1	-1	1	-1	1	z	xy
E	2	0	-2	0	0	(x,y); (R_x,R_y)	(xz,yz)

D_{3d}	E	$2C_3$	$3C_2$	i	$2S_6$	$3\sigma_d$		
A_{1g}	1	1	1	1	1	1		x^2+y^2, z^2
A_{2g}	1	1	-1	1	1	-1	R_z	
E_g	2	-1	0	2	-1	0	(R_x,R_y)	$(x^2-y^2, xy), (xz,yz)$
A_{1u}	1	1	1	-1	-1	-1		
A_{2u}	1	1	-1	-1	-1	1	z	
E_u	2	-1	0	-2	1	0	(x,y)	

D_{4d}	E	$2S_8$	$2C_4$	$2S_8^3$	C_2	$4C_2'$	$4\sigma_d$		
A_1	1	1	1	1	1	1	1		x^2+y^2, z^2
A_2	1	1	1	1	1	-1	-1	R_z	
B_1	1	-1	1	-1	1	1	-1		
B_2	1	-1	1	-1	1	-1	1	z	
E_1	2	$\sqrt{2}$	0	$-\sqrt{2}$	-2	0	0	(x,y)	
E_2	2	0	-2	0	2	0	0		(x^2-y^2, xy)
E_3	2	$-\sqrt{2}$	0	$\sqrt{2}$	-2	0	0	(R_x,R_y)	(xz,yz)

D_{5d}	E	$2C_5$	$2C_5^2$	$5C_2$	i	$2S_{10}^3$	$2S_{10}$	$5\sigma_d$		
A_{1g}	1	1	1	1	1	1	1	1		$x^2 + y^2, z^2$
A_{2g}	1	1	1	-1	1	1	1	-1	R_z	
E_{1g}	2	2 cos 72°	2 cos 144°	0	2	2 cos 72°	2 cos 144°	0	(R_x, R_y)	(xz, yz)
E_{2g}	2	2 cos 144°	2 cos 72°	0	2	2 cos 144°	2 cos 72°	0		$(x^2 - y^2, xy)$
A_{1u}	1	1	1	1	-1	-1	-1	-1		
A_{2u}	1	1	1	-1	-1	-1	-1	1	z	
E_{1u}	2	2 cos 72°	2 cos 144°	0	-2	-2 cos 72°	-2 cos 144°	0	(x, y)	
E_{2u}	2	2 cos 144°	2 cos 72°	0	-2	-2 cos 144°	-2 cos 72°	0		

D_{6d}	E	$2S_{12}$	$2C_6$	$2S_4$	$2C_3$	$2S_{12}^5$	C_2	$6C_2'$	$6\sigma_d$		
A_1	1	1	1	1	1	1	1	1	1		$x^2 + y^2, z^2$
A_2	1	1	1	1	1	1	1	-1	-1	R_z	
B_1	1	-1	1	-1	1	-1	1	1	-1		
B_2	1	-1	1	-1	1	-1	1	-1	1	z	
E_1	2	$\sqrt{3}$	1	0	-1	$-\sqrt{3}$	-2	0	0	(x, y)	
E_2	2	1	-1	-2	-1	1	2	0	0		$(x^2 - y^2, xy)$
E_3	2	0	-2	0	2	0	-2	0	0		
E_4	2	-1	-1	2	-1	-1	2	0	0		
E_5	2	$-\sqrt{3}$	1	0	-1	$\sqrt{3}$	-2	0	0	(R_x, R_y)	(xz, yz)

8. The S_n Groups

S_4	E	S_4	C_2	S_4^3		
A	1	1	1	1	R_z	$x^2 + y^2, z^2$
B	1	-1	1	-1	z	$x^2 - y^2, xy$
E	$\begin{Bmatrix} 1 \\ 1 \end{Bmatrix}$	$\begin{matrix} i \\ -i \end{matrix}$	$\begin{matrix} -1 \\ -1 \end{matrix}$	$\begin{Bmatrix} -i \\ i \end{Bmatrix}$	$(x, y); (R_x, R_y)$	(xz, yz)

S_6	E	C_3	C_3^2	i	S_6^5	S_6		$\varepsilon = \exp(2\pi i/3)$
A_g	1	1	1	1	1	1	R_z	$x^2 + y^2, z^2$
E_g	$\begin{Bmatrix} 1 \\ 1 \end{Bmatrix}$	$\begin{matrix} \varepsilon \\ \varepsilon^* \end{matrix}$	$\begin{matrix} \varepsilon^* \\ \varepsilon \end{matrix}$	$\begin{matrix} 1 \\ 1 \end{matrix}$	$\begin{matrix} \varepsilon \\ \varepsilon^* \end{matrix}$	$\begin{Bmatrix} \varepsilon^* \\ \varepsilon \end{Bmatrix}$	(R_x, R_y)	$(x^2 - y^2, xy); (xz, yz)$
A_u	1	1	1	-1	-1	-1	z	
E_u	$\begin{Bmatrix} 1 \\ 1 \end{Bmatrix}$	$\begin{matrix} \varepsilon \\ \varepsilon^* \end{matrix}$	$\begin{matrix} \varepsilon^* \\ \varepsilon \end{matrix}$	$\begin{matrix} -1 \\ -1 \end{matrix}$	$\begin{matrix} -\varepsilon \\ -\varepsilon^* \end{matrix}$	$\begin{Bmatrix} -\varepsilon^* \\ -\varepsilon \end{Bmatrix}$	(x, y)	

S_8	E	S_8	C_4	S_8^3	C_2	S_8^5	C_4^3	S_8^7		$\varepsilon = \exp(2\pi i/8)$
A	1	1	1	1	1	1	1	1	R_z	$x^2 + y^2, z^2$
B	1	-1	1	-1	1	-1	1	-1	z	
E_1	$\begin{Bmatrix} 1 \\ 1 \end{Bmatrix}$	$\begin{matrix} \varepsilon \\ \varepsilon^* \end{matrix}$	$\begin{matrix} i \\ -i \end{matrix}$	$\begin{matrix} -\varepsilon^* \\ -\varepsilon \end{matrix}$	$\begin{matrix} -1 \\ -1 \end{matrix}$	$\begin{matrix} -\varepsilon \\ -\varepsilon^* \end{matrix}$	$\begin{matrix} -i \\ i \end{matrix}$	$\begin{Bmatrix} \varepsilon^* \\ \varepsilon \end{Bmatrix}$	$(x, y); (R_x, R_y)$	
E_2	$\begin{Bmatrix} 1 \\ 1 \end{Bmatrix}$	$\begin{matrix} i \\ -i \end{matrix}$	$\begin{matrix} -1 \\ -1 \end{matrix}$	$\begin{matrix} -i \\ i \end{matrix}$	$\begin{matrix} 1 \\ 1 \end{matrix}$	$\begin{matrix} i \\ -i \end{matrix}$	$\begin{matrix} -1 \\ -1 \end{matrix}$	$\begin{Bmatrix} -i \\ i \end{Bmatrix}$		$(x^2 - y^2, xy)$
E_3	$\begin{Bmatrix} 1 \\ 1 \end{Bmatrix}$	$\begin{matrix} -\varepsilon^* \\ -\varepsilon \end{matrix}$	$\begin{matrix} -i \\ i \end{matrix}$	$\begin{matrix} \varepsilon \\ \varepsilon^* \end{matrix}$	$\begin{matrix} -1 \\ -1 \end{matrix}$	$\begin{matrix} \varepsilon^* \\ \varepsilon \end{matrix}$	$\begin{matrix} i \\ -i \end{matrix}$	$\begin{Bmatrix} -\varepsilon \\ -\varepsilon^* \end{Bmatrix}$		(xz, yz)

9. The Cubic Groups

T_d	E	$8C_3$	$3C_2$	$6S_4$	$6\sigma_d$		
A_1	1	1	1	1	1		$x^2 + y^2 + z^2$
A_2	1	1	1	-1	-1		
E	2	-1	2	0	0		$(2z^2 - x^2 - y^2, x^2 - y^2)$
T_1	3	0	-1	1	-1	(R_x, R_y, R_z)	
T_2	3	0	-1	-1	1	(x, y, z)	(xy, xz, yz)

O_h	E	$8C_3$	$6C_2$	$6C_4$	$3C_2(=C_4^2)$	i	$6S_4$	$8S_6$	$3\sigma_h$	$6\sigma_d$		
A_{1g}	1	1	1	1	1	1	1	1	1	1		$x^2 + y^2 + z^2$
A_{2g}	1	1	-1	-1	1	1	-1	1	1	-1		
E_g	2	-1	0	0	2	2	0	-1	2	0		$(2z^2 - x^2 - y^2, x^2 - y^2)$
T_{1g}	3	0	-1	1	-1	3	1	0	-1	-1	(R_x, R_y, R_z)	
T_{2g}	3	0	1	-1	-1	3	-1	0	-1	1		(xz, yz, xy)
A_{1u}	1	1	1	1	1	-1	-1	-1	-1	-1		
A_{2u}	1	1	-1	-1	1	-1	1	-1	-1	1		
E_u	2	-1	0	0	2	-2	0	1	-2	0		
T_{1u}	3	0	-1	1	-1	-3	-1	0	1	1	(x, y, z)	
T_{2u}	3	0	1	-1	-1	-3	1	0	1	-1		

10. The Groups $C_{\infty v}$ and $D_{\infty h}$ for Linear Molecules

$C_{\infty v}$	E	$2C_\infty^\Phi$	...	$\infty\sigma_v$		
$A_1 \equiv \Sigma^+$	1	1	...	1	z	$x^2 + y^2, z^2$
$A_2 \equiv \Sigma^-$	1	1	...	-1	R_z	
$E_1 \equiv \Pi$	2	$2\cos\Phi$	...	0	$(x, y); (R_x, R_y)$	(xz, yz)
$E_2 \equiv \Delta$	2	$2\cos 2\Phi$	...	0		$(x^2 - y^2, xy)$
$E_3 \equiv \Phi$	2	$2\cos 3\Phi$	...	0		
...	...	...	...	...		

$D_{\infty h}$	E	$2C_{\infty}^{\Phi}$	...	$\infty\sigma_v$	i	$2S_{\infty}^{\Phi}$	...	∞C_2		
Σ_g^+	1	1	...	1	1	1	...	1		$x^2 + y^2, z^2$
Σ_g^-	1	1	...	−1	1	1	...	−1	R_z	
Π_g	2	2 cos Φ	...	0	2	−2 cos Φ	...	0	(R_x, R_y)	(xz, yz)
Δ_g	2	2 cos 2Φ	...	0	2	2 cos 2Φ	...	0		$(x^2 - y^2, xy)$
...	...	...	...	...	...	...	...	...		
Σ_u^+	1	1	...	1	−1	−1	...	−1	z	
Σ_u^-	1	1	...	−1	−1	−1	...	1		
Π_u	2	2 cos Φ	...	0	−2	2 cos Φ	...	0	(x, y)	
Δ_u	2	2 cos 2Φ	...	0	−2	−2 cos 2Φ	...	0		
...	...	...	...	...	...	...	...	...		

11. The Icosahedral Group

I_h	E	$12C_5$	$12C_5^2$	$20C_3$	$15C_2$	i	$12S_{10}$	$12S_{10}^3$	$20S_6$	15σ		
A_g	1	1	1	1	1	1	1	1	1	1		$x^2+y^2+z^2$
T_{1g}	3	$1/2(1+\sqrt{5})$	$1/2(1-\sqrt{5})$	0	-1	3	$1/2(1-\sqrt{5})$	$1/2(1+\sqrt{5})$	0	-1	(R_x, R_y, R_z)	
T_{2g}	3	$1/2(1-\sqrt{5})$	$1/2(1+\sqrt{5})$	0	-1	3	$1/2(1+\sqrt{5})$	$1/2(1-\sqrt{5})$	0	-1		
G_g	4	-1	-1	1	0	4	-1	-1	1	0		
H_g	5	0	0	-1	1	5	0	0	-1	1		$(2z^2-x^2-y^2, x^2-y^2, xy, yz, zx)$
A_u	1	1	1	1	1	-1	-1	-1	-1	-1		
T_{1u}	3	$1/2(1+\sqrt{5})$	$1/2(1-\sqrt{5})$	0	-1	-3	$-1/2(1-\sqrt{5})$	$-1/2(1+\sqrt{5})$	0	1	(x, y, z)	
T_{2u}	3	$1/2(1-\sqrt{5})$	$1/2(1+\sqrt{5})$	0	-1	-3	$-1/2(1+\sqrt{5})$	$-1/2(1-\sqrt{5})$	0	1		
G_u	4	-1	-1	1	0	-4	1	1	-1	0		
H_u	5	0	0	-1	1	-5	0	0	1	-1		

APPENDIX 4.2

RULES FOR THE EVALUATION OF DIRECT PRODUCTS IN SYMMETRY GROUPS

$A \times A = A$	$B \times A = B$	$E \times A = E$	$T \times A = T$
$A \times B = B$	$B \times B = A$	$E \times B = E$	$T \times B = T$
$A \times E = E$	$B \times E = E$	$E \times E = *$	$T \times E = T_1 + T_2$
$A \times T = T$	$B \times T = T$	$E \times T = T_1 + T_2$	$T \times T = *$
	primes	subscripts	
$g \times g = g$	$' \times ' = '$	$1 \times 1 = 1$	
$g \times u = u$	$' \times '' = ''$	$1 \times 2 = 2$	
$u \times g = u$	$'' \times ' = ''$	$2 \times 1 = 2$	
$u \times u = g$	$'' \times '' = '$	$2 \times 2 = 1$	

* For E.

In some Groups, *e.g.* O, T_d, C_{3v} and D_6:

$$E_1 \times E_1 = E_2 \times E_2 = A_1 + A_2 + E_2$$

$$E_1 \times E_2 = E_2 \times E_1 = B_1 + B_2 + E_1$$

(If there are no subscripts to one of A, B or E, then $A_1 = A_2 = A$, *etc.*)

In C_{4v} and D_4 *etc.*

$$E \times E = A_1 + A_2 + B_1 + B_2.$$

For T.

$$T_1 \times T_1 = T_2 \times T_2 = A_1 + E + T_1 + T_2$$
$$T_1 \times T_2 = T_2 \times T_1 = A_2 + E + T_1 + T_2$$

Examples.

In O_h: $E_g \times T_{1u} = T_{1u} + T_{2u}$, $A_{2u} \times T_{2u} = T_{1g}$,
$E_g \times E_g = A_{1g} + A_{2g} + E_g$.

In D_{4h}: $E_g \times B_{2g} = E_g$, $E_g \times E_u = A_{1u} + A_{2u} + B_{1u} + B_{2u}$.

In C_{3v}: $A_1 \times A_2 = A_2$, $A_2 \times E = E$, $E \times E = A_1 + A_2 + E$.

In D_{3h}: $E' \times A_2' = E'$, $E'' \times E' = A_1'' + A_2'' + E''$.

The complete results for O, are:

O	A_1	A_2	E	T_1	T_2
A_1	A_1	A_2	E	T_1	T_2
A_2	A_2	A_1	E	T_2	T_1
E	E	E	$A_1 + A_2 + E$	$T_1 + T_2$	$T_1 + T_2$
T_1	T_1	T_2	$T_1 + T_2$	$A_1 + E + T_1 + T_2$	$A_2 + E + T_1 + T_2$
T_2	T_2	T_1	$T_1 + T_2$	$A_2 + E + T_1 + T_2$	$A_1 + E + T_1 + T_2$

APPENDIX 4.3

CORRELATION TABLES FOR SELECTED POINT GROUPS

D_{4h}	D_4	$C_2' \to C_2'$ D_{2d}	$C_2'' \to C_2'$ D_{2d}	C_{4v}	C_{4h}	C_2' D_{2h}	C_2'' D_{2h}	C_4	S_4
A_{1g}	A_1	A_1	A_1	A_1	A_g	A_g	A_g	A	A
A_{2g}	A_2	A_2	A_2	A_2	A_g	B_{1g}	B_{1g}	A	A
B_{1g}	B_1	B_1	B_2	B_1	B_g	A_g	B_{1g}	B	B
B_{2g}	B_2	B_2	B_1	B_2	B_g	B_{1g}	A_g	B	B
E_g	E	E	E	E	E_g	$B_{2g} + B_{3g}$	$B_{2g} + B_{3g}$	E	E
A_{1u}	A_1	B_1	B_1	A_2	A_u	A_u	A_u	A	B
A_{2u}	A_2	B_2	B_2	A_1	A_u	B_{1u}	B_{1u}	A	B
B_{1u}	B_1	A_1	A_2	B_2	B_u	A_u	B_{1u}	B	A
B_{2u}	B_2	A_2	A_1	B_1	B_u	B_{1u}	A_u	B	A
E_u	E	E	E	E	E_u	$B_{2u} + B_{3u}$	$B_{2u} + B_{3u}$	E	E

D_{4h} (cont.)	C_2' D_2	C_2'' D_2	C_2, σ_v C_{2v}	C_2, σ_d C_{2v}	C_2' C_{2v}	C_2'' C_{2v}
A_{1g}	A	A	A_1	A_1	A_1	A_1
A_{2g}	B_1	B_1	A_2	A_2	B_1	B_1
B_{1g}	A	B_1	A_1	A_2	A_1	B_1
B_{2g}	B_1	A	A_2	A_1	B_1	A_1
E_g	$B_2 + B_3$	$B_2 + B_3$	$B_1 + B_2$	$B_1 + B_2$	$A_2 + B_2$	$A_2 + B_2$
A_{1u}	A	A	A_2	A_2	A_2	A_2
A_{2u}	B_1	B_1	A_1	A_1	B_2	B_2
B_{1u}	A	B_1	A_2	A_1	A_2	B_2
B_{2u}	B_1	A	A_1	A_2	B_2	A_2
E_u	$B_2 + B_3$	$B_2 + B_3$	$B_1 + B_2$	$B_1 + B_2$	$A_1 + B_1$	$A_1 + B_1$

D_{4h} (cont.)	C_2 C_{2h}	C_2' C_{2h}	C_2'' C_{2h}	C_2 C_2	C_2' C_2	C_2'' C_2	σ_h C_s	σ_v C_s	σ_d C_s	C_i
A_{1g}	A_g	A_g	A_g	A	A	A	A'	A'	A'	A_g
A_{2g}	A_g	B_g	B_g	A	B	B	A'	A''	A''	A_g
B_{1g}	A_g	A_g	B_g	A	A	B	A'	A'	A''	A_g
B_{2g}	A_g	B_g	A_g	A	B	A	A'	A''	A'	A_g
E_g	$2B_g$	$A_g + B_g$	$A_g + B_g$	$2B$	$A + B$	$A + B$	$2A''$	$A' + A''$	$A' + A''$	$2A_g$
A_{1u}	A_u	A_u	A_u	A	A	A	A''	A''	A''	A_u
A_{2u}	A_u	B_u	B_u	A	B	B	A''	A'	A'	A_u
B_{1u}	A_u	A_u	B_u	A	A	B	A''	A''	A'	A_u
B_{2u}	A_u	B_u	A_u	A	B	A	A''	A'	A''	A_u
E_u	$2B_u$	$A_u + B_u$	$A_u + B_u$	$2B$	$A + B$	$A + B$	$2A'$	$A' + A''$	$A' + A''$	$2A_u$

T_d	T	D_{2d}	C_{3v}	S_4	D_2	C_{2v}	C_3	C_2	C_s
A_1	A	A_1	A_1	A	A	A_1	A	A	A'
A_2	A	B_1	A_2	B	A	A_2	A	A	A''
E	E	$A_1 + B_1$	E	$A + B$	$2A$	$A_1 + A_2$	E	$2A$	$A' + A''$
T_1	T	$A_2 + E$	$A_2 + E$	$A + E$	$B_1 + B_2 + B_3$	$A_2 + B_1 + B_2$	$A + E$	$A + 2B$	$A' + 2A''$
T_2	T	$B_2 + E$	$A_1 + E$	$B + E$	$B_1 + B_2 + B_3$	$A_1 + B_1 + B_2$	$A + E$	$A + 2B$	$2A' + A''$

O	T	D_4	D_3	C_4	$3C_2$ D_2	$C_2, 2C_2'$ D_2	C_3	C_2	C_2
A_1	A	A_1	A_1	A	A	A	A	A	A
A_2	A	B_1	A_2	B	A	B_1	A	A	B
E	E	$A_1 + B_1$	E	$A + B$	$2A$	$A + B_1$	E	$2A$	$A + B$
T_1		$A_2 + E$	$A_2 + E$	$A + E$	$B_1 + B_2 + B_3$	$B_1 + B_2 + B_3$	$A + E$	$A + 2B$	$A + 2B$
T_2		$B_2 + E$	$A_1 + E$	$B + E$	$B_1 + B_2 + B_3$	$A + B_2 + B_3$	$A + E$	$A + 2B$	$2A + B$

O_h	O	T_d	T_h	D_{4h}	D_{3d}
A_{1g}	A_1	A_1	A_g	A_{1g}	A_{1g}
A_{2g}	A_2	A_2	A_g	B_{1g}	A_{2g}
E_g	E	E	E_g	$A_{1g} + B_{1g}$	E_g
T_{1g}	T_1	T_1	T_g	$A_{2g} + E_g$	$A_{2g} + E_g$
T_{2g}	T_2	T_2	T_g	$B_{2g} + E_g$	$A_{1g} + E_g$
A_{1u}	A_1	A_2	A_u	A_{1u}	A_{1u}
A_{2u}	A_2	A_1	A_u	B_{1u}	A_{2u}
E_u	E	E	E_u	$A_{1u} + B_{1u}$	E_u
T_{1u}	T_1	T_2	T_u	$A_{2u} + E_u$	$A_{2u} + E_u$
T_{2u}	T_2	T_1	T_u	$B_{2u} + E_u$	$A_{1u} + E_u$

Chapter 5

CHEMICAL BONDING

The study of the nature of bonding is certainly one of the central pursuits of chemistry. In this chapter we have collected problems related to the more important aspects of this fundamental area. We have omitted several more qualitative concepts such as Lewis structures and lone-pair repulsions as these are probably already familiar to students using this book. The problems included are in four main groupings: ionic bonding, valence bond theory, molecular orbital theory, and ligand field theory. Closely related problems may be found in the other chapters. Because this is such a broad topic the list of useful references is virtually endless. However, a selection from the following representative list should be adequate for most self study needs.

1. General Texts
   - (a) C. S. G. Phillips and R. J. P. Williams, *Inorganic Chemistry*, Oxford University Press, Oxford, 1965.
   - (b) F. A. Cotton and G. Wilkinson, *Advanced Inorganic Chemistry*, Interscience, New York, 1966.
   - (c) B. E. Douglas and D. H. McDaniel, *Concepts and Models in Inorganic Chemistry*, Blaisdell Publishing Co., Waltham, Mass., 1965.
2. Bonding and Electronic Structure

(a) L. Pauling, *The Nature of the Chemical Bond*, 3rd ed., Cornell University Press, Ithaca, 1960.

(b) C. A. Coulson, *Valence*, 2nd ed., Oxford University Press, Oxford, 1961.

(c) J. N. Murrell, S. F. A. Kettle and J. M. Tedder, *Valence Theory*, John Wiley and Sons, Inc., New York, 1965.

(d) H. B. Gray, *Electrons and Chemical Bonding*, W. A. Benjamin, Inc., New York, 1965.

(e) L. E. Orgel, *Transition Metal Chemistry*, 2nd ed., John Wiley and Sons, Inc., New York, 1966.

(f) C. J. Ballhausen, *Introduction to Ligand Field Theory*, McGraw-Hill Book Co., New York, 1966.

(g) B. N. Figgis, *Introduction to Ligand Fields*, Interscience, New York, 1966.

(h) F. Basolo and R. G. Pearson, *Mechanisms of Inorganic Reactions*, 2nd ed., John Wiley and Sons, Inc., New York, 1967.

(i) G. W. King, *Spectroscopy and Molecular Structure*, Holt, Rinehart, and Winston, New York, 1964.

The above list is not meant to imply that all the listed books are required. In fact, most of the problems can be solved with reference to one good general text such as 1a, b, or c. In some instances more specific and detailed information is needed such as found in those books listed under 2. Consult Chapter 4 for references on the principles of group theory.

PROBLEMS

5.1 *Ionic bonding--ionic radii.* (Read: Pauling, page 511ff).

(a) Using the data given below, calculate the ionic radii of K^+, Cl^-, Rb^+ and Br^-. Interatomic distances: KCl, 3.14 Å; RbBr, 3.43 Å.

Table 5.1

Screening Constants, S, for Outer Electrons

(L. Pauling, *Proc. Roy. Soc. (A)*, 114, 181 (1927))

Configuration	S
He	0.188
Ne	4.52
Ar	10.87
Kr	26.83
Xe	41.80

(b) Using the values for C_n for the neon core, the argon core, and the krypton core (which can be calculated from the results in part (a)) determine the univalent radii of S^{2-} and Sr^{2+}.

5.2 *Ionic bonding--crystal coordination numbers.*

(a) In each of the following crystal structures, how many nearest neighbors does each cation have?

(1) face-centered cubic (NaCl)
(2) body-centered cubic (CsCl)
(3) zinc blend (ZnS)

(b) In structures (1) and (2) of part (a) above how many second nearest neighbors does each cation have and what are the charges of the second nearest neighbors?

5.3 *Ionic bonding--radius ratio.* (Read: Pauling, page 543ff and general advanced texts). Using the crystal radii in Table 5.2, predict the crystal structures of the following ionic compounds:

(a) CsI, (b) MgO, (c) ZnI_2, (d) TlCl, (e) BeS, (f) KCl.

5.4 *Ionic bonding--lattice energy.* (Read: Pauling, page 505ff, and general advanced texts for this and the next eight problems). Data for these problems were taken from T. C. Waddington, *Adv. in Inorg. Chem. and Radiochem.*, 1, 157 (1959) and references cited therein, or from M. F. Ladd, and W. H. Lee in H. Reiss, ed., *Prog. Solid State Chem.*, Vol. 1, Pergamon Press, London, 1964.

(a) Derive the first seven terms in the coulombic

Table 5.2

Crystal Radii and Univalent Radii of Ions [a]

			H^-	He	Li^+	Be^{++}	B^{3+}	C^{4+}	N^{5+}	O^{6+}	F^{7+}
			2.08		0.60	0.31	0.20	0.15	0.11	0.09	0.07
			(2.08)	(0.93)	(0.60)	(0.44)	(0.35)	(0.29)	(0.25)	(0.22)	(0.19)
C^{4-}	N^{3-}	O^{--}	F^-	Ne	Na^+	Mg^{++}	Al^{3+}	Si^{4+}	P^{5+}	S^{6+}	Cl^{7+}
2.60	1.71	1.40	1.36		0.95	0.65	0.50	0.41	0.34	0.29	0.26
(4.14)	(2.47)	(1.76)	(1.36)	(1.12)	(0.95)	(0.82)	(0.72)	(0.65)	(0.59)	(0.53)	(0.49)
Si^{4-}	P^{3-}	S^{--}	Cl^-	Ar	K^+	Ca^{++}	Sc^{3+}	Ti^{4+}	V^{5+}	Cr^{6+}	Mn^{7+}
2.71	2.12	1.84	1.81		1.33	0.99	0.81	0.68	0.59	0.52	0.46
(3.84)	(2.79)	(2.19)	(1.81)	(1.54)	(1.33)	(1.18)	(1.06)	(0.96)	(0.88)	(0.81)	(0.75)
					Cu^+	Zn^{++}	Ga^{3+}	Ge^{4+}	As^{5+}	Se^{6+}	Br^{7+}
					0.96	0.74	0.62	0.53	0.47	0.42	0.39
					(0.96)	(0.88)	(0.81)	(0.76)	(0.71)	(0.66)	(0.62)
Ge^{4-}	As^{3-}	Se^{--}	Br^-	Kr	Rb^+	Sr^{++}	Y^{3+}	Zr^{4+}	Nb^{5+}	Mo^{6+}	
2.72	2.22	1.98	1.95		1.48	1.13	0.93	0.80	0.70	0.62	
(3.71)	(2.85)	(2.32)	(1.95)	(1.69)	(1.48)	(1.32)	(1.20)	(1.09)	(1.00)	(0.93)	
					Ag^+	Cd^{++}	In^{3+}	Sn^{4+}	Sb^{5+}	Te^{6+}	I^{7+}
					1.26	0.97	0.81	0.71	0.62	0.56	0.50
					(1.26)	(1.14)	(1.04)	(0.96)	(0.89)	(0.82)	(0.77)
Sn^{4-}	Sb^{3-}	Te^{--}	I^-	Xe	Cs^+	Ba^{++}	La^{3+}	Ce^{4+}			
2.94	2.45	2.21	2.16		1.69	1.35	1.15	1.01			
(3.70)	(2.95)	(2.50)	(2.16)	(1.90)	(1.69)	(1.53)	(1.39)	(1.27)			
					Au^+	Hg^{++}	Tl^{3+}	Pb^{4+}	Bi^{5+}		
					1.37	1.10	0.95	0.84	0.74		
					(1.37)	(1.25)	(1.15)	(1.06)	(0.98)		

[a] univalent radii are in parentheses

Table 5.3
Values of Madelung Constants

Structure	A_{r_0}
Sodium Chloride, $M^{1+}X^{1-}$	1.74756
Cesium Chloride, $M^{1+}X^{1-}$	1.76267
Sphalerite, $M^{1+}X^{1-}$	1.63806
Wurtzite, $M^{1+}X^{1-}$	1.64132
Fluroite, $M^{2+}X_2{}^{1-}$	5.03878
Cuprite, $M_2{}^{1+}X^{2-}$	4.11552
Rutile, $M^{2+}X_2{}^{1-}$	4.816
Anatase, $M^{2+}X_2{}^{1-}$	4.800
CdI_2, $M^{2+}X_2{}^{1-}$	4.71
β-Quartz, $M^{2+}X_2{}^{1-}$	4.4394
Corundum, $M_2{}^{3+}X_3{}^{2-}$	25.0312

Values of A_{r_0} are such that the Coulomb energy per stoichiometric molecule is $-A_{r_0}e^2/r_0$ with r_0 the smallest anion-cation distance.

part of the lattice energy expression for sodium chloride in terms of the shortest anion-cation distance. $r_0 = 2.81$ Å.

(b) Derive the first three terms of the coulombic part of the lattice energy expression for cesium chloride in terms of the lattice constant α_0, the length of the edge of the unit cell.

5.5 *Ionic bonding--lattice energy.*

(a) Calculate the lattice energy of sodium chloride. The shortest anion-cation distance r_0 is 2.81 Å. The Madelung constant (see solution to 5.4) A_{r_0} is 1.747 and n = 8.0. ($e^2 = 23.1 \times 10^{-20}$ erg cm)

(b) Calculate the lattice energy of TlCl. The structure is the same as CsCl, $\alpha_0 = 3.85$ Å and $A_{\alpha_0} = 2.035$.

Table 5.4

Values of the Born Exponent n[a]

Ion Type	n
He	5
Ne	7
Ar, Cu^+	9
Kr, Ag^+	10
Xe, Au^+	12

[a] For an ion pair use the average of the n's for the two ions. For example, use 8 for NaCl.

5.6 *Ionic bonding--lattice energies;* theoretical (coulombic) model versus empirical (Born-Haber cycle) model.

(a) Compare the lattice energy calculated in problem 5.5 (b) based on the assumption of pure ionic bonding with the lattice energy derived empirically from the Born-Haber cycle.

Data:

$\Delta H_f°(TlCl)$ = -48.7 kcal/mole
$I^1(Tl)$ = 140.2 kcal/mole
S(Tl) = 45.6 kcal/mole
E(Cl) = 88.2 kcal/mole
$\Delta H_{Dissociation}(Cl_2)$ = 57.8 kcal/mole

(b) Compare the theoretical and empirical lattice energies of NiF_2.

Data:

Structure type: rutile

r_0 = 2.0 Å
n = 8.0
$\Delta H_f°(NiF_2(s))$ = -187.8 kcal/mole
$I(Ni) = I^1 + I^2$ = 593.3 kcal/mole
S(Ni) = 85.0 kcal/mole
E(F) = 79.7 kcal/mole
$\Delta H_{Dissociation}(F_2)$ = 37.6 kcal/mole

5.7 *Ionic bonding--lattice energy and enthalpy of solution.* From the following data calculate the enthalpy of solution of KF at 298°K.

$U_0(KF)$ = 191 kcal/mole
$H(K^+)$ = -84.0 kcal/mole
$H(F^-)$ = -113.2 kcal/mole

5.8 *Ionic bonding--Born-Haber cycle and electron affinity determination.* Calculate the electron affinity of fluorine given in the following data for KF.

Structure type: NaCl

α_0 = 5.33 Å = length of unit cell edge
A_{α_0} = 3.50 for NaCl structure
n = 8.0
$\Delta H_f°(KF(s))$ = -134.5 kcal/mole
I(K) = 99.6 kcal/mole
S(K) = 21.7 kcal/mole
$\Delta H_{Dissociation}(F_2)$ = 37.6 kcal/mole

5.9 *Ionic bonding--Born-Haber cycle and proton affinity determination.*

(a) Calculate the proton affinity of NH_3 from the following data for NH_4Cl:

U_0 = 161.6 kcal/mole
$\Delta H_f°(NH_4Cl(s))$ = -75.4 kcal/mole
$\Delta H_f°(H^+(g))$ = 367.1 kcal/mole = $\Delta H_f°(H(g)) + I_H$
$\Delta H_f°(Cl^-(g))$ = -57 kcal/mole = $1/2\Delta H_D(Cl_2(g)) + E_{Cl}$
$\Delta H_f°(NH_3(g))$ = -11.0 kcal/mole

(b) Compare the result for the proton affinity of NH_3 determined in (a) with that obtained from NH_4F. Propose a reasonable explanation for the difference. (The values from NH_4Br and NH_4I are 210.5 and 208.5 kcal/mole respectively).
Data for NH_4F:

U_0 = 175.2 kcal/mole
$\Delta H_f°(NH_4F(s))$ = 111.6 kcal/mole
$\Delta H_f°(F^-(g))$ = -64.8 kcal/mole
See part (a) for other data.

5.10 *Ionic bonding--Born-Haber cycle and electron affinity determination.* From the data given for CaO, calculate the electron affinity of oxygen. Structure type: NaCl

$\alpha_0 = 4.802$ Å; $r_0 = 2.4$ Å

$n = 8.0$

$\Delta H_f°(CaO(s)) = -151.7$ kcal/mole

$I^1(Ca) + I^2(Ca) = 412.9$ kcal/mole

$S(Ca) = 47.5$ kcal/mole

$1/2\Delta H_{Dissoc.}(O_2) = 59.2$ kcal/mole

5.11 *Ionic bonding--theoretical and empirical values for lattice energies of transition metal fluorides; crystal field stabilization energy.* (Read: Ladd and Lee, *op. cit.*, and D. F. C. Morris, *J. Inorg. and Nuclear Chem.*, 4, 8(1959)). Using the data given below for the second half of the first transition series divalent fluorides:

(a) Calculate the theoretical lattice energies using the Born equation.

(b) Calculate the empirical, apparent lattice energies using the Born-Haber cycle. (Ignore the difference in heat capacities.)

(c) Compare the results of (a) and (b) and from the differences estimate thermodynamic crystal field splitting parameters (Dq's) for FeF_2, CoF_2, NiF_2, and CuF_2.

(d) Estimate spectroscopic Dq's using the empirical parameters given by Figgis (p. 244) and compare these values to the thermodynamic estimates.

Table 5.5
Thermodynamic Data for Transition Metal Fluorides[a]

Salt	Structure type[b]	r_0 (Å)	$-\Delta H_f°$	I^1+I^2	S
MnF_2	rutile	2.12	189	531.9	68.3
FeF_2	rutile	2.06	168	555.1	99.8
CoF_2	rutile	2.04	159	574.4	102.0
NiF_2	rutile	2.02	159.5	594.4	101.6
CuF_2	fluorite	$\alpha_0 = 5.406$	126.9	646.0	81.5
ZnF_2	rutile	2.04	176	630.6	31.2

[a]D. F. C. Morris, *op. cit.* [b]A_{α_0} for fluorite = 11.637.

For F_2, $\Delta H_{Dissoc.} = 37.6$ kcal/mole. For F, E = 82.5 kcal/mole.

The coordination of the metal ions is octahedral in all cases.

5.12 *Valence bond theory--wave functions.* (Read: Coulson, p. 114ff). Write and normalize the first order valence bond wave function for H_2.

5.13 *Valence bond wave functions--ionic structures.* (Read: Coulson, p. 117ff). The second phase of development of the valence bond function for a homonuclear diatomic molecule is the inclusion of the ionic structures

$$\psi_A(1)\psi_A(2) \text{ and } \psi_B(1)\psi_B(2)$$

The wave function then is

$$\psi_{VB} = \psi_A(1)\psi_B(2) \pm \psi_A(2)\psi_B(1) + \lambda[\psi_A(1)\psi_A(2) + \psi_B(1)\psi_B(2)]$$

or

$$\psi_{VB} = \psi_{cov} + \lambda\psi_{ion}$$

or

$$\psi_{VB} = (1-\lambda)\psi_{cov} + \lambda\psi_{ion}$$

Normalize this last function ignoring overlap.

5.14 *Valence bond wave functions--heteronuclear diatomics.* (Read: Coulson, p. 128ff). Write a valence bond wave function including ionic structures for the heteronuclear diatomic molecule A-B. In what respect does this function differ from that for the homonuclear case?

5.15 *Valence bond theory--partial ionic character of covalent bonds.* (Read: Coulson, p. 130ff). Implicit in the formulation of the valence bond wave function is the notion of partial ionic character of covalent bonds. Obviously the degree of ionic character is directly related to the magnitude of the coefficient λ. (See answer to 5.14).

(a) Assuming the following values for λ, calculate the percent ionic character of the bonds in H_2, HCl, and CsF.

H_2	$\lambda = 0.25$
HCl	$\lambda = 0.45$
CsF	$\lambda = 3.17$

Use the form of the wave function

$$\psi_{VB} = \psi_{cov} + \lambda\psi_{ion}$$

(b) What are the values of λ in the wave equations for the species in (a) using the form of the wave equation

$$\psi_{VB} = (1-\lambda)\psi_{cov} + \lambda\psi_{ion}?$$

5.16 *Valence bond wave functions--estimation of* λ. (Read: Coulson, p. 134ff). Determining the value of λ in the heteronuclear diatomic valence bond wave function is not a straight forward problem. Several approaches have been adopted to find approximate values. The simplest is to use the measured dipole moment μ and make some drastic (and in most cases improbable) assumptions about the polarity of "pure" covalent bonds.

Assuming Ψ_{cov} (HCl) is completely non-polar and all the polarity of the HCl molecule derives from Ψ_{ionic}, calculate the percent ionic charcter for the HCl wave function using the following data:

$$\mu = 1.03 \times 10^{-18} \text{ esu} = 1.03 \text{ Debye}$$

$$r_0 = 1.27 \text{ Å} = 1.27 \times 10^{-8} \text{ cm}$$

$$e = 4.802 \times 10^{-10} \text{ esu}$$

$$\psi_{HCl} = \psi_{cov} + \lambda\psi_{ion}$$

5.17 *Valence bond theory--ionic covalent resonance energy and electronegativities.* (Read: Pauling, p. 88ff). Pauling has shown that a consistent scale of electronegativities can be derived from a consideration of the bond energies of diatomic molecules. Using Pauling's method and the bond energies in Table 5.6 calculate the electronegativities of Na, Cl, O, C and H. Compare your answers to those in Table 5.7. Start by assuming (as does Pauling) that $\chi_F = 4.0$.

5.18 *Valence bond theory--ionic covalent resonance energy and the estimation of* λ. (Read: Pauling, p. 79ff and Cotton and Wilkinson, p. 107ff). Pauling has shown that the ionic covalent resonance energy (as determined by the geometric mean Δ') is proportional to the percent ionic character and thus may be used to estimate λ. The empirical relationship is % ionic character = $18|\chi_A-\chi_B|^{1.4}$. Using the values for bond energies in Table 5.5 calculate the percent ionic character and λ for the HF bond.

5.19 *Valence bond theory--electronegativities and bond polarity.* From consideration of electronegativities alone predict the relative polarity of the bonds in $BeCl_2$, BN, and CS_2.

Table 5.6

Energy Values for Single Bonds (Kcal/Mole)

Bond	Bond energy	Bond	Bond energy	Bond	Bond energy
H-H	104.2	P-H	76.4	Si-Cl	85.7
C-C	83.1	As-H	58.6	Si-Br	69.1
Si-Si	42.2	O-H	110.6	Si-I	50.9
Ge-Ge	37.6	S-H	81.1	Ge-Cl	97.5
Sn-Sn	34.2	Se-H	66.1	N-F	64.5
N-N	38.4	Te-H	57.5	N-Cl	47.7
P-P	51.3	H-F	134.6	P-Cl	79.1
As-As	32.1	H-Cl	103.2	P-Br	65.4
Sb-Sb	30.2	H-Br	87.5	P-I	51.4
Bi-Bi	25	H-I	71.4	As-F	111.3
O-O	33.2	C-Si	69.3	As-Cl	68.9
S-S	50.9	C-N	69.7	As-Br	56.5
Se-Se	44.0	C-O	84.0	As-I	41.6
Te-Te	33	C-S	62.0	O-F	44.2
F-F	36.6	C-F	105.4	O-Cl	48.5
Cl-Cl	58.0	C-Cl	78.5	S-Cl	59.7
Br-Br	46.1	C-Br	65.9	S-Br	50.7
I-I	36.1	C-I	57.4	Cl-F	60.6
C-H	98.8	Si-O	88.2	Br-Cl	52.3
Si-H	70.4	Si-S	54.2	I-Cl	50.3
N-H	93.4	Si-F	129.3	I-Br	42.5

5.20 *Valence bond theory--choice of important structures in* Ψ_{ionic}. (Read: Coulson, p. 130).

(a) From the values of ionization potentials and electron affinities of hydrogen and fluorine given below, show that, of the two possible ionic structures, only $\Psi_F(1)\Psi_F(2)$, that is, the structure with both electrons on the fluorine, need be considered.

$$I(H) = 13.60 \text{ eV}$$
$$I(F) = 17.42 \text{ eV}$$
$$E(H) = 0.75 \text{ eV}$$
$$E(F) = 3.45 \text{ eV}$$

(b) Contrast the results for HF with those for the

Table 5.7a

Electronegativities of the Elements
(Values in bold type are calculated using the Allred-Rochow formula; those in italics are estimated by Pauling's method and those in Roman type are calculated by Mulliken's method.)[a]

I	II	III	IV	III	II	I[b]	
H							He
2.20							
Li	Be	B	C	N	O	F	Ne
0.97	**1.47**	**2.01**	**2.50**	**3.07**	**3.50**	**4.10**	
0.98	*1.57*	*2.04*	*2.55*	*3.04*	*3.44*	*3.98*	
0.94	1.46	2.01	2.63	2.33	3.17	3.91	
Na	Mg	Al	Si	P	S	Cl	Ar
1.01	**1.23**	**1.47**	**1.74**	**2.06**	**2.44**	**2.83**	
0.93	*1.31*	*1.61*	*1.90*	*2.19*	*2.58*	*3.16*	
0.93	1.32	1.81	2.44	1.81	2.41	3.00	
K	Ca	Ga	Ge	As	Se	Br	Kr
0.91	**1.04**	**1.82**	**2.02**	**2.20**	**2.48**	**2.74**	
0.82	*1.00*	*1.81*	*2.01*	*2.18*	*2.55*	*2.96*	
0.80		1.95		1.75	2.23	2.76	
Rb	Sr	In	Sn	Sb	Te	I	Xe
0.89	**0.99**	**1.49**	**1.72**	**1.82**	**2.01**	**2.21**	
0.82	*0.95*	*1.78*	*1.96*	*2.05*		*2.66*	
		1.80		1.65	2.10	2.56	
Cs	Ba	Tl	Pb	Bi	Po	At	Rn
0.86	**0.97**	**1.44**	**1.55**	**1.67**	**1.76**	**1.96**	
0.79	*0.89*	*2.04*	*2.33*	*2.02*			
Fr	Ra						
0.86	**0.97**						

Table 5.7b

III	IV	II	II	II	II	II	II	II	II
Sc	Ti	V	Cr	Mn	Fe	Co	Ni	Cu	Zn
1.20	1.32	1.45	1.56	1.60	1.64	1.70	1.75	1.75	1.66
1.36	*1.54*	*1.63*	*1.66*	*1.55*	*1.83*	*1.88*	*1.91*	*1.90*	*1.65*
								1.36	1.49
Y	Zr	Nb	Mo	Tc	Ru	Rh	Pd	Ag	Cd
1.11	1.22	1.23	1.30	1.36	1.42	1.45	1.35	1.42	1.46
1.22	*1.33*		*2.16*			*2.28*	*2.20*	*1.93*	*1.69*
								1.36	1.4
	Hf	Ta	W	Re	Os	Ir	Pt	Au	Hg
	1.23	1.33	1.40	1.46	1.52	1.55	1.44	1.42	1.44
			2.36			*2.20*	*2.28*	*2.54*	*2.00*

[a] Allred-Rochow electronegativities from *J. Inorg. Nucl. Chem.*, 5, 264 (1958); Pauling electronegativities from A. L. Allred, *J. Inorg. Nucl. Chem.*, 17, 215 (1961); Mulliken electronegativities from H. O. Pritchard and H. A. Skinner, *Chem. Rev.*, 55, 745 (1955).

[b] Roman numerals at the top give the oxidation states used for the Pauling electronegativities.

N-Cl bond.

$I(N) = 14.54$ eV
$I(Cl) = 13.01$ eV
$E(N) = -0.1$ eV
$E(Cl) = 3.61$ eV

(c) What structures are important for the H-Cl bond and for the Cl-F bond?

5.21 *Symmetry of σ and π bonds.* (Read: Cotton and Wilkinson, p. 87). Illustrate clearly the overlap of the charge clouds of atomic wave functions (orbitals) to form σ and π bonds as appropriate in the following:

(a) HCl, (b) CO_2, (c) C_2H_2, (d) CO, (e) H_2O

5.22 *Hybrid orbitals.* (Review group theory, Chapter 4). In many molecular species convincing theoretical arguments and experimental evidence indicate that the best atomic wave functions to use in valence bond wave functions are not pure hydrogenic functions, but are better described as hybrid functions (orbitals) consisting of linear combinations of the hydrogenic functions. For example, in the square planar ion $PdCl_4^{2-}$, the hybrid is dsp^2. Considering only the geometry of the following species predict the possible hybrid orbitals which the central atom might use to form σ bonds:

(a) CCl_4, tetrahedral; (b) SF_6, octahedral;

(c) $Fe(CO)_5$, trigonal bipyramidal; (d) BCl_3, trigonal planar; (e) NH_3, trigonal pyramidal.

5.23 *Hybrid orbitals.* (Read: Cotton, Chapter 6; Hochstrasser, Chapter 8). Use the sigma bonds of the tetrahedral molecule, MX_4, as a basis set and construct the reducible representation spanned by this set of bonds. Decompose the reducible representation into its irreducible representation components. Which set of atomic orbitals transform the same way as this set of sigma bonds? On the basis of this information which atomic orbitals will participate in the formation of hybrid orbitals for sigma bonding in SiF_4?

5.24 *Hybrid orbitals.* (Read: Cotton, Chapter 6).

(a) Use group theoretical techniques and construct the trigonal hybrid orbitals which would be used by carbon in the carbonate ion.

(b) Derive analogous expressions for the tetrahedral hybrid orbitals of silicon in SiF_4.

5.25 *Symmetry considerations in orbital overlap.* (Read, for example, Cotton and Wilkinson, p. 69). Evaluate the overlaps indicated below as to whether they are positive, negative, or zero and if finite, as to type (σ, π, or δ). Assume in each case that the internuclear axis is z.

	Atom A orbital	Atom B orbital
(a)	s	p_z
(b)	s	p_x
(c)	p_z	p_z
(d)	$d_{x^2-y^2}$	$d_{x^2-y^2}$
(e)	d_{z^2}	p_z
(f)	p_x	p_x
(g)	p_z	p_x

5.26 *Molecular orbital theory--overlap integral in diatomic molecules.* In the molecular orbital model of the hydrogen molecule ion H_2^+, the overlap integral for two 1s orbitals has the value 0.59. How much of an error would be introduced in a calculation based on functions written with normalization constants that neglect overlap?

5.27 *Molecular orbital theory--comparison to valence bond theory.* (Read: Coulson, p. 155). The valence bond description of a simple homonuclear diatomic molecule can be written:

$$\Psi_{VB} = \psi_A(1)\psi_B(2) + \psi_B(1)\psi_A(2)$$

ignoring ionic forms. The corresponding ground state molecular orbital wave function would be:

$$\Psi_{MO} = [\psi_A(1) + \psi_B(1)][\psi_A(2) + \psi_B(2)].$$

Show that by inclusion of the ionic part

$$\lambda[\psi_A(1)\psi_A(2) + \psi_B(1)\psi_B(2)]$$

in the valence bond model and the mixing in of excited state functions

$$k[\psi_A(1) - \psi_B(1)][\psi_A(2) - \psi_B(2)]$$

with the ground state of the molecular orbital model, the two descriptions become identical, provided

$$\lambda = \frac{1 + k}{1 - k}$$

5.28 *Molecular orbital theory--homonuclear diatomic molecules.* (Read: Gray, Chapter 2).

(a) Depending on the magnitude of the energy difference between the 2s and 2p atomic orbitals two significantly different orders of molecular orbital energies can be conceived. Illustrate each of these. Which is more likely for F_2? Which is more likely for B_2?

(b) Only in certain cases in the second period diatomic molecules can the order referred to in (a) above be checked. B_2 is one of these cases. Illustrate and discuss the necessary experiment for proving the order of levels and the significance of the results.

(c) The unpaired electron in N_2^+ has the properties of a σ electron. What is the order of the molecular orbital energies which this evidence indicates?

5.29 *Molecular orbital theory--bond order in homonuclear diatomics.* (Read: Gray, Chapter 2 and others, *e.g.*, Cotton and Wilkinson, p. 81ff).

(a) Predict the bond order in B_2, C_2, Cl_2 and N_2.

(b) Discuss the relation of the molecular orbital bond orders to the following observed bond lengths:

O_2^+, 1.12 Å

O_2, 1.21 Å

O_2^-, 1.26 Å

$O_2^=$, 1.49 Å

5.30 *Molecular orbital theory--comparison to valence bond theory.* Compare the predictions of the molecular orbital and valence bond models for O_2.

5.31 *Molecular orbital theory--heteronuclear diatomic molecules.* (Read: Gray, Chapter 2). Construct MO energy level diagrams for each of the following diatomic molecules. What are the bond orders and the numbers of unpaired electrons in each?

(a) NO, (b) BN, (c) HCl, (d) ClF, (e) LiI, (f) BO.

5.32 *Molecular orbital theory--triatomic molecules.* (Read: Gray, Chapter 3. Review group theory). Construct molecular orbital energy level diagrams for the following:

(a) BeH_2, (b) CO_2

5.33 *Molecular orbital theory--term symbols.* (Read: Gray, p. 60ff). Derive the term symbols for the molecular orbital ground states of the following molecules:

(a) H_2, (b) O_2, (c) BF, (d) BO, (e) BN, (f) BeO

5.34 *Localized and delocalized molecular orbitals for* H_2O. (Read: King, Chapter 10, and Ballhausen and Gray, Chapter 6).

(a) Using the 2s and the three 2p orbitals on the oxygen construct a set of tetrahedral hybrid orbitals and use these with the two 1s orbitals of the hydrogens to construct a *localized* molecular orbital description of H_2O. (See problem 5.24).

(b) Using as a basis set the 1s, 2s, and three 2p orbitals of the oxygen and the two 1s orbitals of the hydrogens construct a general *delocalized* molecular orbital description for H_2O.

5.35 *Delocalized molecular orbital description of* NH_3. (Read: King, Chapter 10). For the molecule NH_3 construct a delocalized molecular orbital description using as a basis set the 1s, 2s and three 2p orbitals of the nitrogen and the three 1s orbitals of the hydrogens.

5.36 *Ligand field theory--crystal field splittings of d-orbitals.* (Read: Cotton and Wilkinson, Chapter 26 and Orgel, Chapter 2). Using labelled diagrams show qualitatively how the five one-electron d-orbitals are split in ligand fields of the following symmetries:

(a) O_h, (b) T_d, (c) D_{4h}, (d) D_3, (e) D_{3h} (5-coordinate), (f) C_{4v}

5.37 *Ligand field theory--crystal field stabilization energy.* (Read: Cotton and Wilkinson, Chapter 26 and Orgel, Chapter 3; and F. A. Cotton, *J. Chem. Educ.*, 41, 466 (1964)).

Calculate the crystal field stabilization energies of the configurations d^n (n = 1-9) in octahedral and in tetrahedral ligand fields.

5.38 *Ligand field theory--octahedral site preference energy.* A spinel is an oxide with the general formula AB_2O_4. The crystal structure is made up of a close-packed array of oxide ions with the A and B metal ions occupying the holes in the lattice. If A occupies the T_d holes and B the O_h holes, the spinel is said to be *normal*, but, if the A ions are in the O_h holes and B ions in both kinds, the spinel is *inverse*. Use the concept of octahedral site preference energy defined as

$$\text{OSPE} = \text{CFSE}\ (O_h) - \text{CFSE}\ (T_d)$$

and predict whether the following spinels will be normal or inverse:

(a) $NiAl_2O_4$, (b) $MnAl_2O_4$, (c) $FeAl_2O_4$, (d) $MgCr_2O_4$,

(e) Fe_3O_4.

5.39 *Ligand field theory.*

(a) Which first row transition metal ions would have orbitally degenerate ground states in a ligand field of octahedral symmetry?

(b) For which of these can the ground state degeneracy be resolved by:

(1) a trigonal distortion?
(2) a tetragonal distortion?

(c) For which of the first row transition metal ions will the multiplicity, and thus the magnetic moment, not depend on the strength of the *octahedral* ligand field?

5.40 *Ligand field theory.* The chromium(III) ion in a tetrahedral environment has a triply degenerate ground state and consequently is subject to a structural distortion. Which of the following types of distortion is more likely:

(a) elongation along an S_4 axis?

(b) contraction along an S_4 axis?

5.41 *Ligand field theory--crystal field activation energy (CFAE).* (Read: Basolo and Pearson, p. 146).

The *d*-orbital energies in units of D_q for octahedral (O_h), square based pyramidal (C_{4v}), pentagonal bipyramidal (D_{5h}), and octahedral wedge (C_{2v}) structures are given in Table 5.8.

Table 5.8

	$d_{x^2-y^2}$	d_{z^2}	d_{xy}	d_{xz}	d_{yz}
O_h	6.0	6.0	-4.0	-4.0	-4.0
C_{4v}	9.14	0.86	-0.86	-4.57	-4.57
D_{5h}	2.82	4.93	2.82	-5.28	-5.28
C_{2v}[a]	8.79	1.39[b]	-1.51[b]	-2.60[c]	-6.08[c]

[a] square pyramid with two groups above the base

[b] d_{z^2}, d_{xy} hybrids

[c] d_{xz}, d_{yz} hybrids

Calculate the crystal field stabilization energies for the configurations d^0-d^{10} for each structure in both weak and strong crystal fields, and evaluate the crystal field activation energy (CFAE) for the following processes:

(a) $ML_6 \rightarrow ML_5\ (C_{4v}) + L$

(b) $ML_6 + L \rightarrow ML_7\ (D_{5h})$

(c) $ML_6 + L \rightarrow ML_7\ (C_{2v})$

5.42 *Ligand field theory--lability and crystal field activation energy (CFAE).* The hexacyanometallate complexes exhibit the following order of lability:

$V(CN)_6^{3-} > Mn(CN)_6^{3-} \gg Cr(CN)_6^{3-} > Fe(CN)_6^{3-} \sim Co(CN)_6^{3-}$

and

$Cr(CN)_6^{4-} > V(CN)_6^{4-} \sim Mn(CN)_6^{4-} > Fe(CN)_6^{4-}$

Based on CFAE, what is the most likely structure for the intermediate in the ligand exchange reactions?

5.43 *Molecular orbitals in metal complexes.* Review group theory. Read: Gray, Chapter 9; Ballhausen, Chapter 7;

Figgis, Chapter 8) Consider the octahedral complex ion TiF_6^{3-}.

(a) Determine the combinations of fluorine orbitals which can form σ-molecular orbitals with the titanium.

(b) Determine the fluorine orbitals which can form π-molecular orbitals with the titanium.

(c) Construct a molecular orbital energy level correlation diagram for TiF_6^{3-}.

5.44 *Molecular orbital description of SO_2.* In problem 5.34 a delocalized MO model was constructed for the water molecule. The SO_2 molecule is also a bent triatomic structure. However, unlike H_2O, SO_2 may have some π character in its bonding. Derive the symmetries of the delocalized π MO's possible for SO_2. Using the *s* and *p* orbitals of the sulfur, which sulfur orbital can be used exclusively for π bonding? What is the linear combination of oxygen orbitals which can be used in this MO? (See also problem 5.43)

SOLUTIONS

5.1 (a) Pauling assumes that the ionic radius of univalent ions is given by

$$R_1 = C_n/(Z-S),$$

where C_n is a constant for isoelectronic species, Z is the nuclear charge, and S is the reduction in nuclear charge sensed by the valence electrons due to screening by the inner electrons. By choosing salts composed of isoelectronic ions, one obtains the result that the univalent radii R_1 are inversely proportional to the effective nuclear charges $(Z-S)$. Knowing the interionic distance from x-ray diffraction measurements along with the screening constants, permits the direct calculation of R_1's.

Since in both KCl and $RbBr$ the ions are isoelectronic,

$$R_1(K^+)/R_1(Cl^-) = (Z-S)_{Cl^-}/(Z-S)_{K^+}, \text{ and}$$

$$R_1(Rb^+)/R_1(Br^-) = (Z-S)_{Br^-}/(Z-S)_{Rb^+}$$

From x-ray diffraction,

$$R_1(K^+) + R_1(Cl^-) = 3.14 \text{ Å and}$$

$$R_1(Rb^+) + R_1(Br^-) = 3.43 \text{ Å}.$$

It follows then that

$$R_1(K^+) = 1.35 \text{ Å},\ R_1(Cl^-) = 1.79 \text{ Å},\ R_1(Rb^+) = 1.53 \text{ Å},$$

$$\text{and } R_1(Br^-) = 1.90 \text{ Å}.$$

(b) A full account of the relationship between univalent radii and crystal radii for multivalent ions is given by Pauling (Read: Pauling, p. 511ff). The sums of crystal radii obtained by this procedure (with some refinements) agree within at least 0.10 Å with the empirical interionic distances as determined by x-ray diffraction.

From the Born equation,

$$(R_{z+} + R_{z-}) = (nB/Az^2)^{1/(n-1)}$$

$$(R_{1+} + R_{1-}) = (nB/A)^{1/(n-1)} \quad \text{(See Pauling, p. 515)}$$

$$\therefore R_z = R_1[z^{-2/(n-1)}]$$

The technique used in part (a) allows the calculation of the C_n values for the Ar and Kr configurations from which R_1's can be determined and then R_z's.

Ion	C_n	R_1 (Å)	R_z (Å)
S^{2-}	10.97	2.14	1.81
Sr^{2+}	15.56	1.39	1.19

5.2 It is best to get three-dimensional models to study in answering this question. In the figures below the three lattices are pictured in perspective.

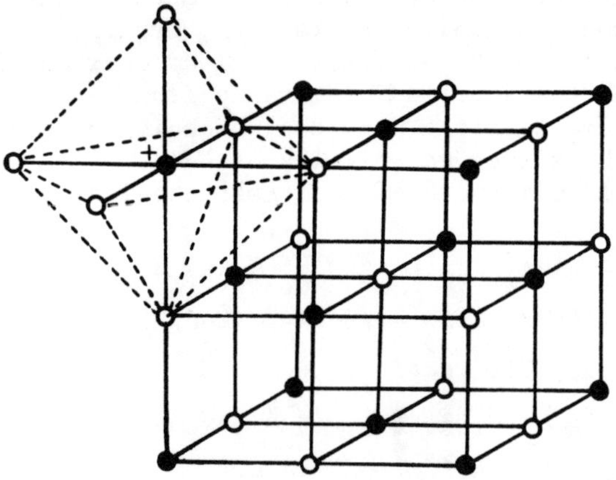

face-centered cubic

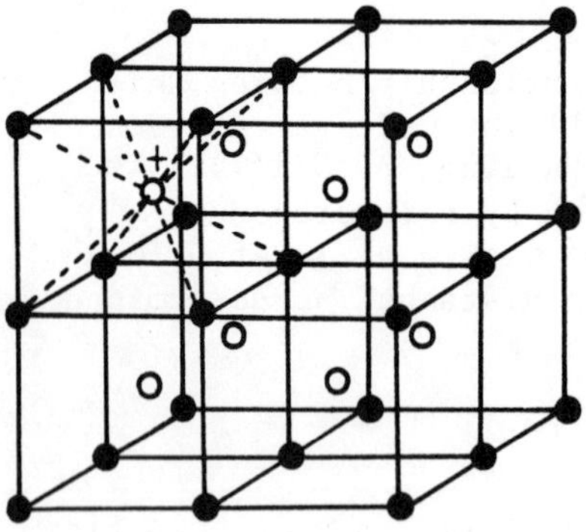

body-centered cubic

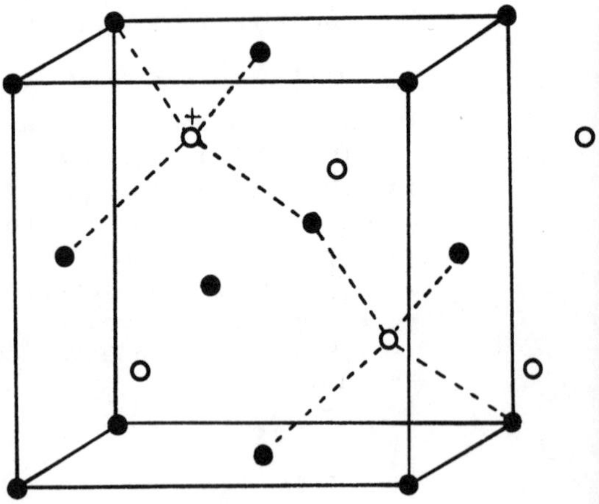

zinc blend

The answers are then:

(a) (1) 6 (b) (1) 12; +
(2) 8 (2) 6; +
(3) 4

5.3 The hard sphere ionic model predicts that maximum stability will obtain when the anion spheres just touch each other and the anion and cation spheres just touch. Such a condition is achieved for different cation coordination numbers and geometries depending on the ratio of anion radius to cation radius r^-/r^+. Some of the more common structure types and their cation coordination numbers and radius ratios are given below.

Cation site geometry	Prototypes	Coord. no.	Maximum radius ratio
cubo-octahedron	K_2SiF_6	12	1.000
cube	$CsCl, CaF_2$	8	1.37
octahedron	$NaCl, TiO_2$	6	2.42
tetrahedron	ZnS, SiO_2	4	4.44

Considering the above and the crystal radii of the ions concerned, the answers are

(a) CsI $r^-/r^+ = 2.16/1.69 = 1.28$ ∴ CsCl structure

(b) MgO $r^-/r^+ = 1.40/0.65 = 2.15$ ∴ NaCl structure

(c) ZnI_2 $r^-/r^+ = 2.16/0.74 = 2.92$ ∴ SiO_2 structure

(d) CsCl structure (See Pauling, p. 543ff)

5.4 (a) Number and charge of ith nearest neighbors of a Na^+ ion in NaCl

i	number	charge
1	6	−
2	12	+
3	8	−
4	6	+
5	24	−
6	24	+
7	12	+

By simple trigonometry the coulombic part is seen to be

$$E = -\frac{e^2|Z|^2}{r_0}\left(6 - \frac{12}{\sqrt{2}} + \frac{8}{\sqrt{3}} - \frac{6}{\sqrt{4}} + \frac{24}{\sqrt{5}} - \frac{24}{\sqrt{6}} - \frac{12}{\sqrt{8}} \cdots\right)$$

Since $e = 4.802 \times 10^{-10}$ esu, $Z = 1$, and
$r_0 = 2.81 \times 10^{-8}$ cm,

$$E = -\frac{23.1 \times 10^{-20}\ \text{erg cm}}{2.81 \times 10^{-8}\ \text{cm}}\left(6 - \frac{12}{\sqrt{2}} + \frac{8}{\sqrt{3}} - \frac{6}{\sqrt{4}} \cdots\right)$$

$$= -8.22 \times 10^{-12}\ \text{erg}\left(6 - \frac{12}{\sqrt{2}} + \frac{8}{\sqrt{3}} - \frac{6}{\sqrt{4}} \cdots\right)$$

The series in parenthesis above converges to give the Madelung constant for the NaCl type lattice in terms of r_0. The series converges very slowly and it is not profitable to evaluate by brute force. A number of sophisticated methods of making the summation have been derived for cubic, as well as more complicated lattices. The Madelung constants for a few structures are given in Table 5.3.

5.5 Considering only the coulomb term and the Born repulsion term and ignoring weak Van der Waals interactions, the correct expression for the lattice energy in terms of the shortest anion-cation distance r_0, is

$$U_0 = (NA_{r_0}Z^2e^2/r_0)(1 - 1/n),$$

where N is Avagadro's number, A_{r_0} is the Madelung constant in terms of the shortest anion cation distance r_0 (see Table 5.3), Z is the largest common factor in the charges on the ions (when the charge ratio is 1:1, 1:2, 1:3,2:3, Z is 1; when 2:2, Z is 2), n is the Born exponent (see Table 5.4), and e is the electronic charge, 4.802×10^{-10} esu.

(a) For NaCl,

$\therefore\ U_0 =$

$$\frac{6.02 \times 10^{23}\ \text{mole}^{-1} \times 1.747 \times 23.1 \times 10^{-20}\ \text{erg cm}}{2.81 \times 10^{-8}\ \text{cm}}(7/8)$$

$U_0 = 8.62 \times 10^{12}$ erg mole^{-1} $\times$ (7/8)

$U_0 = 206.2$ kcal mole^{-1} $\times$ (7/8)

$U_0 = 180.4$ kcal mole^{-1}

5.6 The Born-Haber cycle is an application of Hess' Law of heat summation and relies on the thermodynamic principle that enthalpy differences between states are independent of the path of change. A common form of the cycle for a binary metal halide is given below along with the definition of the terms used. (Waddington, *op.cit.*)

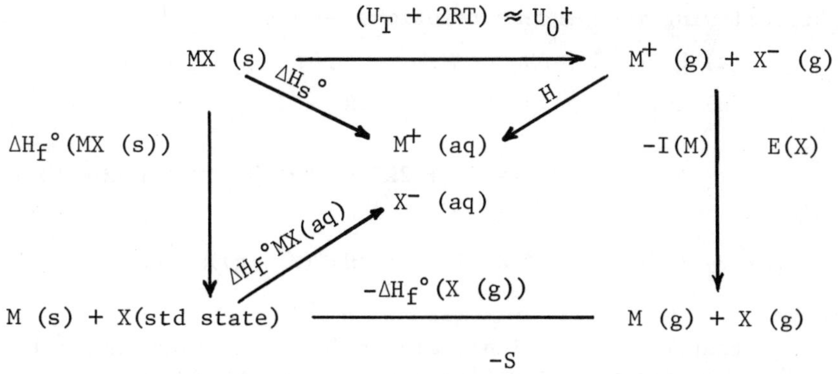

† This is true only if the differences in heat capacities are ignored. Strictly, $U_0 = U_T - [C_p(M^+) + C_p(X^-) - C_p(MX)]\,dT + 2RT$. The difference in heat capacities is generally of the order of a few kcal.

$\Delta H_f°(MX (s))$ is the enthalpy of formation of the crystalline salt.

$\Delta H_f°(MX (aq))$ is the enthalpy of formation of the salt in aqueous solution at unit activity.

$\Delta H_s°$ is the standard enthalpy of solution of the crystalline salt at unit activity.

U_0 is the lattice energy of MX at 0°K.

U_T is the lattice energy of MX at T°K.

$U_T + 2RT$ is the corresponding lattice enthalpy at T°K. (Recall $\Delta H = \Delta E + \Delta nRT$)

H is the enthalpy of hydration of the two gaseous ions to unit activity.

E(X) is the electron affinity of the X radical.

I(M) is the first ionization potential of the metal ion. (The sum of the first two ionization potentials would be used for a divalent metal ion).

$\Delta H_f°(X (g))$ is the standard enthalpy of formation of the X radical in the gas phase. In many simple salts such as oxides and halides $\Delta H_f°\ X(g) = (1/2)\Delta H_{Dissociation}$ of X_2.

S is the enthalpy of sublimation of the metal.

The following independent relations derive:

(a) $U_T = -H + \Delta H_s° - 2RT$

(b) $\Delta H_f°(X\ (g)) - E = U_T + 2RT - I - S + \Delta H_f°(MX)$

or

$\Delta H_f°(X^-\ (g)) = U_T + 2RT - \Delta H_f°(M^+\ (g)) + \Delta H_f°(MX)$

Thus:

$$U_0 \cong U_T + 2RT = \Delta H_f°(X^-\ (g)) + \Delta H_f°(M^+\ (g)) - \Delta H_f°(MX\ (s))$$

Recall that the theoretical calculation of lattice energy by the Born equation or modifications thereof yields U_0.

$$\therefore\ U_0 \simeq \Delta H_f°(X\ (g)) - E + S + I - \Delta H_f°(MX\ (s))$$

The Born-Haber cycle is used primarily for three purposes:

(a) To evaluate electron affinities (E)
(b) To estimate the extent of ionic character of a lattice
(c) To estimate heats of formation and stabilities of as-yet-unknown compounds.

These are illustrated in this and the following five questions:

(a) $U_0 \simeq -\Delta H_f°(TlCl) + I(Tl) + S(Tl) + 1/2\Delta H_{Dissoc.}(Cl_2) - E(Cl)$

$= (48.7 + 140.2 + 45.6 + 28.9 - 88.2)$ kcal mole^{-1}

$= 175.1$ kcal mole^{-1}

This fairly close agreement of the empirical value with that calculated from the Born-Landé equation (see problem 5.5 b) suggests that the lattice forces are principally electrostatic.

5.7 $U_0 \simeq U_T + 2RT = -H + \Delta H_s° = -H(K^+) - H(F^-) + \Delta H_s°(KF)$

$\Delta H_s°(KF) = U_0(KF) + H(K^+) + H(F^-)$

$= 191 - 197.2$

$= -6.2$ kcal mole^{-1}

5.8 $E = -\Delta H_f^\circ(KF) + S(K) + I(K) + 1/2\Delta H_{Dissoc.}(F_2) - U_0(KF)$

$U_0 = \dfrac{NZ^2e^2A_{\alpha\theta}}{\alpha_0}$ $(1-1/n)$

$U_0 = \dfrac{1.39 \times 10^5 \text{ erg cm mole}^{-1} \times 3.50}{5.33 \times 10^{-8} \text{ cm} \times 4.184 \times 10^{10} \text{ erg kcal}^{-1}}$ $(1-1/8)$

   $= 218 \text{ kcal mole}^{-1} \mathbf{x}(1-1/8)$

   $= 191 \text{ kcal mole}^{-1}$

Hence, $E = (134.5 + 21.7 + 99.6 + 18.8 - 191) \text{ kcal mole}^{-1}$

   $= (275 - 191) \text{ kcal mole}^{-1}$

   $= 84 \text{ kcal mole}^{-1}$

5.9   (a) NH_4Cl (s) $\xrightarrow{U_T + 2RT}$ NH_4^+(g) Cl^-(g)

$\Delta H_f^\circ(NH_4Cl(s))$ $\qquad$ $P(NH_3)$

NH_3(g) + H^+(g) + Cl^-(g)

$-I(H) + E(Cl)$

$1/2 N_2 + 2H_2$ $\xleftarrow{-\Delta H_f^\circ(NH_3(g))}$ NH_3(g) + H(g)

$+ 1/2 Cl_2$ $\quad -\Delta H_f^\circ(H(g)) - \Delta H_f^\circ(Cl(g))$ $\quad + Cl$(g)

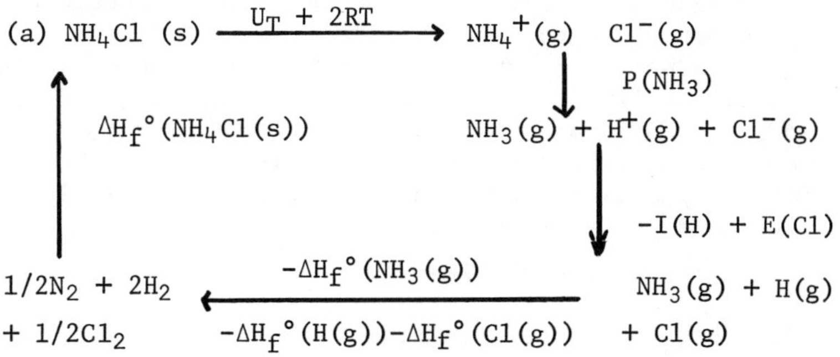

$P(NH_3) = -(U_T + 2RT) - \Delta H_f^\circ(NH_4Cl(s)) + \Delta H_f^\circ(NH_3(g))$

$\quad + \Delta H_f^\circ(H(g)) + \Delta H_f^\circ(Cl(g)) + I(H) - E(Cl)$

$\quad \approx -U_0 - \Delta H_f^\circ(NH_4Cl(s)) + \Delta H_f^\circ(H^+(g)) +$

$\quad \Delta H_f^\circ(Cl^-(g)) + \Delta H_f^\circ(NH_3(g))$

Note: $\Delta H_f^\circ H^+(g) = \Delta H_f^\circ H(g) + I(H) = 1/2\Delta H_{Dissoc.}(H_2) + I(H)$

$\Delta H_f^\circ X^-(g) = \Delta H_f^\circ Cl(g) + E(Cl) = 1/2\Delta H_{Dissoc.} Cl_2$

$\therefore P(NH_3) = 211 \text{ kcal mole}^{-1}$

   (b) $P(NH_3) = 229 \text{ kcal mole}^{-1}$.

The proton affinity calculated from the analogous data for
the other ammonium halides is very close to the value for
the chloride (211 kcal mole^{-1}). NH_4F is probably strongly
hydrogen bonded compared to the other ammonium halides.

5.10 $U_0(CaO) = 848 \text{ kcal mole}^{-1}$

$E(O) = 177 \text{ kcal mole}^{-1}$

   Note: The stability of this crystal in spite of the

negative value of E(0) is owing to the very high lattice energy.

5.11 The answers to these questions are collected in Table 5.9.

The theoretical lattice energies U_2 and the experimental lattice energies U_1 are seen to be in good agreement only for MnF_2 and ZnF_2. The difference in each case should be the crystal field stabilization energy (CFSE) for an octahedral field. The values of the CFSE in terms of the radial parameter Dq are given in column 6 below. Column 7 then gives the thermodynamic Dq determined by equating the values in Column 5 with those in Column 6. In Column 8 are given the values of Dq for each case estimated independently from spectroscopic data.

Another approach to answering this problem is to draw a straight line between the experimental points (U_1) for MnF_2 and ZnF_2, and to take as the thermodynamic CFSE's the vertical distances between this line and the other values for U_1. Compare your results from this graphical method with those obtained below.

The values obtained are

	Difference from Straight Line			Thermodynamic Dq(K)
FeF_2	21.8 kcal	=	7.62 kK	1905
CoF_2	24.8	=	8.67	1083
NiF_2	35.0	=	12.2	1019
CuF_2	27.0	=	9.44	1573

In neither case is the thermodynamic Dq in very good agreement with the spectroscopic value. The agreement is best for CoF_2 and NiF_2.

5.12 $\Psi = c_1\psi_1 + c_2\psi_2$

where $\psi_1 = \psi_A(1)\ \psi_B(2)$

$\psi_2 = \psi_B(1)\ \psi_A(2)$

${c_1}^2 = {c_2}^2 \qquad$ *ie.*, $c_1 = \pm c_2$

$\therefore\ \Psi_+ = \psi_A(1)\ \psi_B(2) + \psi_B(1)\ \psi_A(2)$

$\Psi_- = \psi_A(1)\ \psi_B(2) - \psi_B(1)\ \psi_A(2)$

The normalization requirement is that $\int_{-\infty}^{\infty} \Psi_\pm{}^*\Psi_\pm\ d\tau = 1$.
Assuming the wavefunctions are real, we can write:

5.11

Table 5.9.

Compound	U_1 (exp)[a] (B-H cycle) (kcal/mole)	U_2 (theoret) (Born eqn) (kcal/mole)	U_1-U_2 (kcal/mole)	U_1-U_2[b] (kK)	O_h CFSE	Thermo-dynamic Dq (kK)	Spectroscopic Dq (kK)
MnF_2	662	660	2	0.700	0	----	0.780[d] 0.765[c]
CoF_2	708	686	22	7.69	8Dq	0.961	0.837[c]
NiF_2	728	692	36	12.6	12Dq	1.05	0.801[c]
CuF_2	727	625	102	35.7	6Dq	5.94	1.08[c]
ZnF_2	710	686	24	8.39	0	----	----

[a] Calculated ignoring the difference in heat capacities.

[b] 1 kcal $mole^{-1}$ = 3.4964×10^2 cm^{-1}; 1 kK = 1000 cm^{-1}.

[c] Estimated as described by Figgis, p. 244.

[d] J. W. Stout, *J. Chem. Phys.*, 31, 709 (1959).

$$\int \Psi_+^2 d\tau = \int \psi_A^2(1)\ \psi_B^2(2)\ d\tau + 2\int \psi_A(1)\ \psi_B(1)\ \psi_A(2)\ \psi_B(2)\ d\tau + \int \psi_A^2(2)\ \psi_B^2(1)\ d\tau.$$

Since $\int \psi_A^2(1)\ \psi_B^2(2)\ d\tau = \int \psi_A^2(1)\ d\tau_1 \cdot \int \psi_B^2(2)\ d\tau_2 = 1$,

and $\int \psi_A^2(2)\ \psi_B^2(1)\ d\tau = \int \psi_A^2(2)\ d\tau_2 \cdot \int \psi_B^2(1)\ d\tau_1 = 1$,

and $2\int \psi_A(1)\ \psi_B(1)\ \psi_A(2)\ \psi_B(2)\ d\tau$

$= 2[\int \psi_A(1)\ \psi_B(1)\ d\tau_1 \cdot \int \psi_A(2)\ \psi_B(2)\ d\tau_2]$

$= 2S^2$, where S is called the overlap integral,

then $\int \Psi_+^2\ d\tau = 2 + 2S^2 = 2(1 + S^2)$.

Therefore, in order for $\int \Psi_+^2\ d\tau = 1$, the whole function must be multiplied by $N = [2(1 + S^2)]^{-1/2}$. Thus $\Psi_+ = [2(1 + S^2)]^{-1/2}\ [\psi_A(1)\ \psi_B(2) + \psi_B(1)\ \psi_A(2)]$.

Often the overlap integral is ignored. This leads to substantial error in the H_2 case (See problem 5.26). If overlap is ignored in this case $N = 1/\sqrt{2}$. The general rule, if overlap is ignored, is that $N^{-2} = c_1^2 + c_2^2 ... + c_n^2$, where c_i are the coefficients of the terms in the wave function (assuming each term is separately normalized). In this case

$$N^{-2} = (1/\sqrt{2})^{-2} = 2 = 1^2 + 1^2.$$

5.13 Ignoring overlap the normalizing coefficient is the sum of the squares of the coefficients of the terms raised to the -1/2 power.

$$(1 - \lambda)^2 + \lambda^2 = 1 - 2\lambda + 2\lambda^2$$

$$\therefore\ \Psi_{VB} = (1 - 2\lambda + 2\lambda^2)^{-1/2}\ [(1-\lambda)\psi_{cov} + \lambda\psi_{ion}]$$

5.14 The essential difference is that the two "covalent" structures and the two "ionic" structures have different weights (thus different coefficients) in the wave function for the heteronuclear case owing to the difference in electronegativity of A and B.

$$\Psi_{VB} = c_1\psi_A(1)\ \psi_B(2) + c_2\psi_A(2)\ \psi_B(1) + \lambda\{c_3\psi_A(1)\ \psi_A(2) + c_4\psi_B(1)\ \psi_B(2)\}$$

$c_1 \neq c_2;\ c_3 \neq c_4$

5.15 (a) $\%$ ionic character $= \dfrac{100\lambda^2}{1 + \lambda^2}$

CsF: $\% = 91$

(b) $\%$ ionic character $= \dfrac{100\lambda^2}{(1 - \lambda)^2 + \lambda^2} = \dfrac{100\lambda^2}{1 - 2\lambda + 2\lambda^2}$

H_2: $\% = 5.9$; $\therefore \lambda = 0.20$

CsF: $\% = 91$; $\therefore \lambda = 0.76$

5.16 Since the dipole moment is the product of the charge times the distance, if the structure were totally ionic, the moment would be

$$\mu = 4.802 \times 1.27 \times 10^{-18} \text{ esu cm}$$
$$= 6.1 \text{ Debye}$$

The experimentally determined moment is only 1.03 Debye. This result implies that the percent ionic character is

$$\frac{1.03 \times 100}{6.1} = 17\%.$$

Note that this rather crude method rests on the assumption that the pure covalent bond is totally nonpolar. For various reasons this is highly improbable. (See Cotton and Wilkinson, p. 107ff).

5.17 According to Pauling, the ionic-covalent resonance energy of the A-B bond is given by

$$\Delta = D(AB) - (1/2)\{D(AA) + D(BB)\}$$

or

$$\Delta' = D(AB) - \{D(AA)\cdot D(BB)\}^{1/2}$$

where D(AA), D(AB) and D(BB) are the bond dissociation energies of the A-B, A-A, and B-B bonds and are taken to be roughly equal to the enthalpies of formation. Pauling finds the following empirical relationship between the difference in the electronegativities of A and B, and Δ':

$$|\chi_A - \chi_B| = 0.18\sqrt{\Delta'}$$

Starting with this equation and a definition of $\chi_F = 4.0$, he proceeds as follows then to set up a complete scale of electronegativities:

$$\Delta'OF = D(OF) - \{D(OO) \cdot D(FF)\}^{1/2}$$
$$= 44.2 - \{(33.2)(36.6)\}^{1/2}$$
$$= 9.3$$
$$\therefore |\chi_O - \chi_F| = 0.18\sqrt{9.3}$$
$$= 0.5$$
$$\therefore \chi_O = 3.5$$

By the same method using CO, χ_C is found to be 2.5. Likewise using CCl and HCl and NaH, χ_{Cl}, χ_H and χ_{Na} are 3.0, 2.1 and 1.7, respectively.
(Note that a knowledge of the trends in electronegativity is required to assess the sign of $\Delta\chi$).

5.18 For HF

$$\Delta' = 134.6 - (104.2 \times 36.6)^{1/2} = 72.8 \text{ kcal/mole}$$
$$\therefore \% \text{ ionic character} = 18(0.18\sqrt{72.8})^{1.4}$$
$$= 32\%$$

Note: The value of $0.18\sqrt{\Delta'}$ of 1.5 here is less than usually accepted for $|\chi_F - \chi_H|$. Using the more generally accepted value of 1.8, the percent ionic character is 41%.

Pauling developed this relationship from the assumption that the polarity of a bond and thus the dipole moment of a diatomic molecule was directly proportional to the % ionic character. He compared measured dipole moments of a few alkali halides to the moments calculated by assuming 100% ionic character (that is, a (+) and a (-) charge separated by the bond length). As Pauling admits (Ref. 2.a, p. 98) expansion of this treatment to include other compounds gives a great deal of scatter of the data from the equation calculated for the alkali halides. Implicit in the whole argument also is the assumption that the purely covalent (0% ionic) part of the bond between the two different atoms A and B has zero dipole moment. The reasons why this is not so are discussed by Cotton and Wilkinson (p. 107ff). In short, the argument is to be taken as extremely approximate--in fact, as Cotton and Wilkinson point out, there really is no *intrinsic* connection between dipole moment and % ionic character.

5.19 Considering the polarity of the bond to be at least roughly proportional to the difference in electronegativity, the order of increasing polarity is $CS_2 < BN < BeCl_2$.

5.20 (a) For the structure $\psi_F(1)\ \psi_F(2)$ (that is, both electrons on the fluorine) we consider the energy required to take an electron from the hydrogen to the fluorine. For the structure $\psi_H(1)\ \psi_H(2)$ the energy to take an electron from the fluorine to the hydrogen is calculated.

$$H + F \rightarrow H^+ + F^- \qquad \Delta H = I(H) - E(F)$$
$$= 13.60 - 3.61 = 9.99 \text{ eV}$$

$$H + F \rightarrow H^- + F^+ \qquad \Delta H = I(F) - E(H)$$
$$= 17.42 - 0.75 = 16.67 \text{ eV}$$

The large difference in energy between these processes indicates that the two structures will not interact to any appreciable extent and only the lower energy structure $\psi_F(1)\ \psi_F(2)$ need be considered.

(b) Energy for $\psi_N(1)\ \psi_N(2) = I(Cl) - E(N) =$
$13.01 + 0.1 = 13.11$ eV

Energy for $\psi_{Cl}(1)\psi_{Cl}(2) = I(N) - E(Cl) =$
$14.54 - 3.61 = 10.93$ eV

The closeness of the energies of these two structures indicates both will be important in the "best" wave function for the N-Cl bond.

5.21 (a)

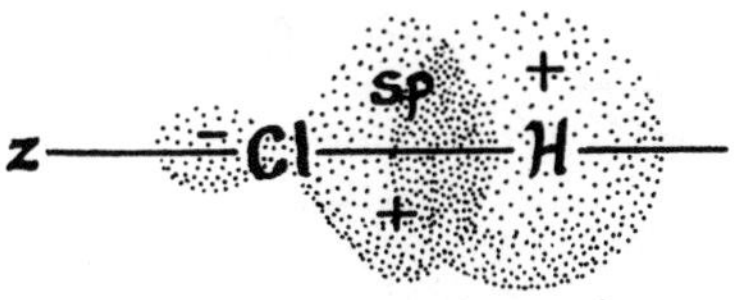

H-Cl
1σ bond
(no π bonding possible)

(b)

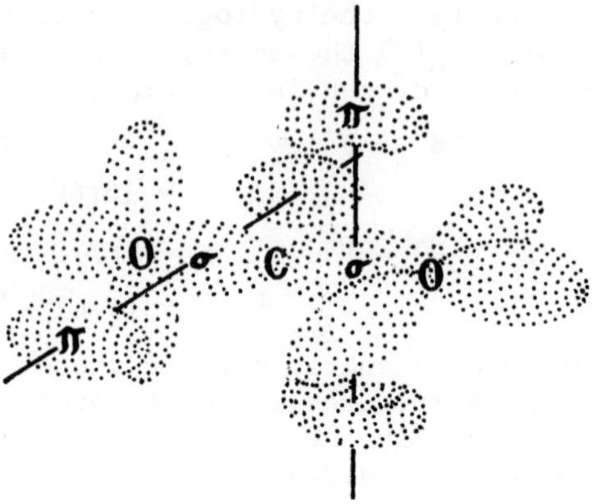

O=C=O
1σ and 1π bond between each oxygen and the carbon

(c)

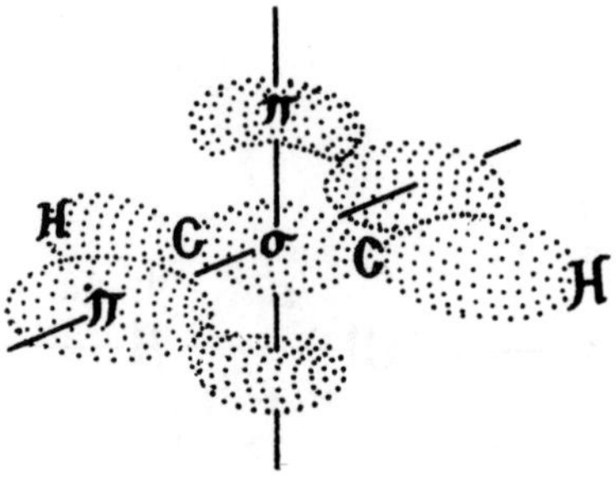

C_2H_2
1σ and 2π bonds

5.22 Each of these problems can be solved by application of simple group theory. The approach is to find the irreducible representation(s) spanned by the set of σ bonds in the complex and then to choose a collection of *s*, *p*, and *d* orbitals which transform as the same irreducible representation(s).

(a) For a tetrahedral molecule the representation spanned by the σ bonds is

T_d	E	$8C_3$	$3C_2$	$6S_4$	$6\sigma_d$
Γ_σ	4	1	0	0	2

By standard techniques (see Chapter 4) Γ_σ can be decomposed to $A_1 + T_2$. Reference to the character table for T_d shows that the collection $2s + 2p_x + 2p_y + 2p_z$ transforms as $A_1 + T_2$. Therefore, the hybrid is sp^3.

(b)

O_h	E	$8C_3$	$6C_2$	$6C_4$	$3C_2$	i	$6S_4$	$8S_6$	$3\sigma_h$	$6\sigma_d$
Γ_σ	6	0	0	2	2	0	0	0	4	2

$= A_{1g} + E_g + T_{1u}$.

A possible combination is $3d_{z^2} + 3d_{x^2-y^2} + 3s + 3p_x + 3p_y + 3p_z$; that is, d^2sp^3.

(c)

D_{3h}	E	$2C_3$	$3C_2$	σ_h	$2S_3$	$3\sigma_v$
Γ_σ	5	2	1	3	0	3

$= 2A_1' + E' + A_2''$

5.23 The four sigma bonds form a basis for the reducible representation:

T_d	E	$8C_3$	$3C_2$	$6S_4$	$6\sigma_d$
Γ_σ	4	1	0	0	2

$= A_1 + T_2$

From the character table we find that the s-orbital transforms as A_1, that the set p_x, p_y, p_z transforms as T_2, and, also, that the set (d_{xz}, d_{yz}, d_{xy}) transforms as T_2. Thus, assuming d-orbital participation, the silicon sigma bonding orbitals might be considered a mixture of sp^3 and sd^3 hybrids.

5.24 This is a rather long problem which can be "solved" much more quickly by some simple algebraic manipulations, but going through it systematically is valuable since it illustrates techniques which are required for more difficult problems.

First determine the symmetry properties of the trigonal hybrid orbitals. Let the reducible representation spanned by the set of three sigma bonds be $\Gamma_{\sigma_{1,2,3}}$.

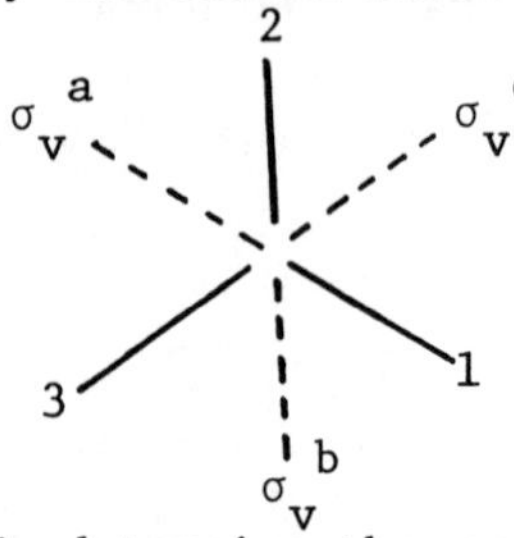

D_{3h}	E	$2C_3$	$3C_2$	σ_h	$2S_3$	$3\sigma_v$
$\Gamma_{\sigma_{1,2,3}}$	3	0	1	3	0	1

$= A_1' + E'$

To determine the symmetry orbitals, we first construct a transformation table. Then for each row of the table, take the following sum:

$\sum_s D_s^{(\alpha)}\sigma_s$, where the sum is over the s elements of the group. The factor $D_s^{(\alpha)}$ is the character of the operation s of the irreducible representation, and σ_s is the corresponding entry in each row of the transformation table. Here α is A_1' and E". The transformation table is

	E	C_3^1	C_3^2	C_2^a	C_2^b	C_2^c	σ_h	S_3^1	S_3^2	σ_v^a	σ_v^b	σ_v^c
σ_1	σ_1	σ_2	σ_3	σ_1	σ_3	σ_2	σ_1	σ_2	σ_3	σ_1	σ_3	σ_2
σ_2	σ_2	σ_3	σ_1	σ_3	σ_2	σ_1	σ_2	σ_3	σ_1	σ_3	σ_2	σ_1
σ_3	σ_3	σ_1	σ_2	σ_2	σ_1	σ_3	σ_3	σ_1	σ_2	σ_2	σ_1	σ_3
A_1'	1	1	1	1	1	1	1	1	1	1	1	1
E'	2	-1	-1	0	0	0	2	-1	-1	0	0	0

For A_1', $\sum_s D_s^{(A_1')}\sigma_s$ gives us

$4(\sigma_1 + \sigma_2 + \sigma_3)$ from the first row,

$4(\sigma_1 + \sigma_2 + \sigma_3)$ from the second row, and

$4(\sigma_1 + \sigma_2 + \sigma_3)$ from the third row.

Consequently the normalized A_1' symmetry orbital is $(1/\sqrt{3})(\sigma_1 + \sigma_2 + \sigma_3)$. For E' we get

$2(2\sigma_1 - \sigma_2 - \sigma_3)$ from the first row,

$2(2\sigma_2 - \sigma_3 - \sigma_1)$ from the second row, and

$2(2\sigma_3 - \sigma_2 - \sigma_1)$ from the third row.

These three expressions are not orthogonal, so we choose one, the first, and take a linear combination of the other two. The normalized symmetry orbitals are

$$(1/\sqrt{6})(2\sigma_1 - \sigma_2 - \sigma_3) \text{ and } (1/\sqrt{2})(\sigma_2 - \sigma_3).$$

We can summarize in matrix notation,

$$\begin{pmatrix} 1/\sqrt{3} & 1/\sqrt{3} & 1/\sqrt{3} \\ 2/\sqrt{6} & -1/\sqrt{6} & -1/\sqrt{6} \\ 0 & 1/\sqrt{2} & -1/\sqrt{2} \end{pmatrix} \begin{pmatrix} \sigma_1 \\ \sigma_2 \\ \sigma_3 \end{pmatrix} = \begin{pmatrix} b_1 \\ b_2 \\ b_3 \end{pmatrix}$$

where $\sigma_{1,2,3}$ are the hybrid orbitals and $b_{1,2,3}$ are a set of orthonormal functions which also have the symmetry of the group. Such a set is a set of atomic orbitals which transform as $A_1' + E'$, *e.g.*, s and p_x, p_y. To find $\sigma_{1,2,3}$ in terms of linear combinations of the s, p_x and p_y orbitals we need to find $\mathbb{A}^{-1}$. In general $\mathbb{A}^{-1}$ = cofactor $\mathbb{A}^{\dagger *}/|\mathbb{A}|$, but since we have used orthonormal sets of functions, in this case $\mathbb{A}$ is unitary, and thus, $\mathbb{A}^{-1} = \mathbb{A}^{\dagger *}$. Further, since the functions are all real, $\mathbb{A}^{-1} = \mathbb{A}^{\dagger}$. That is,

$$\mathbb{A}^{-1} = \begin{pmatrix} 1/\sqrt{3} & 2/\sqrt{6} & 0 \\ 1/\sqrt{3} & -1/\sqrt{6} & 1/\sqrt{2} \\ 1/\sqrt{3} & -1/\sqrt{6} & -1/\sqrt{2} \end{pmatrix}$$

So multiplying from the left we have,

$$\begin{pmatrix} 1/\sqrt{3} & 2/\sqrt{6} & 0 \\ 1/\sqrt{3} & -1/\sqrt{6} & 1/\sqrt{2} \\ 1/\sqrt{3} & -1/\sqrt{6} & -1/\sqrt{2} \end{pmatrix} \begin{pmatrix} 1/\sqrt{3} & 1/\sqrt{3} & 1/\sqrt{3} \\ 2/\sqrt{6} & -1/\sqrt{6} & -1/\sqrt{6} \\ 0 & 1/\sqrt{2} & -1/\sqrt{2} \end{pmatrix} \begin{pmatrix} \sigma_1 \\ \sigma_2 \\ \sigma_3 \end{pmatrix} =$$

$$\begin{pmatrix} 1/\sqrt{3} & 2/\sqrt{6} & 0 \\ 1/\sqrt{3} & -1/\sqrt{6} & 1/\sqrt{2} \\ 1/\sqrt{3} & -1/\sqrt{6} & -1/\sqrt{2} \end{pmatrix} \begin{pmatrix} s \\ p_x \\ p_y \end{pmatrix}$$

$$\begin{pmatrix} \sigma_1 \\ \sigma_2 \\ \sigma_3 \end{pmatrix} = \begin{pmatrix} 1/\sqrt{3} & 2/\sqrt{6} & 0 \\ 1/\sqrt{3} & -1/\sqrt{6} & 1/\sqrt{2} \\ 1/\sqrt{3} & -1/\sqrt{6} & -1/\sqrt{2} \end{pmatrix} \begin{pmatrix} s \\ p_x \\ p_y \end{pmatrix} =$$

$$\begin{matrix} (1/\sqrt{3})s + (2/\sqrt{6})p_x \\ (1/\sqrt{3})s - (1/\sqrt{6})p_x + (1/\sqrt{2})p_y \\ (1/\sqrt{3})s - (1/\sqrt{6})p_x - (1/\sqrt{2})p_y \end{matrix}$$

5.25 The simplest most reliable qualitative criterion for bond strength is that of overlap. Overlaps can be positive, in which case accumulation of electron density between the nuclei is indicated, negative in which case electron density between the nuclei is reduced, or *exactly* zero in which case no net interaction occurs. The magnitude of a finite net overlap depends sensitively on the internuclear separation and on the individual radial functions. However, it is very easy to tell if the overlap is zero or non-zero by simply looking at the symmetry of the angular functions.

(a)

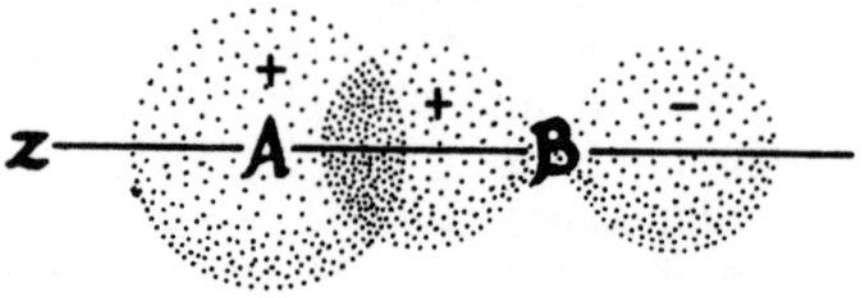

overlap finite; σ bond

(b)

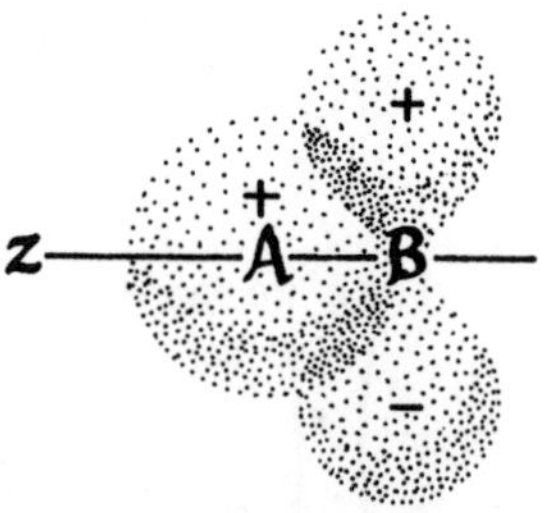

overlap exactly zero

(c)

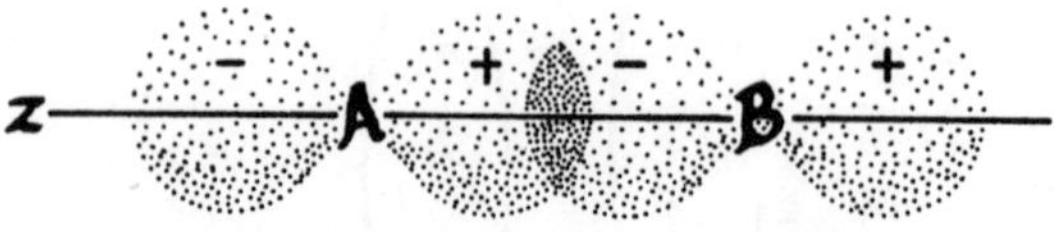

overlap finite; σ bond

(d)

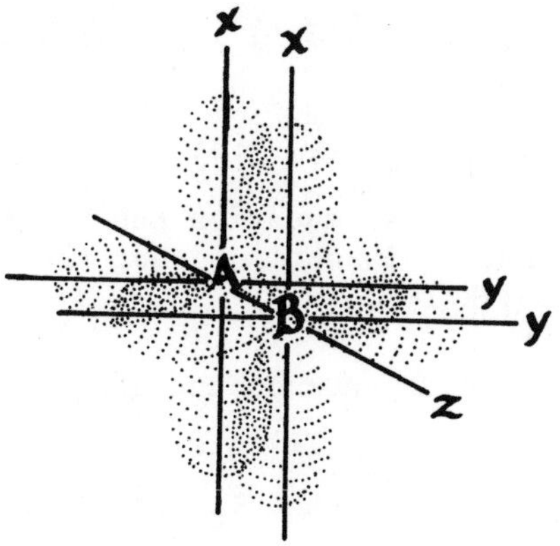

overlap finite;
δ bond

(f)

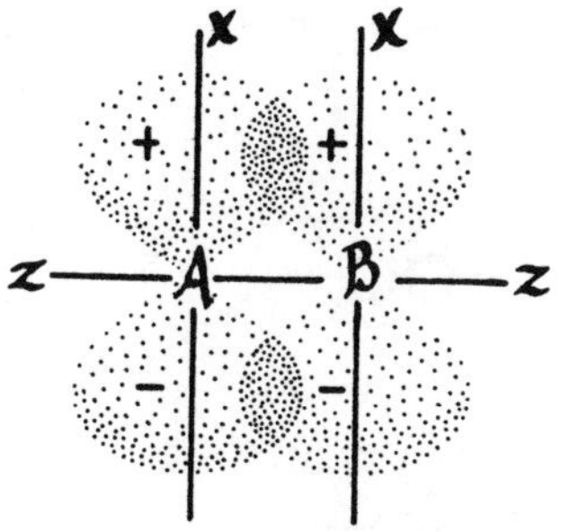

overlap finite;
π bond

5.26 For the function $\Psi = \psi_A + \psi_B$ the normalization constant N is $(2 + 2S)^{-1/2} = 0.56$. If overlap is neglected N is $(2)^{-1/2} = 0.71$. Thus in this case there is a substantial error incurred in neglecting the overlap integral. However, in most cases (with larger atoms) S is very small and can be neglected without introducing significant error.

5.27 Let the valence bond function be

$$\Psi_{VB} = (ab + ba) + \lambda(aa + bb)$$

and the molecular orbital function be

$$\Psi_{MO} = (a + b)(a + b) + k(a - b)(a - b)$$

in a short-hand notation. Then expanding each function

$$\Psi_{VB} = ab + ba + \lambda aa + \lambda bb = 2ab + \lambda(aa + bb)$$

$$\Psi_{MO} = aa + 2ab + bb + kaa - 2kab + kbb$$
$$= aa\ (1 + k) + bb\ (1 + k) + 2ab\ (1 - k)$$
$$= (aa + bb)(1 + k) + 2ab\ (1 - k)$$
$$= \{2ab + [(1+k)/(1-k)](aa + bb)\}\ (1 - k)$$

Since (1 - k) in the last equation can be absorbed into the normalizing constant it is obvious that the two functions are equal if only $\lambda = (1+k)/(1-k)$.

5.28 (a) The difference depends on the gap between the 2*s* and 2*p* atomic energies. If this is relatively small, the interaction between the σ^b_{2s} and σ^b_{2p} orbitals may push the σ^b_{2p} orbital to higher energy than the π^b_{2p}. This apparently happens in the case of B_2, but in F_2 the *s*-*p* separation is probably too great.

(1)

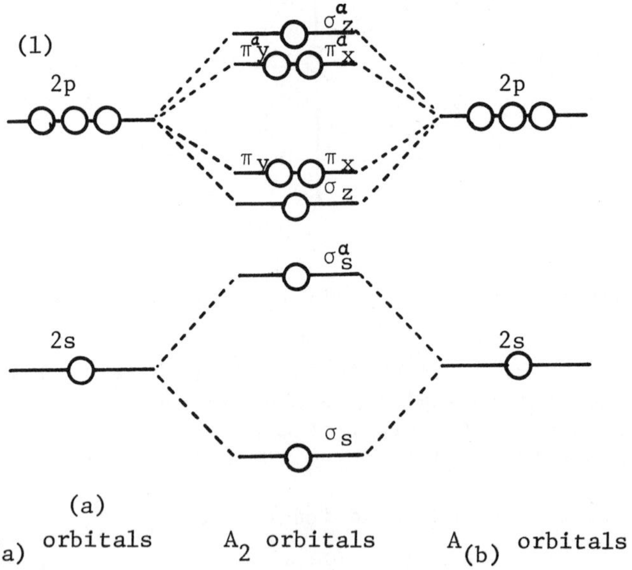

(a)

F_2-type

(2)

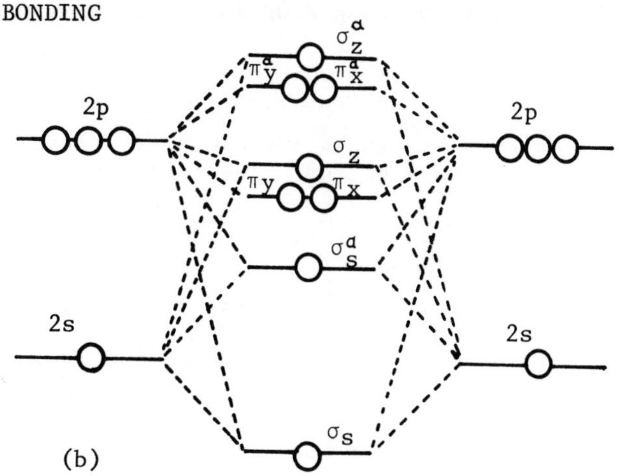

$A_{(a)}$ orbitals A_2 orbitals $A_{(b)}$ orbitals

B_2-type

(b) The B_2 molecule has ten electrons. If the order is as in (2) above the distribution will be $(\sigma_{1s}{}^{b})^2\ (\sigma_{1s}{}^{a})^2\ (\sigma_{2s}{}^{b})^2\ (\sigma_{2s}{}^{a})^2\ (\pi_{x,y}{}^{b})^2$. On the other hand if the order of levels is as in (1), the configuration will be $(\sigma_{1s}{}^{b})^2\ (\sigma_{1s}{}^{a})^2\ (\sigma_{2s})^2\ (\sigma_{2s}{}^{a})^2\ (\sigma_z{}^{b})^2$. The electrons in the degenerate set $(\pi_{x,y})^2$ will be unpaired and thus, if the order is as in (2), B_2 will be paramagnetic. All electrons in the configuration arising from level scheme (1) will be paired. Thus, magnetic susceptibility measurement should establish the correct order. (Experimentally, B_2 is found to be paramagnetic, indicating that scheme (2) is correct.)

5.29 In the molecular orbital model bond order is the number of pairs of electrons in bonding orbitals minus the number of pairs of electrons in antibonding orbitals.

(a)(1) The configuration of B_2 is given in 5.28 b. The bond order is one.

(2) For C_2 the configuration is $KK(\sigma_{2s}{}^{b})^2(\sigma_{2s}{}^{a})^2(\pi_{2p_{x,y}}{}^{b})^4$. Thus the bond order is two.

(3) For Cl_2 the bond order is one.

(b) The bond length is inversely proportional to the bond order, as expected. The configuration of O_2 is $KK(\sigma_{2s}{}^{b})^2\ (\sigma_{2s}{}^{a})^2\ (\pi_{2p_{x,y}}{}^{b})^4\ (\sigma_{p_z}{}^{b})^2\ (\pi_{2p_{x,y}}\ )^2$. Thus the bond orders are

O_2^{+}, 2 1/2; O_2, 2; O_2^{-}, 1 1/2; $O_2^{=}$, 1.

5.31 In the heteronuclear diatomic case the energy differences of the corresponding atomic orbitals must be considered as well as their occupancy. One of the criteria for molecular orbital interaction is that the atomic orbital energies be close to each other. Furthermore, as in the homonuclear diatomic cases, the 1s orbitals of the second period atoms are not close enough to each other to overlap appreciably owing to the repulsion between the 2s electrons. That is to say, the 1s electrons in the second period elements are essentially non-bonding. Energy level differences, and established covalencies indicate that other non-bonding orbitals are also formed in some cases. (See for example (c) below). The bond orders in these molecules are given by the same formula as for the homonuclear cases; the non-bonding electrons do not contribute.

(a) Bond order is 2 1/2.

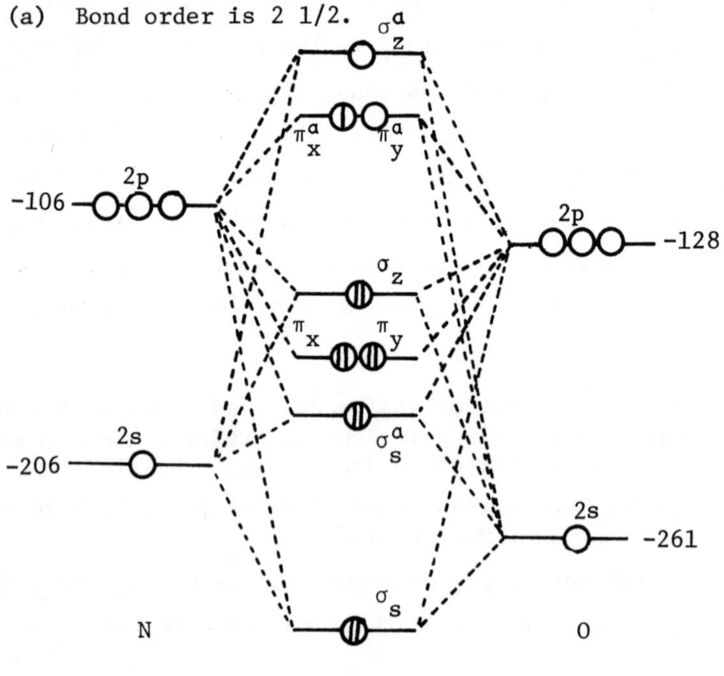

(b) Bond order is 2.

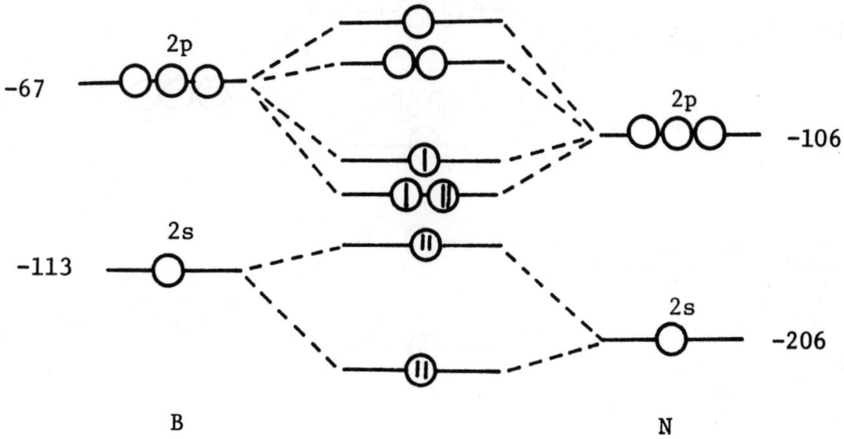

(c)

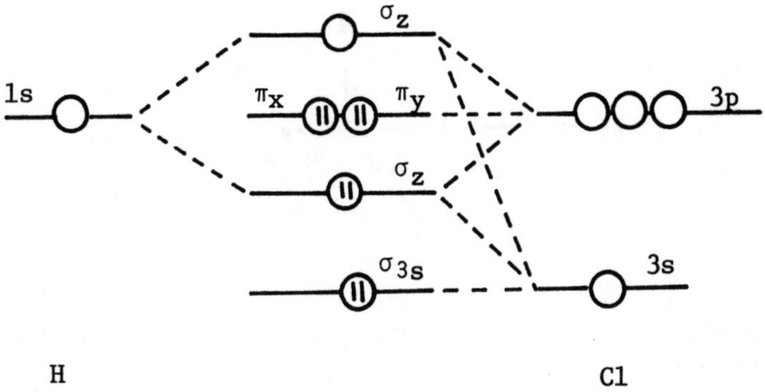

(d)

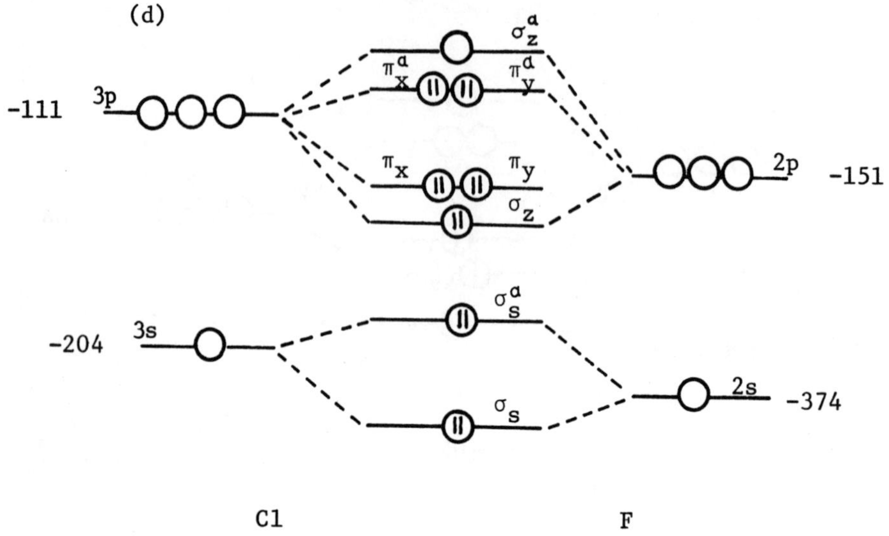

5.32 Linear symmetric triatomic molecules belong to the point group $D_{\infty h}$. The useful coordinate system is

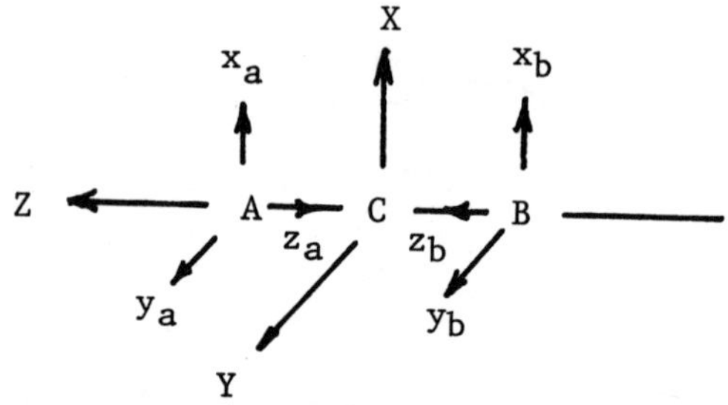

The central atom has available for σ bonding its s and p_z orbitals which transform as Σ_g^+ and Σ_u^- respectively. The linear combination of A and B orbitals used for σ MO's are

$$\psi_{\Sigma_g^+} = (1/\sqrt{2})[\phi_{sA} + \phi_{sB}]$$

$$\psi_{\Sigma_g^+} = (1/\sqrt{2})[\phi_{p_zA} + \phi_{p_zB}]$$

$$\psi_{\Sigma_u^-} = (1/\sqrt{2})[\phi_{sA} - \phi_{sB}]$$

$$\psi_{\Sigma_u^-} = (1/\sqrt{2})[\phi_{p_zA} - \phi_{p_zB}]$$

For π bonding the central atom and the other two can use their p_x and p_y orbitals which transform as Π_u. The $A + B$ LCAO's are

$$\psi_{\Pi_{u_x}} = 1/\sqrt{2}\ [\phi_{p_xA} + \phi_{p_xB}]$$

$$\psi_{\Pi_{u_y}} = 1/\sqrt{2}\ [\phi_{p_yA} + \phi_{p_yB}]$$

The other combinations $\phi_{p_xA} - \phi_{p_xB}$ and $\phi_{p_yA} - \phi_{p_yB}$ give identically zero overlap and, in fact, do not transform as Π_u, but as Π_g. They are non-bonding ligand LCAO's.

The MO wave functions are of the form:

$$\psi_{\Gamma_i}{}^b = N^b[\lambda\psi_{\Gamma_i}(C) + (1-\lambda)\psi_{\Gamma_i}(AB)]$$

$$\psi_{\Gamma_i}{}^a = N^a[\lambda\psi_{\Gamma_i}(C) - (1-\lambda)\psi_{\Gamma_i}(AB)]$$

(a) Only σ bonding is possible in BeH_2 since H has no available Π-type orbitals and only s orbitals are available for combination with the beryllium s and p orbitals.

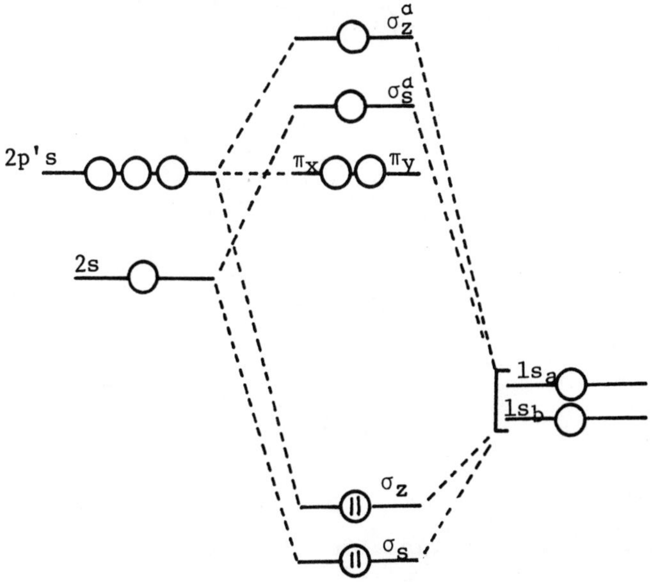

(b) In CO_2 both *s* and *p* orbitals are available on all atoms and thus σ and π MO's must be included. To simplify the diagram only the oxygen *p* orbitals have been used in bonding.

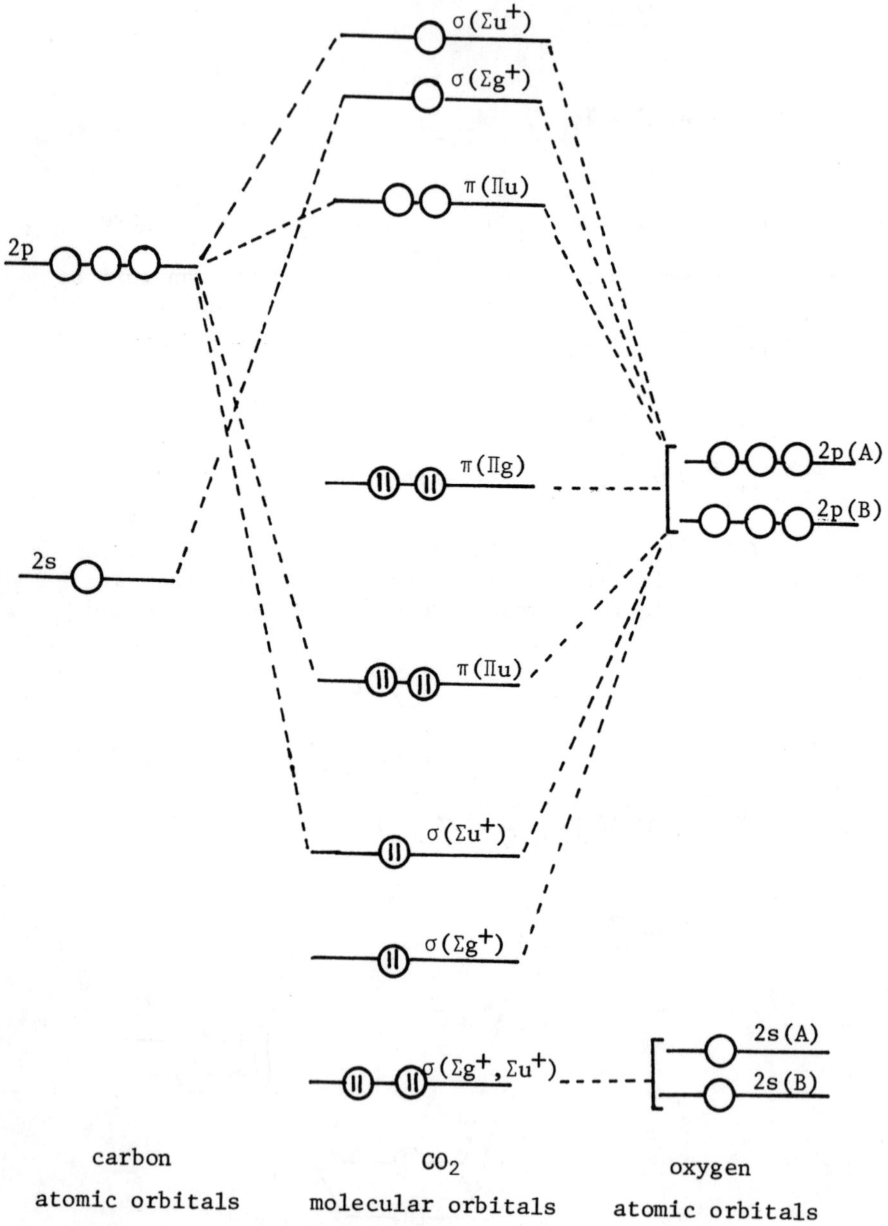

5.33 The term symbols are of the form $^{(2S+1)}X$, where X is a symbol determined by the magnitude of $M_L = \sum_i m_{\ell i}$ for all i electrons. The m_ℓ values corresponding to the several types of molecular orbitals in linear molecules are given in Table 5.10

Table 5.10

Molecular orbitals	m_ℓ	Atomic orbitals
σ	0	s, p_z, d_{z^2}
π	±1	p_x, p_y, d_{xz}, d_{yz}
δ	±2	d_{xy}, $d_{x^2-y^2}$

Since linear molecules belong either to point group $C_{\infty v}$ or $D_{\infty h}$, the molecular orbital states can be classified as irreducible representations of one of these special point groups. The term symbols corresponding to the several values of M_L are Σ, Π, Δ, and Φ for M_L = 0, ±1, ±2, and ±3 respectively. The procedure for handling degenerate configurations is entirely analogous to the procedures for sorting out atomic term symbols. Hund's rules apply to assignment of ground terms.

(a) Both electrons in H_2 are in the $\sigma_s{}^b$ orbital. Therefore, $M_L = m_{\ell a} + m_{\ell b} = 0 + 0 = 0$. The electron spins must be paired so $M_s = 0$ and therefore $S = 0$. The term symbol is $^1\Sigma$.

(b) Since for any filled orbital $\Sigma m_\ell = \Sigma m_s = 0$, only partially filled orbitals need be considered in O_2. This is because the term symbol for a state will be the direct product of the representations of all the electrons and $^1\Sigma \times \Gamma_n = \Gamma_n$. The only incompletely filled orbital in O_2 is $\pi_{2p_{x,y}}{}^a$. The two electrons can be arranged several ways. A mnemonic device similar to that in Chapter 3 is used.

M_L \ M_S	+1	0	-1
2		$(\overset{+}{\pi}_1\overset{-}{\pi}_1)$	
1			
0	$(\overset{+}{\pi}_1\overset{+}{\pi}_{-1})$	$(\overset{+}{\pi}_1\overset{-}{\pi}_{-1})(\overset{-}{\pi}_1\overset{+}{\pi}_{-1})$	$(\overset{-}{\pi}_1\overset{-}{\pi}_{-1})$
-1			
-2		$(\overset{+}{\pi}_{-1}\overset{-}{\pi}_{-1})$	

The states are thus $^3\Sigma$, $^1\Sigma$, and $^1\Delta$. Hund's rule requires that $^3\Sigma$ be the lowest. This is consistent with the measured paramagnetism.

(c) There are no partially filled orbitals in BF and therefore the ground term symbol can be seen immediately to be $^1\Sigma$.

(d) The interaction between the $\sigma_{2s}{}^a$ and $\sigma_{2p_z}{}^b$ in BO probably pushes the latter above the $\pi_{2p_{x,y}}{}^b$ orbital causing the unpaired electron to be in the $\sigma_{2p_z}{}^b$ level. Therefore the term symbol is $^2\Sigma$.

(e) The closeness of the $\pi_{2p_{x,y}}{}^b$ and $\sigma_{2p_z}{}^b$ orbitals in BN causes a population of both in the lowest state. The term symbol is then $^3\Pi \times {}^3\Sigma = {}^3\Pi$.

5.34 (a) The tetrahedral hybrid orbitals of the oxygen are

$\Psi_1 = (1/2)\,[2s + 2p_x + 2p_y + 2p_z]$

$\Psi_2 = (1/2)\,[2s - 2p_x - 2p_y + 2p_z]$

$\Psi_3 = (1/2)\,[2s - 2p_x + 2p_y - 2p_z]$

$\Psi_4 = (1/2)\,[2s + 2p_x - 2p_y - 2p_z]$

(See problems 5.24 and 5.44)

Choosing any two of the above orbitals, the two localized (two-center) molecular orbitals are

$\phi_1 = \alpha\Psi_1 + \beta\sigma(H_1)$

$\phi_2 = \alpha\Psi_2 + \beta\sigma(H_2)$

Qualitatively the energies will be given

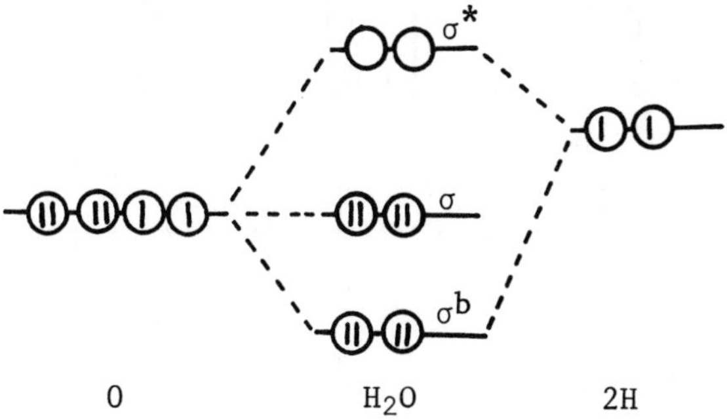

(b) The first step is to determine the symmetry properties of each of the oxygen orbitals and of the set of two equivalent hydrogen 1s orbitals in the point group C_{2v}.

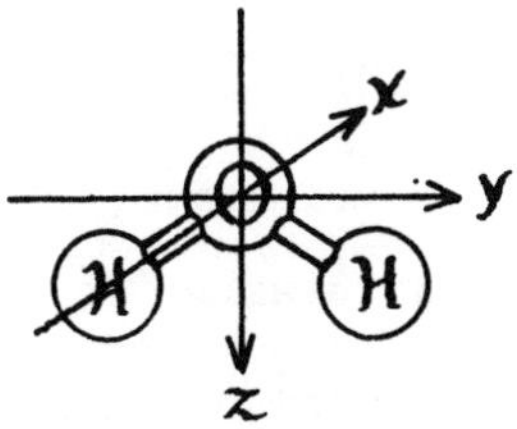

The representations spanned by each of the orbitals are found by determining their transformation properties under the operations of the point group.

C_{2v}	E	C_2	$\sigma_v(xz)$	$\sigma_v(yz)$	
$\chi 1s$	1	1	1	1	$= A_1$
$\chi 2s$	1	1	1	1	$= A_1$
$\chi 2p_x$	1	-1	1	-1	$= B_1$
$\chi 2p_y$	1	-1	-1	1	$= B_2$
$\chi 2p_z$	1	1	1	1	$= A_1$

The matrices of the representation for which the set of two 1s hydrogen orbitals forms a basis are

E	C_2	$\sigma_v(xz)$	$\sigma_v(yz)$
$\begin{bmatrix} 1 & 0 \\ 0 & 1 \end{bmatrix}$	$\begin{bmatrix} 0 & 1 \\ 1 & 0 \end{bmatrix}$	$\begin{bmatrix} 0 & 1 \\ 1 & 0 \end{bmatrix}$	$\begin{bmatrix} 1 & 0 \\ 0 & 1 \end{bmatrix}$
$\chi = 2$	0	0	$2 = A_1 + B_2$

The linear combinations which transform as A_1 and B_2 are

A_1: $(1s_a + 1s_b)$

B_2: $(1s_a - 1s_b)$

Six molecular symmetry orbitals of A_1 and B_2 symmetry and a non-bonding orbital $2p_{xO}$ of B_1 symmetry are formed.

$$na_1 = C_{i_1}(1s_a + 1s_b) + C_{i_2}1s_O + C_{i_3}2s_O + C_{i_4}2p_{zO}$$

$$(n = i = 1,2,3,4)$$

$$mb_2 = C_{i_5}(1s^a - 1s^b) + C_{i_6}2p_{yO}$$

$$(i = 5,6;\ m = i-4 = 1,2)$$

The orbital energy diagram as determined by the calculation of F. O. Edison and H. Shull, *J. Chem. Phys.*, 23, 2348 (1955), is given below.

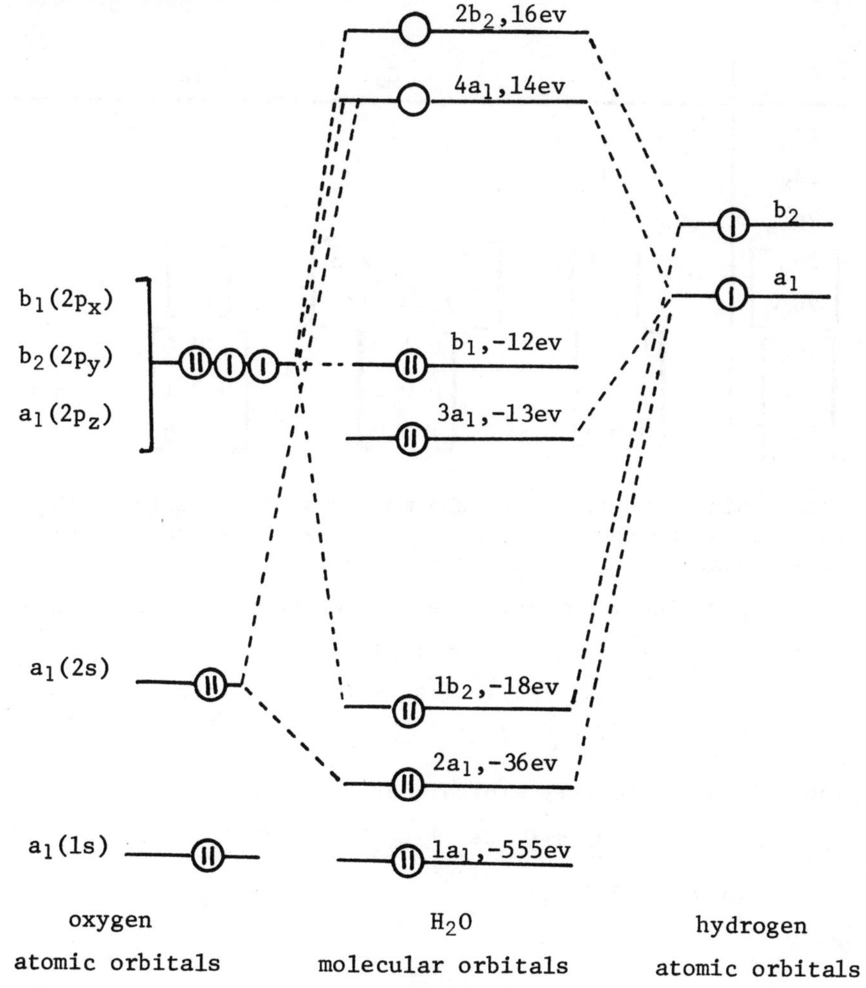

5.35 The point group is C_{3v}. The representations spanned by the basis set are

C_{3v}	E	C_3	$\sigma_v(a)$	
$1s_N$	1	1	1	$= A_1$
$2s_N$	1	1	1	$= A_1$
$2p_{zN}$	1	1	1	$= A_1$
$\begin{bmatrix} 2p_{xN} \\ 2p_{yN} \end{bmatrix}$	$\begin{bmatrix} 1 & 0 \\ 0 & 1 \end{bmatrix}$	$\begin{bmatrix} -1/2 & -1/2\sqrt{3} \\ 1/2\sqrt{3} & -1/2 \end{bmatrix}$	$\begin{bmatrix} 1 & 0 \\ 0 & -1 \end{bmatrix}$	$= E$
$\begin{bmatrix} 1s_a \\ 1s_b \\ 1s_c \end{bmatrix}$	$\begin{bmatrix} 1 & 0 & 0 \\ 0 & 1 & 0 \\ 0 & 0 & 1 \end{bmatrix}$	$\begin{bmatrix} 0 & 1 & 0 \\ 0 & 0 & 1 \\ 1 & 0 & 0 \end{bmatrix}$	$\begin{bmatrix} 1 & 0 & 0 \\ 0 & 0 & 1 \\ 0 & 1 & 0 \end{bmatrix}$	$= A_1 + E$

Note: Since the traces of the matrices of elements in the same class are the same, these three matrices are sufficient to establish the identity of the representations.

The proper linear combinations of the hydrogen orbitals are

$$H_z = (1/\sqrt{3})(1s_a + 1s_b + 1s_c) \qquad = A_1$$

$$\left.\begin{aligned} H_x &= (\sqrt{2/3})1s_a - 1/\sqrt{6}\,(1s_b + 1s_c) \\ H_y &= (1/\sqrt{2})(1s_b - 1s_c) \end{aligned}\right\} = E$$

The molecular symmetry orbitals are

$$na_1 = C_{i_1}1s_N + C_{i_2}2s_N + C_{i_3}2p_{zN} + C_{i_4}H_z$$

$$(n = i = 1,2,3,4)$$

and

$$me = \left\{ \begin{aligned} C_{i_5}2p_{xN} + C_{i_6}H_x \\ C_{i_5}2p_{xN} + C_{i_6}H_y \end{aligned} \right\}$$

$$(i = 5,6;\ m = i-4 = 1,2)$$

The orbital energy level diagram as calculated by H. Kaplan, *J. Chem. Phys.*, 26, 1704 (1957) is given below.

nitrogen NH_3 hydrogen

atomic orbitals   molecular orbitals   atomic orbitals

5.36   (a) O_h

$$\left\{ \begin{array}{l} d_{x^2-y^2} \\ d_{z^2} \end{array} \right.$$

$$\Delta_0 = 10 \ Dq \ (oct)$$

$$\left\{ \begin{array}{l} d_{xz} \\ d_{yz} \\ d_{xy} \end{array} \right.$$

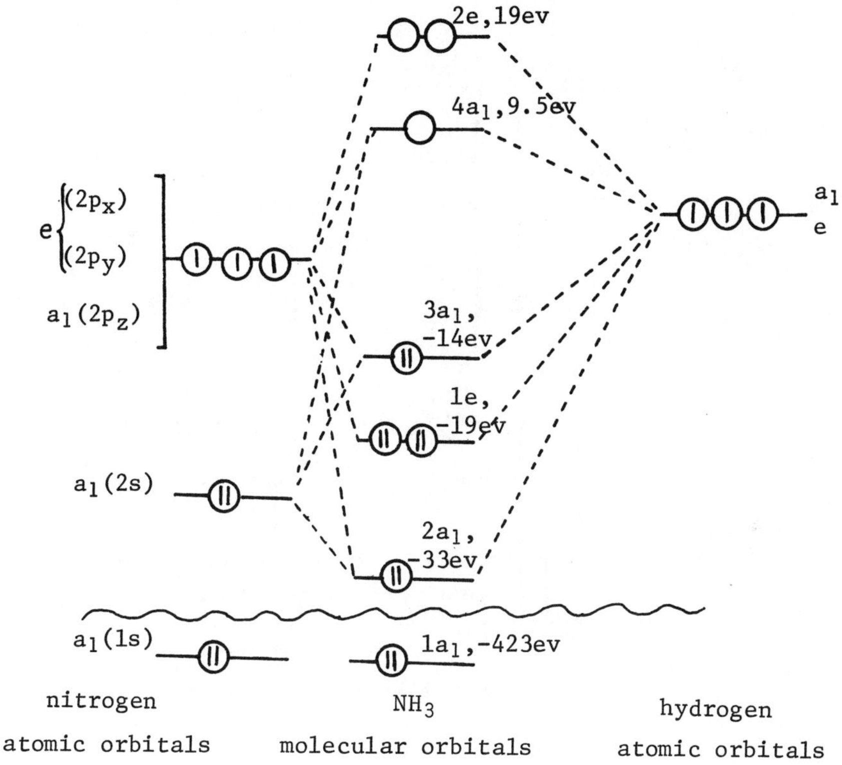

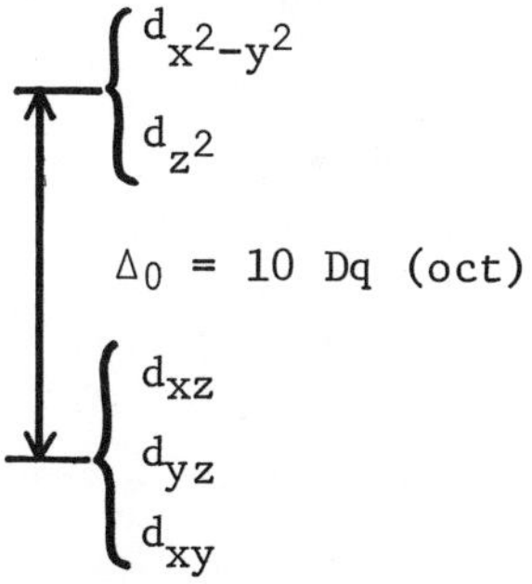

$\{ \}$ indicates degeneracy

(b) T_d

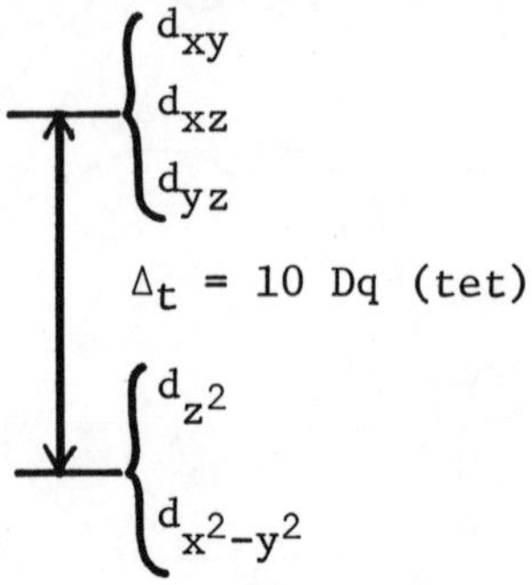

(c) D_{4h}

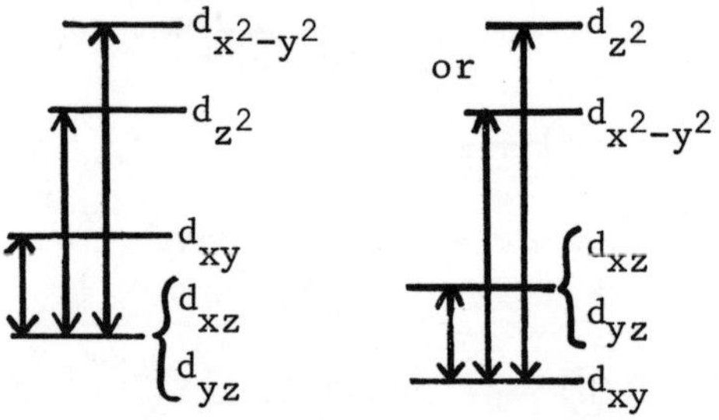

elongated along four-fold axis — compressed along four-fold axis

(d) D_3

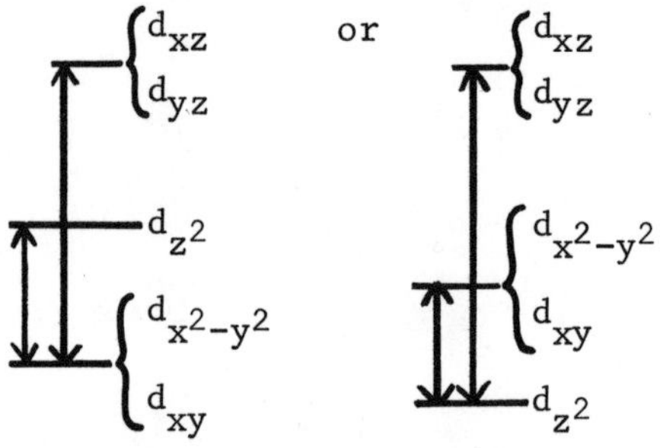

compression along three-fold axis — elongation along three-fold axis

5.37 The most general answer to this question proceeds, as Cotton (F. A. Cotton, *J. Chem. Educ.*, 41, 466 (1964)) has pointed out, by avoiding the "center of gravity" concept entirely. For a series of transition metal ions having the configurations d^n; n = 0, 1-10, in an octahedral field the d-like orbitals are found to be split into two sets, the e_g

antibonding orbitals, and the t_{2g} non-bonding orbitals. As n increases from 0 to 10 the populations of these two sets increase from 0 to 4 and from 0 to 6 respectively. This would be an increase of 2/5 electron for the e_g and 3/5 electron for the t_{2g} per integral change in n *if* the populations of the e_g level and the t_{2g} level increased *uniformly*. In fact they do not. If the populations expected for uniform increase are compared to the actual populations, the differences yield the stabilization in terms of the octahedral (t_{2g}-e_g) splitting Δ_0. The data are tabulated in Table 5.11.

5.38 (a) $NiAl_2O_4$ -- Using the method of 5.37, we see that for Ni(II) CFSE(T_d) = $(4/5)\Delta_t$ and CFSE(O_h) = $(6/5)\Delta_0$. Since $\Delta_t = (4/9)\Delta_0$, then OSPE $= (6/5)\Delta_0 - (4/5)\Delta_t$

$$= (6/5)\Delta_0 - (4/5)(4/9)\Delta_0$$

$$= 0.84\Delta_0$$

With such a large OSPE, we predict $NiAl_2O_4$ to be an inverse spinel. In fact, $NiAl_2O_4$ is about 75% inverse; the site-occupation formula is $Ni_{0.25}Al_{0.75}[Ni_{0.75}Al_{1.25}]O_4$, where the brackets indicate the octahedral sites. See D. S. McClure, *J. Phys. Chem. Solids*, 3, 311 (1957).

(b) $MnAl_2O_4$ -- Since OSPE for both ions = 0 the spinel should be normal.

(c) $FeAl_2O_4$ -- OSPE for Fe(II) $= (2/5)\Delta_0 - (3/5)(4/9)\Delta_0 = 0.13\Delta_0$. It is difficult to predict on the basis of such a small OSPE. (In fact, the spinel is normal).

(d) $MgCr_2O_4$ -- OSPE for Cr(III) = $0.84\Delta_0$; predict normal structure.

Table 5.11

Actual and "Uniform" Orbital Populations in an Octahedral Ligand Field[a]

Total "d" electrons		0	1	2	3	4	5	6	7	8	9	10
"Uniform" populations	e_g	0	$\frac{2}{5}$	$\frac{4}{5}$	$\frac{6}{5}$	$\frac{8}{5}$	2	$\frac{12}{5}$	$\frac{14}{5}$	$\frac{16}{5}$	$\frac{18}{5}$	4
	t_{2g}	0	$\frac{3}{5}$	$\frac{6}{5}$	$\frac{9}{5}$	$\frac{12}{5}$	3	$\frac{18}{5}$	$\frac{21}{5}$	$\frac{24}{5}$	$\frac{27}{5}$	6
Actual populations	e_g	0	0	0	0	1	2	2	2	2	3	4
	t_{2g}	0	1	2	3	3	3	4	5	6	6	6
Population differences		0	$\frac{2}{5}$	$\frac{4}{5}$	$\frac{6}{5}$	$\frac{3}{5}$	0	$\frac{2}{5}$	$\frac{4}{5}$	$\frac{6}{5}$	$\frac{3}{5}$	0
Stabilizations (CFSE's)		0	$\frac{2}{5}\Delta_0$	$\frac{4}{5}\Delta_0$	$\frac{6}{5}\Delta_0$	$\frac{3}{5}\Delta_0$	0	$\frac{2}{5}\Delta_0$	$\frac{4}{5}\Delta_0$	$\frac{6}{5}\Delta_0$	$\frac{3}{5}\Delta_0$	0

[a] The populations given here are for spin-free cases.

5.39 (a)

Configuration	Ground State	Examples
(1) d^1	$^2T_{2g}$	Ti(III) V(IV)
(2) d^2	$^3T_{1g}$	V(III)
(3) d^4 high spin	^5E_g	Cr(II)
(4) d^4 low spin	$^3T_{1g}$	Cr(II)
(5) d^5 low spin	$^2T_{2g}$	Mn(II) Fe(III)
(6) d^6 high spin	$^5T_{2g}$	Fe(II) Co(III)
(7) d^7 high spin	$^4T_{1g}$	Co(II)
(8) d^7 low spin	^2E_g	Co(II)
(9) d^9	^2E_g	Cu(II)

(b)

(1) (1), (2), (4), (5), (6), (7)

(2) (1)-(9)

5.40 Using the coordinate system shown below, the energy level diagrams indicate that elongation along the z axis will remove the degeneracy of the t_{2g} level and give rise to a structure with a non-degenerate ground state. According to the Jahn-Teller theorem, this is the more likely distortion. Note that the flattening distortion still leaves the ground state degenerate.

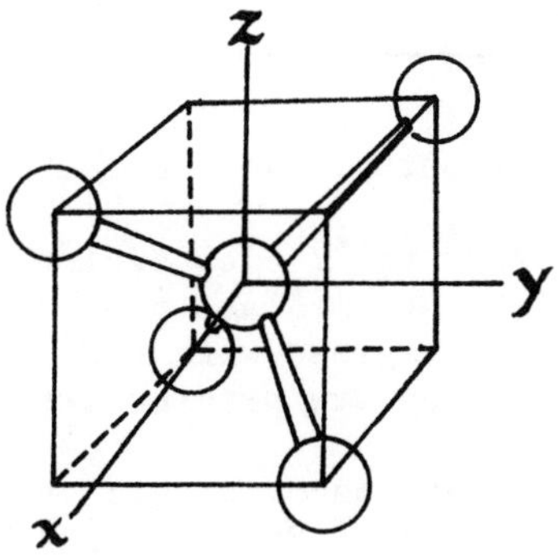

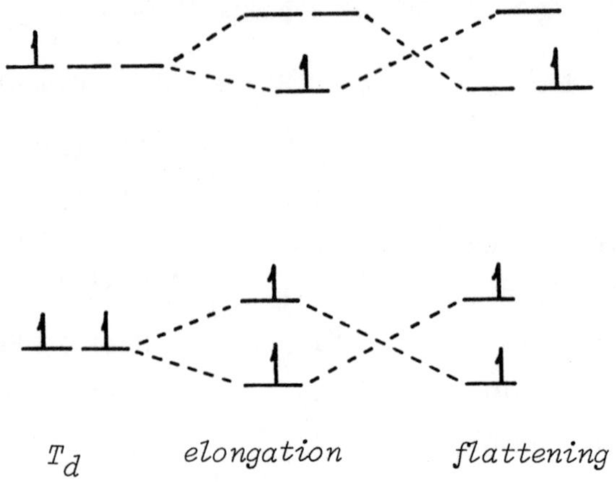

5.41

Strong Fields

Configuration	CFSE[a] O_h	C_{4v}	CFAE[a] $O_h \rightarrow C_{4v}$	CFSE[a] D_{5h}	CFAE[a] $O_h \rightarrow D_{5h}$	CFSE[a] C_{2v}	CFAE[a] $O_h \rightarrow C_{2v}$
d^0	0	0	0	0	0	0	0
d^1	4	4.57	-.57	5.28	-1.28	6.08	-2.08
d^2	8	9.14	-1.14	10.56	-2.56	8.68	-0.68
d^3	12	10.00	2.00	7.74	4.26	10.20	1.80
d^4	16	14.57	1.43	13.02	2.98	16.26	-0.26
d^5	20	19.14	0.86	18.30	1.70	18.86	1.14
d^6	24	20.00	4.00	15.48	8.52	20.37	3.63
d^7	18	19.14	-1.14	12.66	5.34	18.98	-0.98
d^8	12	10.00	2.00	7.74	4.26	10.20	1.80
d^9	6	9.14	-3.14	4.93	1.07	8.79	-2.79
d^{10}	0	0	0	0	0	0	0

[a] In units of Dq

See Basolo and Pearson, p. 146 for the weak field values.

5.42 From the answer to Problem 5.41 we see that CFAE for an octahedral wedge intermediate predicts the lability order $d^2 > d^4 > d^5 > d^3 > d^6$. This agrees with the observed orders except for d^5 and d^3, but they are predicted to be close together.

5.43 (a) The best approach to this problem is through group theory. From Problem 5.22 b we see that the set of σ orbitals in an octahedral complex span the irreducible representations $A_{1g} + E_g + T_{1u}$ of the point group O_h. The collection of fluorine orbitals pointing along the six fluorine-titanium internuclear axes have the same symmetry. The most direct way to find what linear combinations of the fluorine orbitals span each of the separate irreducible representations is to note from the character table for O_h (Appendix 4.1) how the metal bonding *s*, *p* and *d* orbitals transform and by inspection combine fluorine orbitals to get linear combinations transforming the same way.

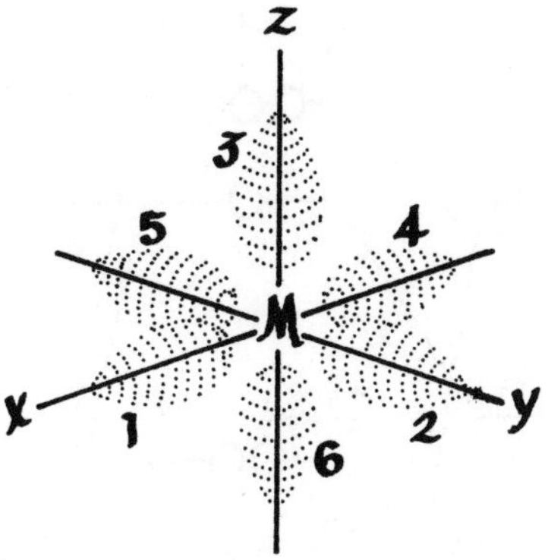

The totally symmetric representation A_{1g} corresponds to the *s*-orbital of the metal and to the sum of all six ligand orbitals. The normalized ligand wave function ignoring overlap is

$$\Psi A_{1g} = (1/\sqrt{6})[\phi_{\sigma_1} + \phi_{\sigma_2} + \phi_{\sigma_3} + \phi_{\sigma_4} + \phi_{\sigma_5} + \phi_{\sigma_6}]$$

The metal orbitals transforming as E_g are d_{z^2} and $d_{x^2-y^2}$. The corresponding ligand LCAO's are

$$\Psi E_g{}^a = (1/\sqrt{12})(2\phi_{\sigma_3} + 2\phi_{\sigma_6} - \phi_{\sigma_1} - \phi_{\sigma_2} - \phi_{\sigma_4} - \phi_{\sigma_5})$$

$$\Psi E_g{}^b = (1/2)(\phi_{\sigma_1} + \phi_{\sigma_4} - \phi_{\sigma_2} - \phi_{\sigma_5})$$

Corresponding to the metal *p* orbitals, which together transform as T_{1u}, are the following ligand combinations

$$\Psi T_{1u}{}^a = (1/\sqrt{2})(\phi_{\sigma_1} - \phi_{\sigma_4})$$

$$\Psi T_{1u}{}^b = (1/\sqrt{2})(\phi_{\sigma_2} - \phi_{\sigma_5})$$

$$\Psi T_{1u}{}^c = (1/\sqrt{2})(\phi_{\sigma_3} - \phi_{\sigma_6})$$

If only σ bonding is considered we need go no further before setting up our correlation diagram. The metal d_{xy}, d_{yz}, and d_{xz} orbitals are non-bonding in the σ-bond model since they do not point along the internuclear directions. Together they transform as T_{2g}. The σ-bond MO diagram is given below.

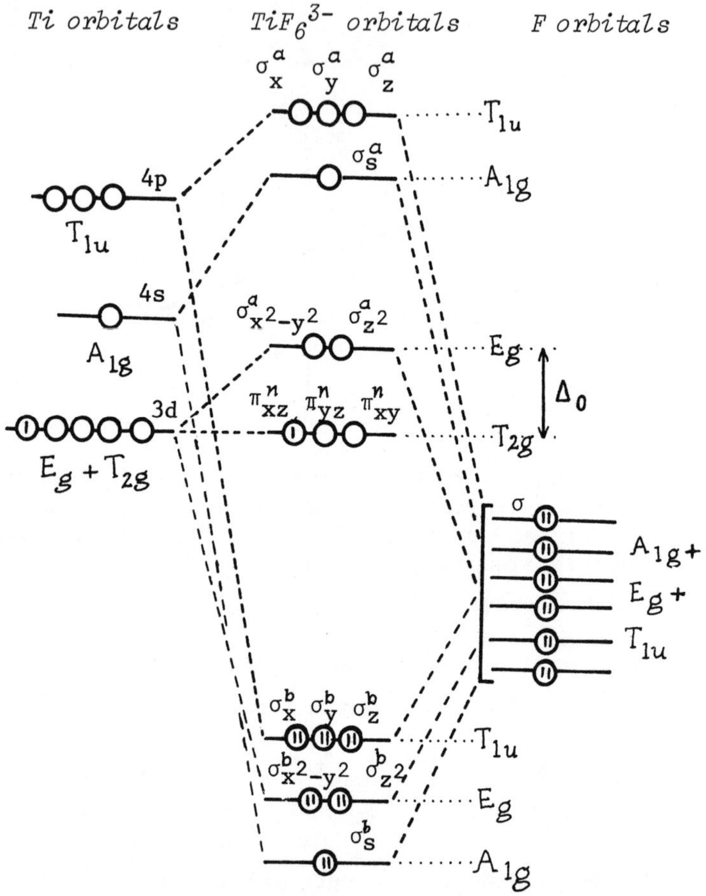

A point to be noticed is that the spacing between the t_{2g} non-bonding MO and the e_g antibonding orbital is analogous to the crystal field splitting Δ_0 in the ionic model (see Problem 5.36).

(b) The correspondence between the ionic model and the MO model pointed out above does not extend to the consideration of π bonding since the ionic model does not include this concept. However, in the MO model essentially the same procedure used above for finding the ligand σ-LCAO's may be used for constructing the ligand π-LCAO's. The orientation of the coordinate system of each of the ligands is given below.

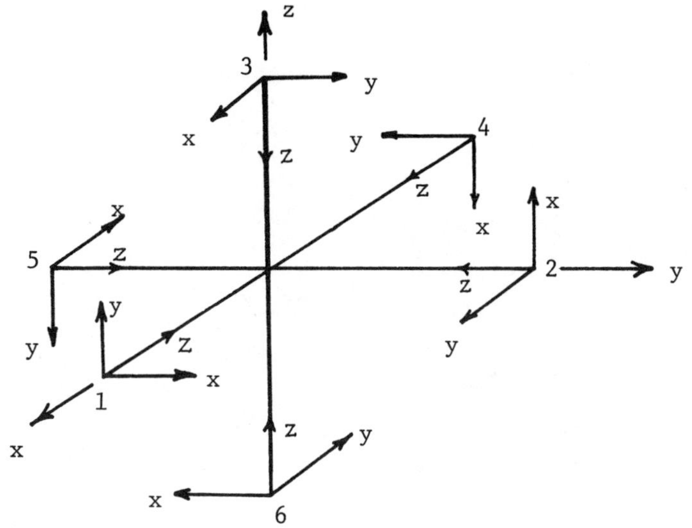

The considerations for choosing this particular set of coordinates are reviewed by Ballhausen (p. 153). By operating on the set of ligand x and y coordinates (which have the same symmetry as the sets of p_x and p_y ligand orbitals) a 12-fold degenerate representation of O_h can be formed using the technique covered in earlier problems.

	E	$8C_3$	$6C_2$	$6C_4$	$3C_2$	i	$6S_4$	$8S_6$	$3\sigma_h$	$6\sigma_d$
Γ_π	12	0	0	0	-4	0	0	0	0	0

By the usual method this can be decomposed to T_{1g} + T_{1u} + T_{2g} + T_{2u}. A glance at the O_h character table (Appendix 4.1) shows us that the metal orbitals which can form π MO's are d_{xz}, d_{yz}, d_{xz} (T_{2g}) and p_x, p_y, p_z (T_{1u}). Since the metal p orbitals have already been used in part (a) above for σ bonding we will consider only the T_{2g} set for simplicity. By comparison to the d_{xz}, d_{yz} and d_{xy} transformation properties the correct ligand LCAO's are

$$\Psi T_{2g} = \begin{cases} \Psi_{xz} = \phi_{x_3} + \phi_{y_1} + \phi_{x_4} + \phi_{y_6} \\ \Psi_{yz} = \phi_{x_2} + \phi_{y_3} + \phi_{x_6} + \phi_{y_5} \\ \Psi_{xy} = \phi_{x_1} + \phi_{y_2} + \phi_{x_5} + \phi_{y_4} \end{cases}$$

The resulting MO correlation diagram is

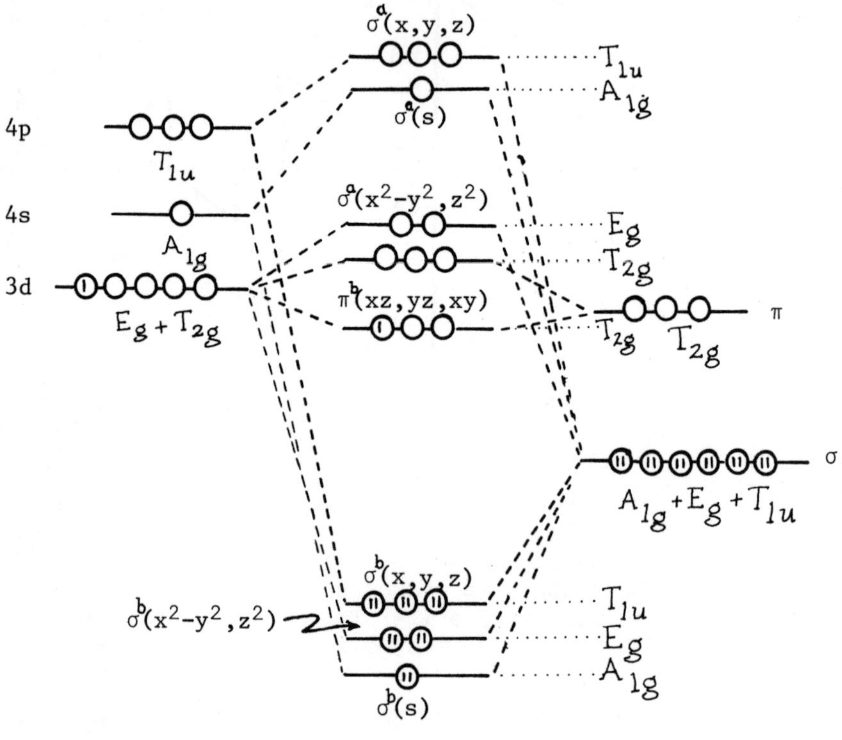

Note that the effect of π bonding when the ligand π orbitals are filled is to effectively *lower* Δ_0. The opposite effect obtains when the ligand π orbitals are empty. Of course the more rigorous methods of problem 5.24 can also be used to derive the form of the ligand orbitals.

Chapter 6

VIBRATIONAL SPECTROSCOPY

Infrared spectroscopy is mainly concerned with the vibrational motions of molecules. Detailed information about chemical bonds can be obtained for relatively simple molecules from a careful study of their vibrational and rotational spectra. For more complicated molecules the inherent complexities preclude detailed analysis; nevertheless, in such cases much relative information can be obtained from a comparison of the infrared spectra of a structurally related series of compounds.

Excellent text books and reference books on infrared spectroscopy and its application to inorganic chemistry are available. Some especially useful ones are the following:

1. G. Herzberg, *Infrared and Raman Spectra*, D. Van Nostrand, New York, 1945.
2. E. B. Wilson, J. C. Decius, and P. C. Cross, *Molecular Vibrations*, McGraw-Hill, New York, 1955.
3. K. Nakamoto, *Infrared Spectra of Inorganic and Coordination Compounds*, John Wiley, New York, 1963.
4. F. A. Cotton, *Chemical Applications of Group Theory*, Interscience, New York, 1963.
5. R. S. Drago, *Physical Methods in Inorganic Chemistry*, Reinhold, New York, 1965, Chapter 7.

PROBLEMS

6.1 *Some basic concepts.*

(a) The region of the electromagnetic spectrum in which most vibrational motions occur is divided into the far infrared, the infrared, and the near infrared. Define the span of these regions in terms of wave numbers.

(b) Infrared absorption bands are reported in terms of wave lengths (angstroms, millimicrons or microns) or in terms of frequency (Kaysers, wave numbers or cm^{-1}). Which quantity is directly proportional to absorbed energy?

(c) How many fundamental vibrations does a non-linear molecule exhibit? A linear molecule?

(d) Vibrational frequencies can be calculated from the harmonic oscillator approximation and the equation

$$\Delta E = (1/2\pi)(k/\mu)^{1/2} \qquad \begin{aligned} k &= \text{force constant} \\ \mu &= \text{reduced mass} \\ &= m_1 m_2/(m_1+m_2) \end{aligned}$$

Generally speaking what do you expect to happen to the frequency as the force constant increases? What will happen to the energy as the reduced mass of the bonded atom or group increases?

6.2 *Group frequencies.* (Read: Nakamoto, Chapter 2). To a good approximation many bands in infrared and Raman spectra may be associated with the vibrations of certain functional groups. Efficient use of vibrational spectroscopy requires a familiarity with many group frequencies. Toward this goal construct a table containing vibrational frequencies or ranges of frequencies for the following:

(a) OH^-, (b) CN^-, (c) CO, (d) N_2, (e) O_2, (f) NO, (g) NO^+, (h) C-N in NCS^-(ionic), (i) C-S in NCS^- (ionic), (j) N-N in N_3^- (ionic), (k) U-O in $[UO_2]^{2+}$, (l) P-S, (m) P-O, (n) M-C in metal carbonyls, (o) Br-O, (p) Ge-Cl, (q) N-Br, (r) O-S-O (bending), (s) S-H.

6.3 *Selection rules.* (Read: Cotton, Chapter 9). The intensities of absorption bands arising from the excitation of vibrational motions are proportional to integrals of the type $\int \Psi_{e.s.} \mu_{x,y,z} \Psi_{g.s.} d\tau$ where $\Psi_{e.s.}$ and $\Psi_{g.s.}$ are the wave functions for the excited and ground states, respectively, and μ_x, μ_y, and μ_z are the components of the electric dipole of the radiation. The components of the electric dipole transform as x, y, and z; $\Psi_{g.s.}$ is always totally symmetric, and for vibrational states in which only one fundamental mode is excited, $\Psi_{e.s.}$ corresponds to the symmetry of the vibrational mode. This leads to the conclusion that the integral will be non-zero only if the fundamental mode transforms the same way as x, y, or z.

For Raman scattering the components of the polarizability operator transform as the squares and binary products of x, y, and z. Therefore, Raman active fundamental modes must transform the same way as x^2, y^2, z^2, xy, xz, yz, or some combination such as x^2-y^2.

Refer to the character tables in Appendix 4.1 and decide which of the following fundamental modes of vibration are infrared active and which are Raman active:

(a) C_{2h}: A_g, A_u, B_u

(b) C_{3v}: A_1, E

(c) T_d: A_1, E, T_2

(d) O_h: $A_{1g} + E_g + T_{2g} + T_{1u} + T_{2u}$

6.4 *Number of normal modes.* (Read: Cotton, Chapter 9). The infrared spectrum of $H_3B\text{-}PH_3$ has been interpreted in terms of a staggered ethane type configuration with symmetry C_{3v}. How many normal modes of vibration does this molecule exhibit, to which irreducible representations of the point group do they correspond, and which are infrared active?

6.5 *Application of selection rules.* (Read: Cotton, Chapter 9). The fundamental bands of the Raman and infrared spectra of nickel tetracarbonyl are tabulated below. Use group theoretical techniques, assume tetrahedral symmetry about the central nickel atom, and suggest assignments for the observed bands. The energies of the bands in cm^{-1} are given.

Infrared	---	2039	---	459	422	---	not studied
Raman	2128[a]	2037	600[a]	461	421	380[a]	78[b]

[a] strongly polarized; [b] very broad

6.6 *Symmetry labels and ν notation.* (Read: Drago, Chapter 7). Determine the symmetry of the normal modes of vibration of the methane molecule, and assign the symmetry labels to the following observed bands for the fundamental vibrations: ν_1 = 2914.2; ν_2 = 1526; ν_3 = 3020.3; and ν_4 = 1306.2 cm^{-1}.

6.7 *Vibrational wave functions.* If the vibrational wave function for the tetrahedral molecule $TiCl_4$ is written as $\Psi(\nu_1, \nu_2, \nu_3, \nu_4)$, then the ground state is $\Psi(0,0,0,0)$ and transforms as A_1. To which irreducible representations do the following wave functions belong?

(a) $\Psi(1,0,0,0)$; (b) $\Psi(0,1,0,0)$; (c) $\Psi(0,0,1,0)$.

6.8 *Application of selection rules.* The infrared and Raman spectra of $Fe(CO)_5$ exhibit the following bands:

	metal-carbon region	carbonyl region
I.R.	472, 377 cm^{-1}	2028, 1994 cm^{-1}
Raman	492, 414, 377	2114, 2031, 1984

By means of selection rule arguments and using these data, decide on the most probable structure for $Fe(CO)_5$.

6.9 *Spectral trends.* (Read: Nakamoto, Chapter 3). The infrared spectra of some salts of $Co(NO_2)_6{}^{3-}$ are shown in Figure 6.1. Assume that the absorptions arise from the Co-N stretching mode and account for the variations in the spectra.

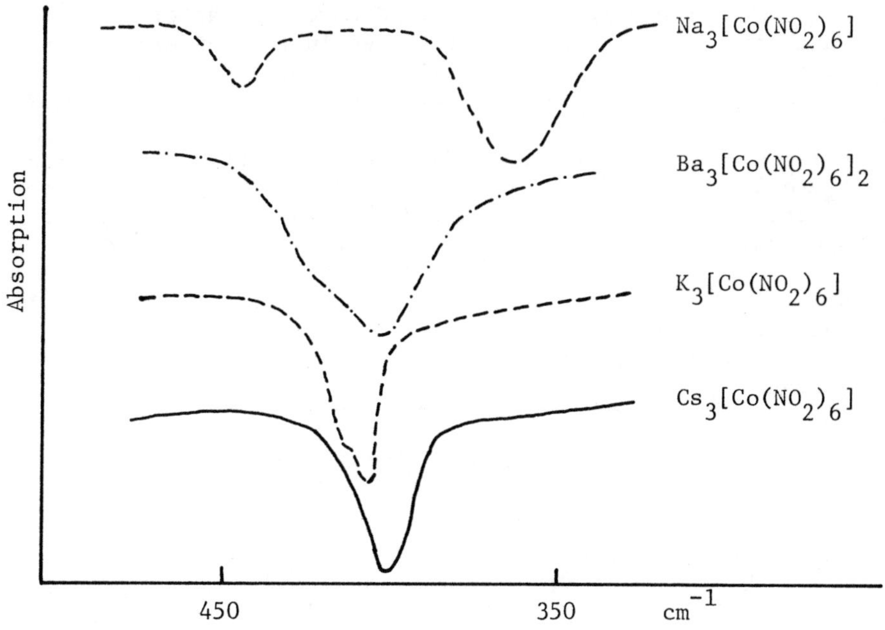

Figure 6.1

6.10 *Application of selection rules.* Under irradiation with ultraviolet light tertiary phosphines react with dimanganesedecacarbonyl, $Mn_2(CO)_{10}$, to yield compounds with the formula $[Mn(CO)_4L]_2$ which retain the Mn–Mn bond. In the carbonyl stretching region, the infrared spectra of the compounds exhibit two bands; one very sharp band at 1960 cm^{-1} with a shoulder at 1980 cm^{-1}. Predict the structure of $[Mn(CO)_4L]_2$.

6.11 *Application of selection rules.* For the compounds $Fe(CO)_4X_2$, where X = Cl, Br, or I, the infrared spectra in the range 1800-2200 cm^{-1} and 200-350 cm^{-1} show the following absorptions:

	1800-2200 cm^{-1}	200-350 cm^{-1}
$Fe(CO)_4Cl_2$	2084, 2108, 2124, 2164	294, 318
$Fe(CO)_4Br_2$	2075, 2099, 2108, 2150	217, 237
$Fe(CO)_4I_2$	2063, 2081, 2086, 2132	---

Deduce the stereochemistry of the compounds using this data.

6.12 *Symmetry effects.* (Read: Nakamoto, Chapter 2). What effect should one expect to see in the infrared spectrum if one Y atom of the trigonal planar XY_3 molecule is substituted with a Z atom to give the planar XY_2Z molecule?

6.13 *Coordinated ligands.* The local symmetry of the nitrate ion is reduced to C_{2v} when it is covalently bonded to a metal ion. What effect should this have on the infrared spectrum?

6.14 *Coordinated ligands.* The infrared spectra in the sulfate absorption region for three complexes of cobalt(III) are shown in Figure 6.2. Account for the spectral differences.

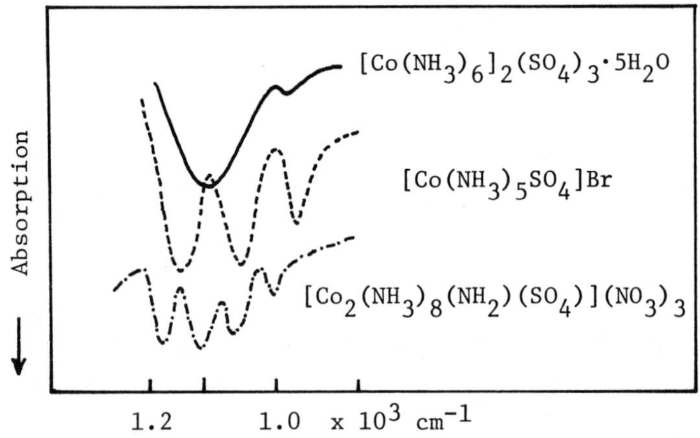

Figure 6.2

6.15 *Isotope effects.* The infrared spectrum of HCl shows a maximum at 2886 cm^{-1} and the corresponding band in DCl occurs at 2091 cm^{-1}. An absorption at 3312 cm^{-1} in HCN is found to shift to 2629 cm^{-1} on deuteration. Assume that these bands arise from the X-H and X-D modes of vibration, respectively, (X = Cl or CN), calculate the expected shift, and comment on any difference between theory and experiment.

6.16 *Isotope effects.* The infrared spectrum of $Mn(CO)_5Cl$ in the metal-halogen stretching region shows a maximum at 295 cm^{-1} with a shoulder at 289 cm^{-1}. Comment on the origin of the shoulder.

6.17 *Identification of unknown compounds.* Commercial ruthenium trichloride was stirred in a nitrogen atmosphere for about 15 hours in an aqueous solution containing an excess of hydrazine. The resulting deep orange-red solution was filtered and treated with a large excess of a salt of one of the anions Cl^-, Br^-, I^-, PF_6^-, or BF_4^-. The diamagnetic orange-yellow precipitates have in each case the elemental composition $RuN_7H_{15}X_2$ (X is the precipitating anion), and exhibit an unexpected band in the infrared at 2112-2167 cm^{-1}. Upon treatment with acid, a gas is evolved and the acido-pentaammine ruthenium complex results. Comment on the probable nature of the orange-yellow products.

6.18 *Distinction of geometrical isomers.* A blue solid forms when ethanolic solutions of copper(II) perchlorate hexahydrate and tri-(2-pyridyl)amine (*tripam*) are mixed. Elemental analysis confirms the composition $Cu(tripyam)_2(ClO_4)_2$. When the blue solid is dissolved in hot ethanol for purification by recrystallization, a yellow-green solid with the same elemental composition is obtained. The infrared spectra in the region 900-1200 cm^{-1}, where Cl-O stretching vibrations occur, are shown in Figure 6.3. Suggest an explanation.

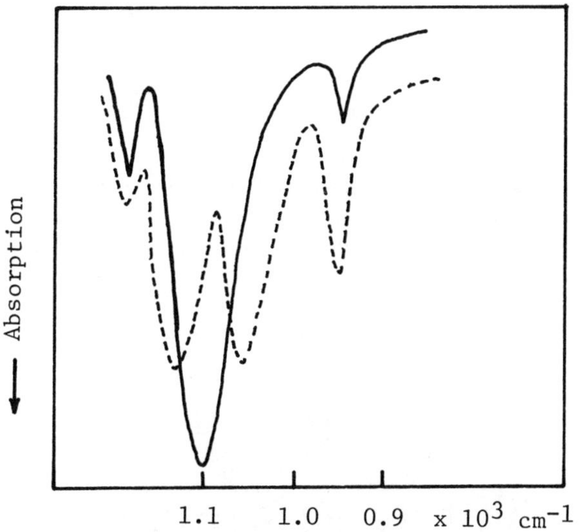

Figure 6.3 Infrared spectra of yellow-green _ _ _ _ and blue ______ $Cu(tripyam)_2(ClO_4)_2$.

6.19 *Verification of structure.* The infrared spectrum of $Os_3(CO)_{12}$ is given in Figure 6.4. Show how this spectrum is consistent with the structure of the molecule.

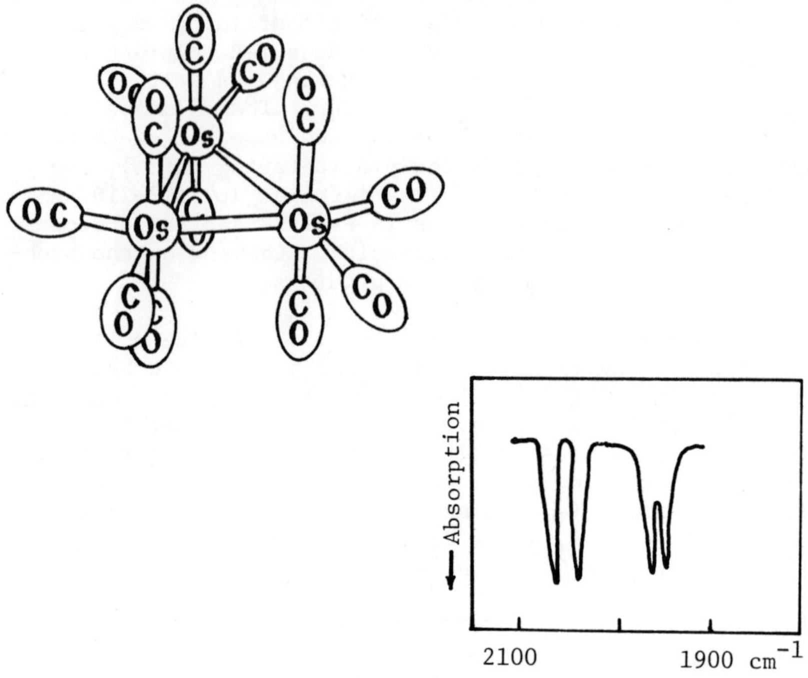

Figure 6.4 Spectrum and structure of $Os_3(CO)_{12}$.

6.20 *Verification of structure.* Show how the infrared spectrum of $Fe_2(CO)_9$ is consistent with the structure, which has symmetry D_{3h}. The spectrum and structure are shown in Figure 6.5.

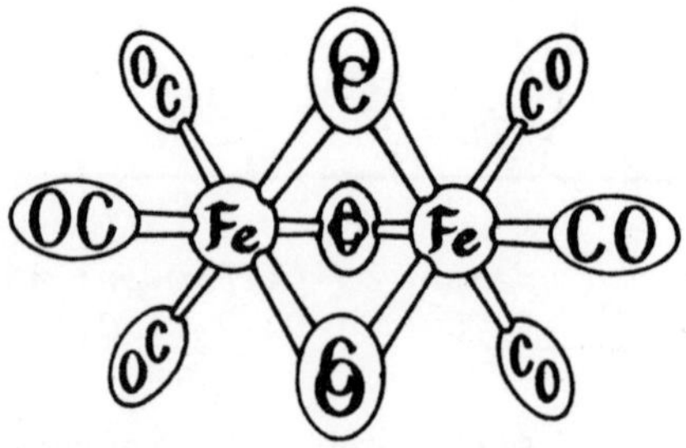

Figure 6.5a Structure of $Fe_2(CO)_9$

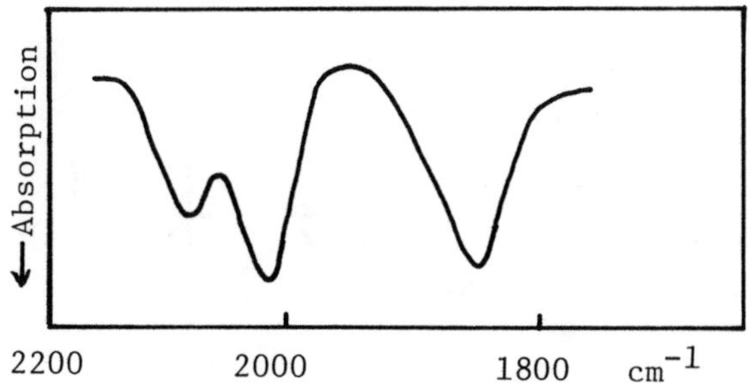

Figure 6.5b Spectrum of $Fe_2(CO)_9$

6.21 *Information about chemical bonds.* The Raman spectra in the region of the C≡O and M–C stretching vibrations of the isostructural series $Ni(CO)_4$, $Co(CO)_4^-$, and $Fe(CO)_4^{2-}$ exhibit the following bands:

	C≡O region	M–C region
$Fe(CO)_4^{2-}$	1788 cm^{-1}	550, 464 cm^{-1}
$Co(CO)_4^-$	1918, 1883	532, 439
$Ni(CO)_4$	2121, 2039	422, 381

Suggest an explanation for these data.

6.22 *Infrared as a diagnostic tool.* The hydroxide ion forms a series of complexes with heavy atoms. How could infrared spectroscopy be used as a diagnostic test for a bound hydroxide as opposed to a free ion as, for example, in NaOH?

6.23 *Symmetries of normal modes.* Four of the eight normal modes of vibration of a trigonal bipyramidal molecule are illustrated in Figure 6.6. To which irreducible representations of the D_{3h} point group do they correspond?

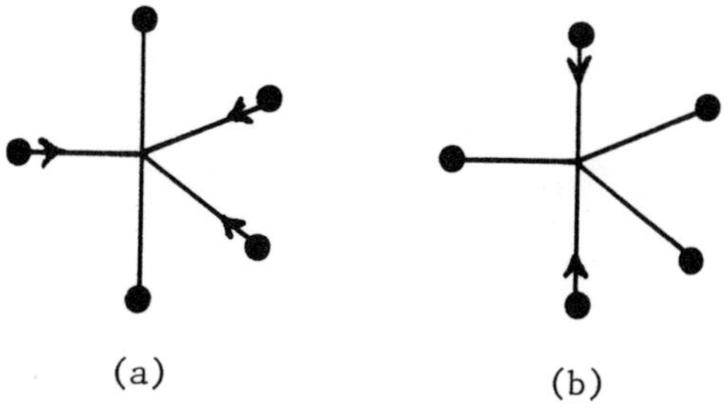

Figure 6.6 Some normal modes of a trigonal bipyramidal structure. *(Continued next page.)*

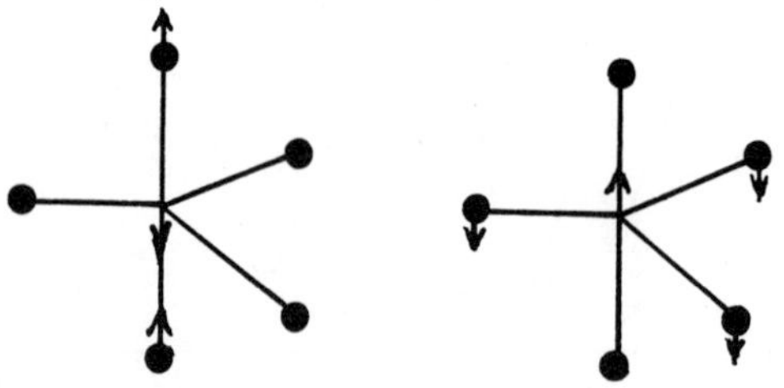

(c) (d)

Figure 6.6 Some normal modes of a trigonal bipyramidal structure.

6.24 *Verification of structure.* A Raman spectrum of solid XeF_4 exhibited three bands at 534, 502, and 235 cm^{-1}, respectively. Show that this experimental observation is consistent with a square planar structure.

6.25 *Comparison of infrared and Raman spectra.* The infrared and Raman spectra of $XeOF_4$ are shown in Figure 6.7. Show how symmetry arguments may be used in the assignments of the observed bands.

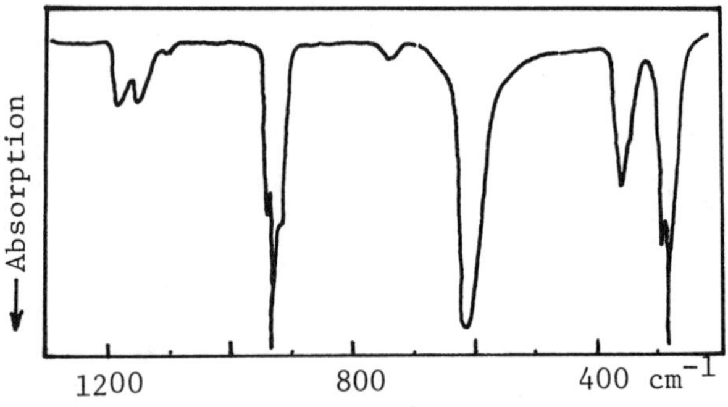

Infrared spectrum of $XeOF_4$ vapor.

Figure 6.7a

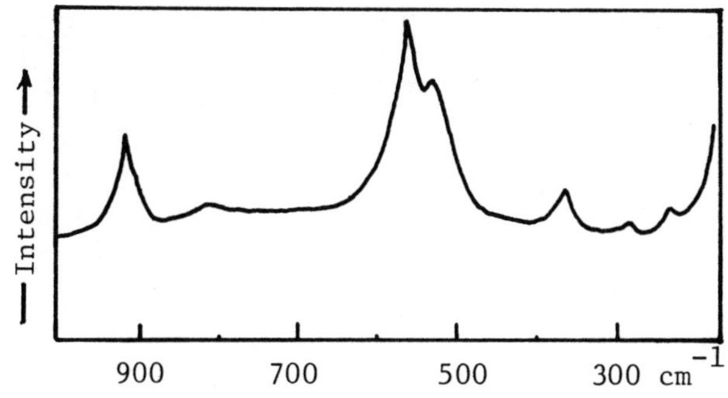

Raman spectrum of liquid $XeOF_4$.

Figure 6.7b

6.26 *Band assignments--overtones and combinations.* The infrared spectrum of RhF_6 is shown in Figure 6.8 with the band assignments. Show that the assignments are self consistent.

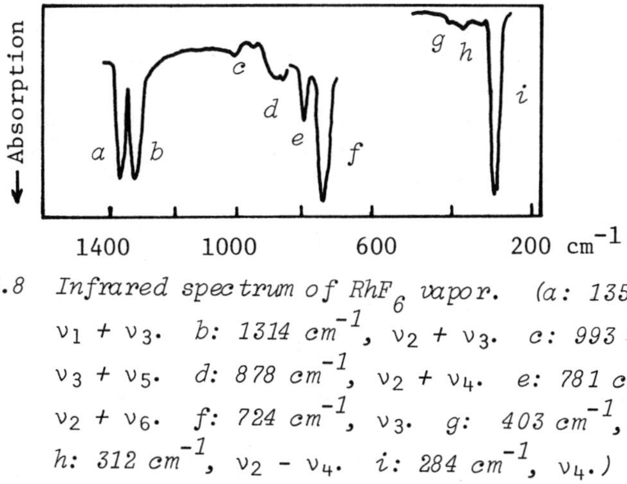

Figure 6.8 *Infrared spectrum of RhF_6 vapor.* (*a*: 1356 cm^{-1}, $\nu_1 + \nu_3$. *b*: 1314 cm^{-1}, $\nu_2 + \nu_3$. *c*: 993 cm^{-1}, $\nu_3 + \nu_5$. *d*: 878 cm^{-1}, $\nu_2 + \nu_4$. *e*: 781 cm^{-1}, $\nu_2 + \nu_6$. *f*: 724 cm^{-1}, ν_3. *g*: 403 cm^{-1}, $\nu_2 - \nu_6$. *h*: 312 cm^{-1}, $\nu_2 - \nu_4$. *i*: 284 cm^{-1}, ν_4.)

6.27 *Band assignments--overtones and combinations.* The infrared spectrum of RuF_6 is shown in Figure 6.9. Recall the assignments of the bands in the spectrum of RhF_6, and provide assignments for the bands in the infrared spectrum of RuF_6.

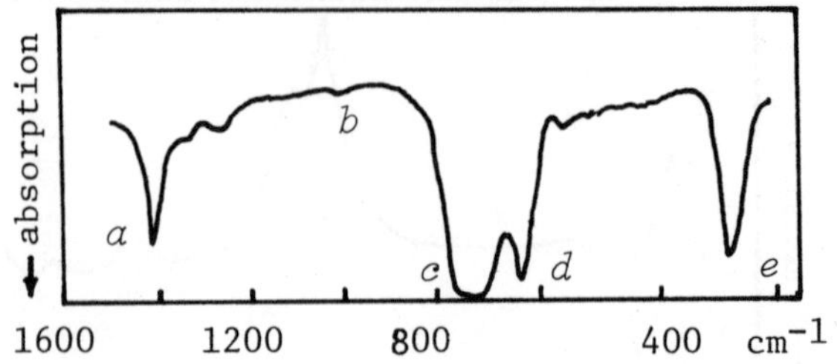

Figure 6.9 *Infrared spectrum of RuF_6 vapor. (a: 1410 cm^{-1}. b: 997 cm^{-1}. c: 735 cm^{-1}. d: 650 cm^{-1}, this band arises from an impurity in the cell. e: 275 cm^{-1}.)*

6.28 *Band assignments.* Although the Raman spectrum of a gaseous sample of SO_2 consists of one line at 1150.5 cm^{-1}, two additional lines, at 525 and 1336 cm^{-1} are seen in the spectrum of a liquid sample. The infrared spectrum of a gaseous sample exhibits bands at 519, 606, 1151, 1361, 1871, 2305 and 2499 cm^{-1}, respectively. Provide assignments for these bands. Consider the xz plane the molecular plane.

6.29 *Selection rules.* The Raman spectrum of SbF_5 exhibits bands at 716, 667, 491, 264, 228, and 90 cm^{-1}. With what structure are these data consistent?

6.30 *Selection rules.* The infrared and Raman spectra of BrF_5 exhibit the following bands:

I.R.	690	645	503	---	---	418	(not studied)		
Raman	683	626	572	536	481	415	365	315	244

With what structure are these data consistent?

6.31 *Functional group frequencies as structural aids.* A triphenylphosphine-containing iridium complex (I) of empirical formula $C_{37}H_{30}IrClOP_2$ reacts with hydrogen chloride affording a compound $C_{37}H_{31}IrCl_2OP_2$ (II). In the 1800-2300 cm^{-1} region the infrared spectrum of (I) shows an absorption at 1944 cm^{-1} and (II) at 2045 and 2245 cm^{-1}. Both (I) and (II) react with mercuric chloride yielding a complex (III) of empirical formula $C_{37}H_{30}HgIrCl_3OP_2$. Treatment of (III) with chlorine yields $C_{37}H_{30}IrCl_3OP_2$ (IV). Reaction of (III) with hydrogen chloride and hydrogen yields (II).

Suggest structural formulas for (I)-(IV), and comment on the above observations.

6.32 *Functional group frequencies.* Carbon-carbon bond stretching modes of vibration occur at ca. 2230 cm^{-1} in free acetylenes, whereas ethylenic groups absorb at ca. 1600 cm^{-1}. In complexes of the type $Pt(PPh_3)_2[RC{\equiv}CR']$ the carbon-carbon stretching vibration is found to occur at about 1700 cm^{-1}. What does this indicate about the bonding in these complexes?

SOLUTIONS

6.1 (a) far infrared - 50-667 cm^{-1}; infrared - 667-4,000 cm^{-1}; near infrared - 4,000-12,500 cm^{-1}.

(b) $\Delta E = h\bar{\nu}$ where h has units of erg-sec. and the frequency, ν, has units of sec^{-1}. Therefore, $\Delta E = hc\bar{\nu}$ where c is the speed of light and $\bar{\nu}$ is in cm^{-1}. Thus, wave numbers are directly proportional to energy.

(c) Non-linear molecules have $3n - 6$ fundamental vibrations, while linear molecules have $3n - 5$, where n is the number of atoms in the molecule.

(d) The frequency will increase as k increases if μ remains constant. The energy will decrease (as will the frequency) as μ increases if k remains constant.

6.2 (a) OH^-, 3500 - 3700 cm^{-1}

(b) CN^-, 2050 - 2250

(c) $^{12}C^{16}O$, 2143.16; $^{13}C^{16}O$, 2096.07

(d) N_2, 2331 (e) O_2, 1555

(f) $^{14}N^{16}O$, 1876.11; $^{15}N^{16}O$, 1843

(g) NO^+, 2220 (h) C-N in $^{14}N^{12}C^{32}S^-$, 2041

(i) C-S in $^{14}N^{12}C^{32}S^-$, 747

(k) U-O in $UO_2{}^{2+}$, 860

6.3 (a) IR active, A_u, B_u; Raman active, A_g.

(b) A_1 and E are both IR and Raman active.

(d) IR active, T_{1u}; Raman active, A_{1g}, E_g, T_{2g}; T_{2u} is inactive in both IR and Raman.

6.4 The structure of the molecule with the set of Cartesian displacement coordinates for one of the atoms is shown below.

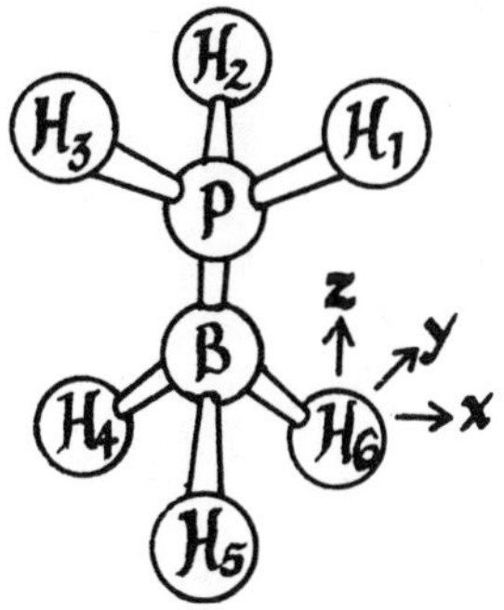

The number of normal modes of vibration equals 3n-6 = 18. Since all motions of the atoms (including rotations and translations) can be described in terms of displacement coordinates for each of the atoms, and since each fundamental mode transforms as an irreducible representation of the point group, we need to determine only the reducible representation spanned by the set of displacement coordinates in order to describe the motions of the molecule in terms of their symmetry properties.

By agreeing that the nuclei are fixed and that the displacement coordinates are moved by a symmetry operation, we can write down the 24 dimensional reducible representation by counting the number of displacement coordinates that are not moved off their original nuclei by the symmetry operation. The reducible representation is

C_{3v}	E	$2C_3$	$3\sigma_v$
$\Gamma_{d.c.}$	24	0	4

Two complicating features arise in the construction of this reducible representation. Let us first consider the C_3 operation. We note immediately that the displacement coordinates on all six hydrogen atoms are moved off their original nuclei by the rotation and contribute nothing to the trace of the representation matrix. However, after the rotation, the x vector on the phosphorus atom, for example, can be described in terms of the original set of displacement vectors. To consider this in detail let us isolate the phosphorus atom and perform the C_3 operation on the set of displacement vectors. The situation can be illustrated as follows:

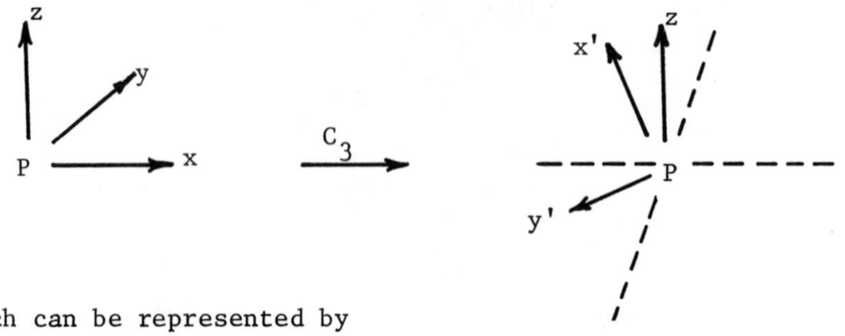

which can be represented by

$$C_3 \begin{pmatrix} x \\ y \\ z \end{pmatrix} = \begin{pmatrix} -1/2 & \sqrt{3}/2 & 0 \\ -\sqrt{3}/2 & -1/2 & 0 \\ 0 & 0 & 1 \end{pmatrix} \begin{pmatrix} x \\ y \\ z \end{pmatrix}$$

The trace of the representation matrix is zero. Obviously the same result obtains for the boron atom and the trace of the 24 x 24 representation matrix equals zero even though some of the vectors remain on their original atoms.

For the reflection through the vertical mirror plane, let us choose the plane containing hydrogen atoms 1 and 4, the phosphorus atom and the boron atom. From the figure we see that the displacement coordinates on hydrogen atoms 2 and 3 and on hydrogen atoms 5 and 6 will be interchanged, will contribute nothing to the trace of the representation matrix, and need not be considered further. In order to clarify the discussion, the operation as it affects the atoms in the plane is illustrated below.

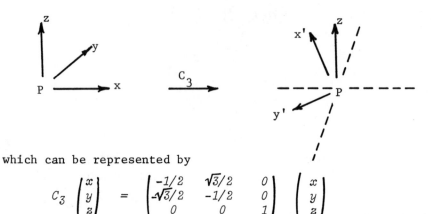

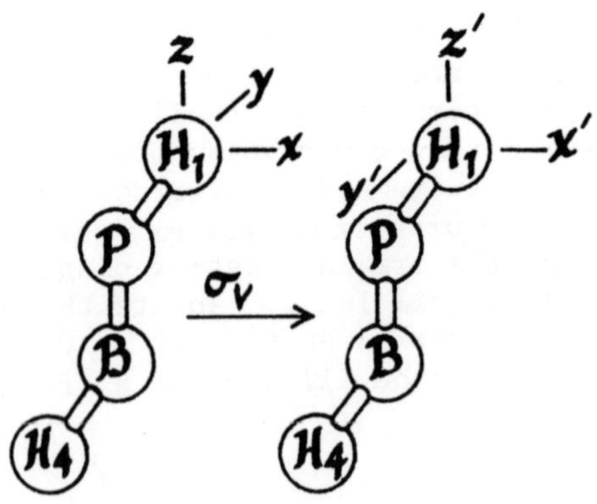

This can be represented as

$$\sigma_v \begin{bmatrix} Hx \\ y \\ z \\ Px \\ y \\ z \\ Bx \\ y \\ z \\ Hx \\ y \\ z \end{bmatrix} = \begin{bmatrix} 1 &&&&&&&&&&& \\ & -1 &&&&&&&&&& \\ && 1 &&&&&&&&& \\ &&& 1 &&&&&&&& \\ &&&& -1 &&&&&&& \\ &&&&& 1 &&&&&& \\ &&&&&& 1 &&&&& \\ &&&&&&& -1 &&&& \\ &&&&&&&& 1 &&& \\ &&&&&&&&& 1 && \\ &&&&&&&&&& -1 & \\ &&&&&&&&&&& 1 \end{bmatrix} \begin{bmatrix} Hx \\ y \\ z \\ Px \\ y \\ z \\ Bx \\ y \\ z \\ Hx \\ y \\ z \end{bmatrix}$$

The trace of the representation matrix is equal to 4.

The reducible representation spanned by the set of displacement coordinates is decomposed by the usual means to yield

$$\Gamma_{d.c.} = 6A_1 + 2A_2 + 8E$$

Since translational motions transform as the Cartesian coordinates x, y, z and rotational motions as rotations about x, y, and z (r_x, r_y, and r_z), we use the character table for C_{3v} (Appendix 4.1) and find

$$\Gamma_{trans} = A_1 + E$$

$$\Gamma_{rot} = A_2 + E$$

which leaves

$$\Gamma_{vib} = 5A_1 + A_2 + 6E$$

From the character table we see that the A_1 and E normal modes give rise to infrared active vibrations and that A_2 is not infrared-active. Consequently, there should be eleven infrared-active motions. Reference: R. W. Rudolph, R. W. Parry, C. F. Farran, *Inorg. Chem.*, 5, 723 (1966).

6.5 In nickel tetracarbonyl the nine atoms each with three coordinates give rise to the following reducible representation:

T_d	E	$8C_3$	$3C_2$	$6S_4$	$6\sigma_d$
$\Gamma_{d.c.}$	27	0	-1	-1	5

which decomposes to $2A_1 + 2E + 2T_1 + 5T_2$. From the character table we see

$$\Gamma_{trans} = T_2 \text{ and } \Gamma_{rot} = T_1$$

Thus, the genuine vibrations are $2A_1 + 2E + T_1 + 4T_2$. Considering just the various sets of chemically equivalent bonds and the angles between them we can obtain reducible representations for the stretching and bending modes of vibration. These are

T_d	E	$8C_3$	$3C_2$	$6S_4$	$6\sigma_d$	
ΓM-C stretch	4	1	0	0	2	$= A_1 + T_2$
ΓC-O stretch	4	1	0	0	2	$= A_1 + T_2$
ΓC-M-C bend	6	0	2	0	2	$= A_1 + E + T_2$
ΓM-C-O bend	8	-1	0	0	0	$= E + T_1 + T_2$

Note that there are three A_1 modes in this group although only two are allowed by symmetry. One is redundant. This is the C-M-C bending vibration since the change in the sixth angle is fixed if the other five are altered. For an A_1 bending mode all angles would have to simultaneously increase or decrease, and this is impossible. We should expect to observe the following modes of vibration:

vibrational mode	infrared active	Raman active
C-O stretch	T_2	A_1 T_2
M-C stretch	T_2	A_1 T_2
M-C-O bend	T_2	E T_2
C-M-C bend	T_2	E T_2

From considerations of the masses of the atoms the absorptions are expected to occur in the following regions:

C-O stretch	about 2000 cm^{-1}
M-C-O bend	about 600
M-C stretch	about 400
C-M-C bend	about 100

The 2037/2039 cm^{-1} band is the T_2 mode since it occurs in both the infrared and Raman spectra. Similarly the

2128 cm^{-1} band is the A_1 mode since it occurs only in the Raman spectrum. Using the same reasons, the 421/422 cm^{-1} band is assigned to the T_2 M-C stretching vibration and the 380 band to the A_1 M-C stretching vibration. The 600 cm^{-1} band must be the M-C-O E-mode of vibration since it does not occur in the infrared spectrum and the 459/461 band is probably the T_2 bending vibration of M-C-O. It is possible that the assignments of the 421/422 and 459/461 cm^{-1} bands could be reversed. Also, since these vibrations have the same symmetry and similar energies, considerable coupling between the two may be anticipated. Finally, the only band not assigned occurs at 78 cm^{-1} in the Raman spectrum (Recall that the infrared spectrum was not recorded in this region.). It is suggested that the E and T_2 modes lie under the very broad band envelope.

6.6 The reducible representation spanned by the displacement coordinates is decomposed as follows:

T_d	E	$8C_3$	$6\sigma_d$	$6S_4$	$3C_2$
$\Gamma_{d.c.}$	15	0	3	-1	-1

$$= A_1 + E + T_1 + 3T_2$$

From the character table we find that translations transform as T_2 and rotations as T_1, so for vibrations we have

$$\Gamma_{vib} = A_1 + E + 2T_2$$

By convention the ν_n symbolism is assigned first in order of highest symmetry, then in order of highest energy within each symmetry species. This gives $\nu_1 = A_1$; $\nu_2 = E$; $\nu_3 = T_2$ and $\nu_4 = T_2$.

6.7 Since in each case only one fundamental vibration is excited, then (a) is A_1; (b) is E.

6.8 Since molecules with coordination number five usually exhibit a trigonal bipyramidal structure with symmetry D_{3h} or a tetragonal pyramidal structure with symmetry C_{4v}, we need only determine the number of C≡O stretching vibrations and simultaneously the metal-carbon vibrations that are infrared active in order to make a structural assignment.

First consider the trigonal bipyramidal structure:

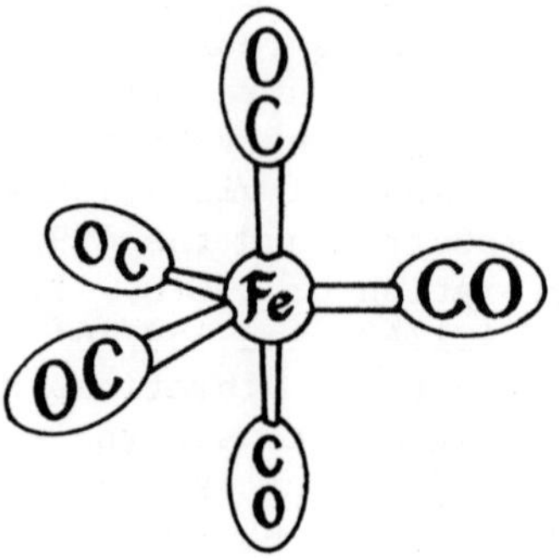

Using the five C≡O bonds as a basis for a five-dimensional representation of the group D_{3h}, we arrive at the reducible representation

D_{3h}	E	$2C_3$	$3C_2$	σ_h	$2S_3$	$3\sigma_v$
$\Gamma_{C\equiv O}$	5	2	1	3	0	3

Once again, to construct the reducible representation we make use of the rule that any bond shifted by the symmetry operation makes no conbribution to the trace of the representation matrix, and consequently makes no contribution to the character of the reducible representation. The reducible representation decomposes to $\Gamma_{C\equiv O} = 2A_1' + E' + A_2''$. Recalling that a fundamental vibration will be infrared active only if the normal mode belongs to the same representation as one or more of the Cartesian coordinates, we can see from the character table that the selection rules are

A_1', inactive
E' and A_2'', active.

Since only two fundamentals will give rise to absorption bands, we conclude that the structure may be a trigonal bipyramid.

For the tetragonal pyramidal structure

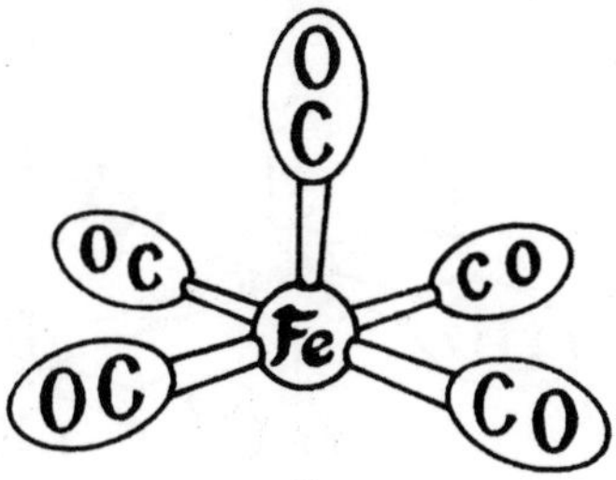

the C≡O bonds form a basis for the representation

C_{4v}	E	$2C_4$	C_2	$2\sigma_v$	$2\sigma_d$
$\Gamma_{C\equiv O}$	5	1	1	3	1 = $E + 2A_1 + B_1$

Since the selection rules are A_1 and E, active; B_1 inactive, three bands are expected, and thus the infrared spectrum is not consistent with the tetragonal pyramidal structure.

6.9 It appears that the splitting between the bands found in the sodium salt decreases as the size of the cation increases, and finally is undetectable in the cesium salt. Since in octahedral symmetry only one Co-N stretch should be observed, all but the cesium salt appear to have lower symmetry--perhaps tetragonal.

6.10 To answer the question posed by this problem, we will draw models for possible structures and by standard methods determine the number of infrared active bands expected for each structure. The structure with two infrared active bands will very likely be correct. Since there are only two bands we need consider only high symmetry structures.

Possible structures are

I.

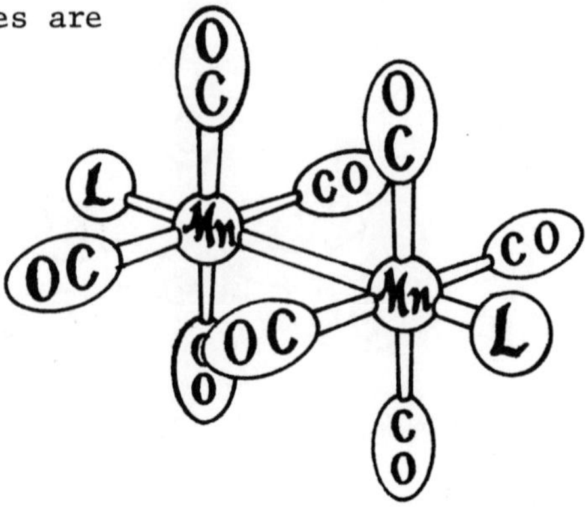

D_{4h}	E	$2C_4$	C_2	$2C_2'$	$2C_2''$	i	$2S_4$	σ_h	$2\sigma_v$	$2\sigma_d$
$\Gamma_{C\equiv O}$	8	0	0	0	0	0	0	0	4	0

$$= A_{1g} + B_{1g} + E_g + A_{2u} + B_{2u} + E_u$$

II.

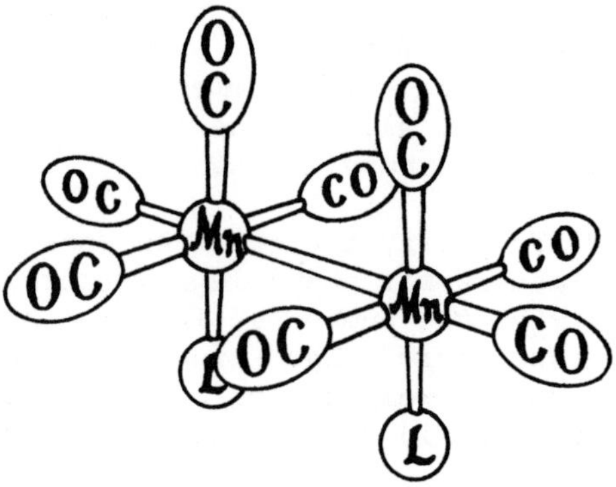

C_{2v}	E	C_2	$\sigma_v(xz)$	$\sigma_v'(yz)$
$\Gamma_{C\equiv O}$	8	0	4	0

$$= 3A_1 + A_2 + 3B_1 + B_2$$

III.

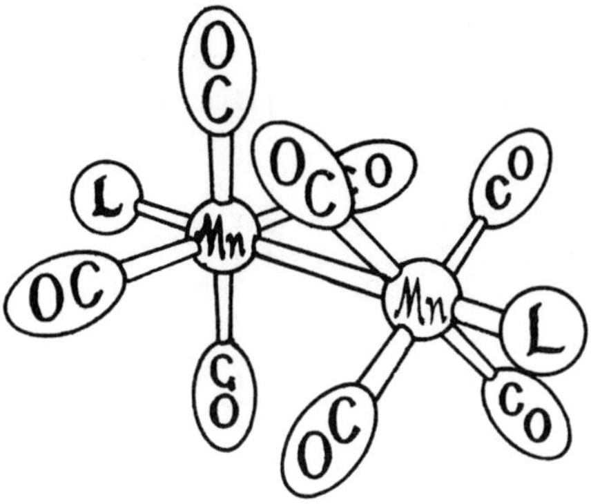

D_{4d}	E	$2S_8$	$2C_4$	$2S_6{}^3$	C_2	$4C_2'$	$4\sigma_d$
$\Gamma_{C\equiv O}$	8	0	0	0	0	0	2

$$= A_1 + B_2 + E_1 + E_2 + E_3$$

IV.

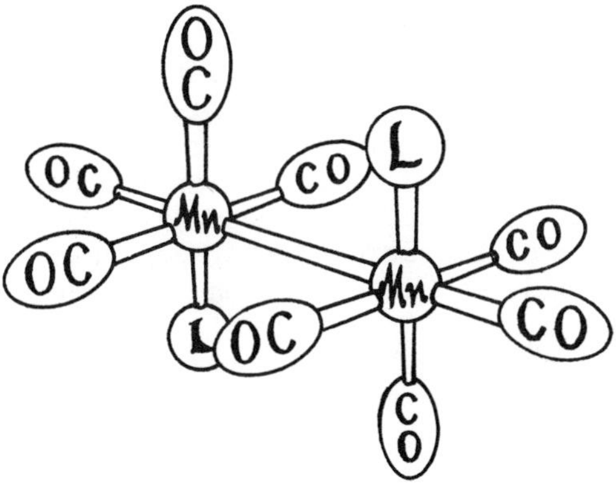

C_{2h}	E	C_2	i	σ_h
$\Gamma_{C\equiv O}$	8	0	0	4

$$= 3A_g + B_g + A_u + 3B_u$$

By reference to the character tables we see that structures I and III should exhibit two infrared active carbonyl stretching modes while IV should have four and II, seven bands. Although we can not make a choice between structures I and III based on infrared evidence, x-ray experiments have shown structure III to be the correct one.

6.11 Assuming octahedral coordination about the iron atom, the structures might be either *cis* or *trans*:

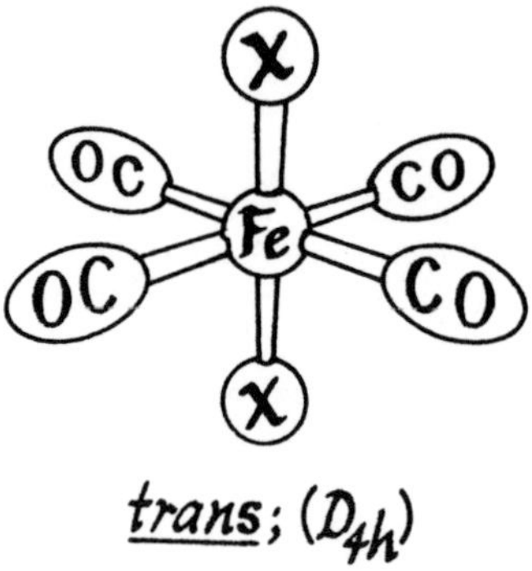

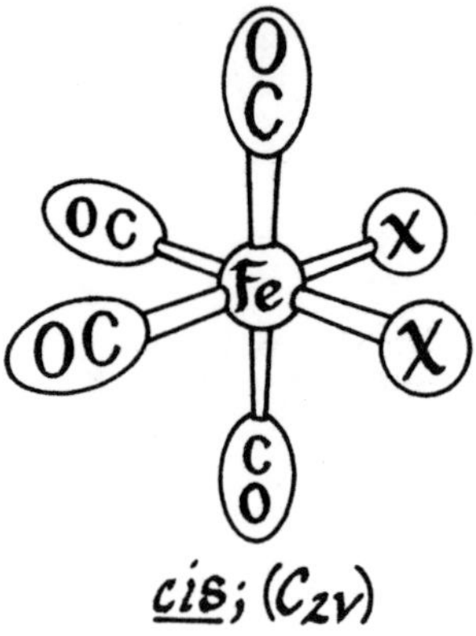

The region 1800-2200 cm^{-1} is the carbonyl stretching region and the region 200-300 cm^{-1} is where the metal-halogen stretching vibrations may be expected to occur. The fact that no bands are observed in the latter region for $Fe(CO)_4I_2$ coupled with the trend in frequencies on changing from Cl to Br implies that the metal-iodine stretching frequencies lie below 200 cm^{-1}. This would be anticipated on a mass-effect basis.

Since the observed absorptions are primarily stretching vibrations, we need consider only the stretchings of the M-X and C-O bonds. There are two M-X and four C-O bonds in each case.

First consider *cis*-$Fe(CO)_4X_2$:

C_{2v}	E	C_2	σ_v	σ_v'	
Γ_{C-O}	4	0	2	2	$= 2A_1 + B_1 + B_2$
Γ_{M-X}	2	0	0	2	$= A_1 + B_2$

From the character table we see that these are all infrared-active modes of vibration and therefore we should expect four infrared carbonyl stretching bands and two bands arising from the metal-halogen stretching vibrations.

Now consider the *trans*-$Fe(CO)_4X_2$.

D_{4h}	E	$2C_4$	C_2	$2C_2'$	$2C_2''$	i	$2S_4$	σ_h	$2\sigma_v$	$2\sigma_d$	
Γ_{C-O}	4	0	0	2	0	0	0	4	2	0	$= A_{1g} + B_{1g} + E_u$
Γ_{M-X}	2	2	2	0	0	0	0	0	2	2	$= A_{1g} + A_{2u}$

From the character table we see that only E_u and A_{2u} are infrared active. Consequently, only one infrared-active carbonyl stretching vibration and one metal-halogen stretching vibration should be observed for this configuration.

Symmetry arguments then indicate that the complexes will have the structure with the halogen atoms mutually *cis* since there are four carbonyl bands and two metal-halogen bands in the infrared spectra.

6.12 Planar XY_3 belongs to group D_{3h}. Its normal modes are derived as follows:

D_{3h}	E	$2C_3$	$3C_2$	σ_h	$2S_3$	$3\sigma_v$
$\Gamma_{d.c.}$	12	0	-2	4	-2	2

$$= A_1' + 3E' + 2A_2'' + E'' + A_2'$$

Since $\Gamma_{trans} = A_2'' + E'$ and $\Gamma_{rot} = A_2' + E''$, we have left $\Gamma_{vib} = A_1' + 2E' + A_2''$. The point group of XY_2Z is C_{2v}, in which there are no two-dimensional irreducible representations. Thus we expect the degeneracy of the E' bands to be lifted. Six bands should be observed. The normal modes are $3A_1$, B_1, and $2B_2$. (Under C_{2v}, z transforms as A_1, x as B_1, and y as B_2.)

6.14 The structural formulas tell us that the local symmetry of the sulfate ion decreases from T_d in $[Co(NH_3)_6]_2(SO_4)_3 \cdot 5H_2O$ to C_{3v} in $[Co(NH_3)_5SO_4]Br$ where it functions as a unidentate ligand to C_{2v} in the third compound in which it functions as a bidentate bridging group. Recalling that the fundamental mode ν_1 is infrared-inactive and that ν_3, which belongs to T_2, is infrared-active, it is seen that the triply degenerate mode is split in the lower symmetry environments and that ν_1 becomes an allowed transition. The deductions from experiment are confirmed by the entries in the correlation table given below. The selection rules for allowed transitions are given in parentheses. (I = infrared-active; R = Raman active)

Correlation Table

point group	ν_1	ν_2	ν_3	ν_4
T_d	A_1(R)	E(R)	T_2(I,R)	T_2(I,R)
C_{3v}	A_1(I,R)	E(I,R)	A_1(I,R) E(I,R)	A_1(I,R) E(I,R)
C_{2v}	A_1(I,R)	A_1(I,R) A_2(R)	A_1(I,R) B_1(I,R) B_2(I,R)	A_1(I,R) B_1(I,R) B_2(I,R)

6.15 The stretching frequency of a diatomic molecule is given by

$$\bar{\nu}(cm^{-1}) = (1/2\pi c)\sqrt{k/\mu}$$

where c is the velocity of light, k is the force constant, and μ is the reduced mass. It is seen that

$$\bar{\nu}(HCl)/\bar{\nu}(DCl) \sim \sqrt{\mu(DCl)/\mu(HCl)} \sim \sqrt{M_D/M_H} = 1.41$$

The calculated ratio $\bar{\nu}$HCl/$\bar{\nu}$DCl is 1.41 whereas the experimental value is 1.38. The expected frequency shift is 844 cm^{-1}.

In the case of HCN and DCN we can again approximate the ratio to be 1.26, yielding an expected shift of 969 cm^{-1}. The significant difference between the expected and observed values in this instance probably reflects the coupling between the C-H (and C-D) and the C-N vibrations.

6.16 This is an example of the isotopic splitting effect. The 295 cm^{-1} band is due to the Mn-^{35}Cl vibration and the shoulder at lower energy arises from the Mn-^{37}Cl vibration. The relative abundances of the two isotopes 35:37 are 0.75:0.25. This explains the lower intensity of the Mn-^{37}Cl absorption.

6.17 A band about 2000 cm^{-1} suggests a triple bond such as in CO, N_3^- or N_2. The elemental analysis supports the formulation of the product as $[Ru(NH_3)_5N_2]X_2$.

6.20 From the structure we expect bands in the ketonic region, about 1800 cm^{-1}, due to bridging carbonyl groups, and at 2000-2100 cm^{-1} owing to terminal carbonyl groups. The spectrum bears out the expectations.

Since $Fe_2(CO)_9$ has symmetry D_{3h}, we find $\Gamma_{(C=O)} = A_1' + E'$ and $\Gamma_{(C\equiv O)} = A_1' + E' + A_2'' + E''$. Of these only E' and A_2'' are infrared-active, and must account for the three bands in the spectrum; one in the bridging carbonyl region and two in the terminal carbonyl region.

6.21 The decrease in the C≡O stretching frequency coupled with the increase in the metal-carbon stretching frequency suggests an increase in the back donation (dative π-bonding) of electrons from the metal in the order Fe > Co > Ni.

6.23 To answer this question, we must determine the effect of the symmetry operations of D_{3h} on the sets of atomic motions indicated by the arrows. If a set of arrows is transformed into an equivalent set by an operation, then the character under that operation is 1. A set of arrows transformed into its exact negative has a character of -1 for that operation. All others are zero. Following this procedure it can be shown that the symmetries of the normal modes are (a) A_1', (b) A_1', (c) A_2''.

6.24 An analysis of the displacement coordinates yields the reducible representation

D_{4h}	E	$2C_4$	C_2	$2C_2'$	$2C_2''$	i	$2S_4$	σ_h	$2\sigma_v$	$2\sigma_d$
$\Gamma_{d.c.}$	15	1	-1	-3	-1	-3	-1	5	3	1

This may be decomposed into

$$A_{1g} + A_{2g} + B_{1g} + B_{2g} + E_g + 2A_{2u} + B_{2u} + 3E_u.$$

Since translations transform as $A_{2u} + E_u$ and rotations as $A_{2g} + E_g$, this leaves the following symmetry species for vibration:

$$A_{1g} + B_{1g} + B_{2g} + A_{2u} + B_{2u} + 2E_u.$$

From the character table we find that A_{1g}, B_{1g}, and B_{2g} are Raman active. Since three bands are allowed by the selection rules, and three bands are observed, the square planar structure is consistent with the observation.

6.25 The set of basis vectors composed of the 18 displacement coordinates spans the reducible representation

C_{4v}	E	$2C_4$	C_2	$2\sigma_v$	$2\sigma_d$
$\Gamma_{d.c.}$	18	2	-2	4	2

which contains the irreducible representations

$$4A_1 + A_2 + 2B_1 + B_2 + 5E$$

We see in the character table that translations transform as A_1 + E and rotations as A_2 + E. This leaves $3A_1 + 2B_1 + B_2 + 3E$ for vibrational modes. The selection rules are

	IR	Raman
A_1	✓	✓
B_1	x	✓
B_2	x	✓
E	✓	✓

✓ = allowed

x = forbidden

There should be three bands ($2B_1 + B_2$) in the Raman spectrum which are not in the infrared spectrum. Comparison of the spectra shows bands at 818, 530 and 231 cm^{-1} which could be these three. (However, it is believed that the band at 818 cm^{-1} may not be a fundamental mode). There should be six coincidences due to $3A_1 + 3E$, but only four are found: 919 (R), 928 (IR); 566 (R), 578 (IR); 364 (R), 362 (IR); 286 (R), 288(IR). The other bands in the infrared spectrum are due to combinations or differences.

6.26 With the assignment of the band at 284 cm^{-1} to ν_4, we see from $\nu_2 - \nu_4 = 312$ that $\nu_2 = 596$ cm^{-1}. Therefore

$$\nu_2 - \nu_6 = 403; \; \nu_6 = 193 \text{ cm}^{-1}$$

$$\nu_2 + \nu_6 = 596 + 193 = 789 \text{ cm}^{-1}$$

$$\nu_2 + \nu_4 = 596 + 284 = 880 \text{ cm}^{-1}$$

$$\nu_2 + \nu_3 = 596 + 724 = 1320 \text{ cm}^{-1}$$

There are no independent checks on ν_1 and ν_5, but we will evaluate them anyway.

$$\nu_1 + \nu_3 = 1358; \quad \nu_1 = 1358 - 724 = 634\ cm^{-1}$$
$$\nu_3 + \nu_5 = 993; \quad \nu_5 = 993 - 724 = 269\ cm^{-1}$$

6.28 By the usual procedures we find

$$\Gamma_{d.c.} = 3A_1 + A_2 + 3B_1 + 2B_2$$

Subtracting $\Gamma_{trans} = A_1 + B_1 + B_2$ and $\Gamma_{rot} = A_2 + B_1 + B_2$, we find that $\Gamma_{vib} = 2A_1 + B_2$. Using the internal coordinates it is seen that the angle deformation transforms as A_1 and bond stretching modes as $A_1 + B_1$. Thus, the two A_1 modes will consist of both bond stretching and angle deformation, while the B_1 will be the asymmetric bond stretching vibration only. Since bending vibrations usually occur at lower energies than stretching vibrations, and asymmetric stretching vibrations frequently occur at higher energies than symmetric stretches, then reasonable assignments are the following:

ν_1	A_1	1151 cm^{-1}
ν_2	A_1	519
ν_3	B_1	1361

The additional bands in the infrared spectrum are due to overtones and combinations. The band at 2305 cm^{-1} is just about two times the energy of ν_1, and is safely assigned as the first overtone of ν_1. Overtones of the other fundamental modes are not seen in the spectrum, although they are allowed by the selection rules.

The band at 1871 cm^{-1} is due to the simultaneous excitation of ν_2 and ν_3 ($\nu_2 + \nu_3 = 1880\ cm^{-1}$), and the band at 2499 cm^{-1} is the combination of $\nu_1 + \nu_3$.

The only band left for assignment is at 606 cm^{-1}. This is a difference band, and arises from $\nu_1 - \nu_2$. Note that $\nu_1 - \nu_2 = 632\ cm^{-1}$. The differences between calculated and observed values are caused both by experimental error and the anharmonicity of the vibrational modes.

6.29 Trigonal bipyramidal since six Raman bands are expected.

6.31 Compound (I). From the empirical formula this probably contains two molecules of triphenylphosphine [$(C_6H_5)_3P$] and the infrared absorption at 1944 cm^{-1} indicates the presence of a carbonyl group. Thus (I) may be written as a four coordinate iridium (I) complex $[(C_6H_5)_3P]_2Ir(CO)Cl$. The two possible structures for this compound are those with the phosphine ligands *cis* or *trans* to each other as illustrated below.

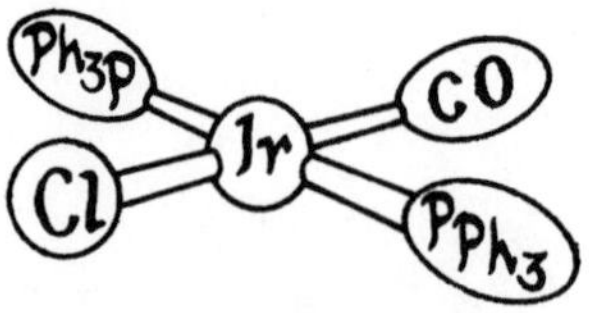

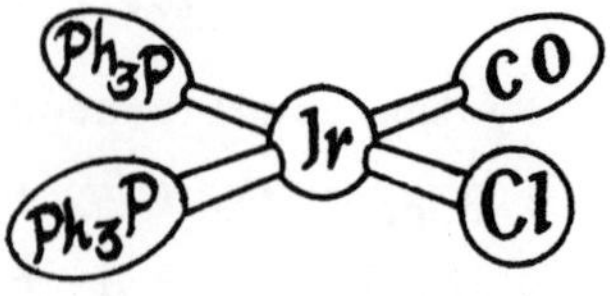

The data given here does not enable us to distinguish between these two structures.

Compound (II). The increase in carbonyl stretching frequency on going from I to II indicates that oxidation has taken place. From the empirical formula this reaction involves a net addition of HCl and the data can be explained by assuming a change from a 4-coordinate Ir(I) complex to a 6-coordinate Ir(III) species.

One possible structural formula for (II) is

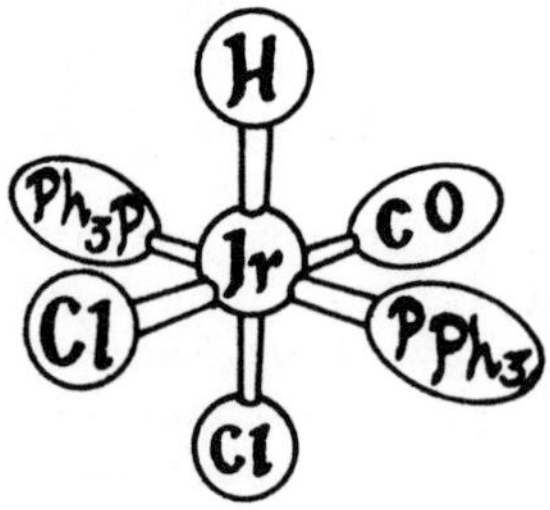

There are obviously several structural isomers of this compound.

The observed infrared absorption at 2245 cm^{-1} corresponds to the Ir-H stretching mode.

Compound (III). Treatment of (I) with $HgCl_2$ results in the net addition of $HgCl_2$ to the molecule. Similarly (II) is converted into (III) with addition of HgCl and loss of one H. Compound (III) is a species containing a metal-metal bond.

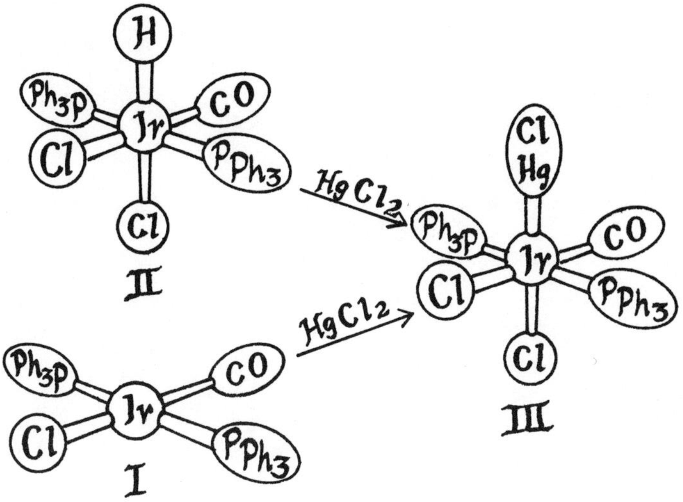

Compound (IV). Treatment of III with chlorine results in rupture of the metal-metal bond and replacement of HgCl by Cl.

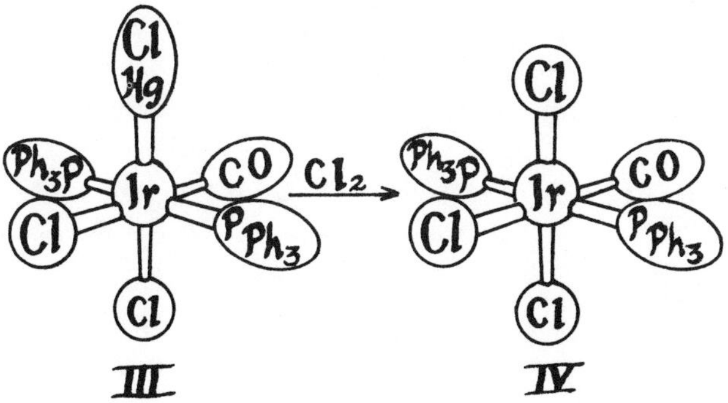

Treatment of III with molecular hydrogen and hydrogen chloride also breaks the metal-metal bond, replacing HgCl by H and so forming (II).

CHAPTER 7

ELECTRONIC SPECTROSCOPY

Electronic absorption spectroscopy is concerned with the transitions of electrons from occupied orbitals in the ground state to some appropriate orbital in the excited state. This powerful tool coupled with crystal field and ligand field theories has stimulated much interest in the properties of transition metal compounds. However, the importance of electronic spectroscopy in the study of the chemistry of the representative elements must not be minimized. With the increasing success of molecular orbital theory, the features of absorption spectra are more readily understood and much progress has been made in this area in recent years.

The relative energies and intensities of the bands in the absorption spectrum, the number of such bands, the temperature dependence and shape of the band envelope, all combine to yield information about the bonding and geometry of molecules. Problems in this chapter will be based on these concepts.

Important references include:

1. C. K. Jørgensen, *Absorption Spectra and Chemical Bonding in Complexes*, Pergamon Press, London, 1962.

2. C. J. Ballhausen, *Introduction to Ligand Field Theory*, McGraw-Hill Book Co., New York, 1962.

3. J. S. Griffith, *The Theory of Transition Metal Ions*, Cambridge University Press, 1961.

4. B. N. Figgis, *Introduction to Ligand Fields*, Interscience Publishers, Inc., New York, 1966.

5. C. J. Ballhausen and H. B. Gray, *Molecular Orbital Theory*, W. A. Benjamin, Inc., New York, 1964.

6. T. M. Dunn, D. S. McClure, and R. G. Pearson, *Some Aspects of Crystal Field Theory*, Harper and Row, Publishers, New York, 1965.

7. N. S. Hush and R. J. M. Hobbs, *Progress in Inorganic Chemistry*, 10, 259 (1968).

8. G. W. King, *Spectroscopy and Molecular Structure*, Holt, Reinhart and Winston, Inc., New York, 1964.

9. G. Herzberg, *Electronic Spectra and Electronic Structure of Polyatomic Molecules*, D. Van Nostrand Co., Inc., Princeton, 1966.

PROBLEMS

7.1 *Ions with a* d^1 *configuration in* O_h *and* T_d *crystal fields.* (Read: Dunn, McClure and Pearson, Chapter 1; Figgis, Chapter 2; or Cotton, Chapter 8.)

(a) What is the Russell-Saunders term symbol for a free ion with the configuration d^1?

(b) Into what crystal field states does the free ion term split in a complex with octahedral (O_h) symmetry? What are the states in a tetrahedral (T_d) complex?

(c) From qualitative considerations underlying crystal field theory, what is the ground state for the d^1 ion in an octahedral complex and in a tetrahedral complex?

(d) What *d-d* transitions are anticipated for octahedral and tetrahedral d^1 complexes?

7.2 *The* d^9 *configuration in* O_h *and* T_d *crystal fields.* Answer each question in 7.1 for the d^9 ion.

7.3 *The* d^2 *ion in* O_h *crystal field.* (Read: Cotton, Chapter 8).

(a) What Russell-Saunders states will arise for a free ion with a d^2 configuration?

(b) Into what crystal field states do the free ion terms split in a complex with O_h symmetry? (Weak field approximation)

(c) The configuration for the ground state of the ion in an octahedral complex may be written $(t_{2g})^2$. Excited state configurations are $(t_{2g})(e_g)$ and $(e_g)^2$. Determine the symmetry labels for the states which may arise from these configurations. (Strong field approximation)

(d) Construct a correlation diagram for the d^2 ion in an octahedral field. Plot the weak octahedral field states which arise from the free ion terms on the left hand abscissa and the states arising from the three strong field configurations $(t_{2g})^2$, (t_{2g})-

(e_g) and $(e_g)^2$ on the right hand. There must be a one to one correspondence between states arising from the free ion terms and the states on the right. Further, two lines representing states of the same symmetry never cross. From these two considerations the multiplicities of the strong field states may be assigned.

(e) Compare the correlation diagram with the Tanabe-Sugano diagram (Appendix 7.1).

7.4 *Tanabe-Sugano diagrams.* (Read: Jørgensen, Chapter 5; Figgis, Chapter 4).

(a) Ions with a d^4 configuration may form both spin-free and spin-paired octahedral complexes. Refer to the appropriate Tanabe-Sugano diagram (Appendix 7.1) and list the spin allowed transitions for both cases.

(b) List the spin allowed and spin forbidden transitions for the high spin and low spin d^6 configuration.

7.5 *Use of Tanabe-Sugano diagrams.* Match the following complex molecules with the proper ground state symbol:

(a) $[Cr(NH_3)_6]^{3+}$		(1) $^6A_{1g}$
(b) $[CoF_6]^{3-}$	(high spin)	(2) $^1A_{1g}$
(c) $[Mn(H_2O)_6]^{2+}$	(high spin)	(3) $^3T_{1g}$
(d) $[Ni(NH_3)_6]^{2+}$	(paramagnetic)	(4) $^4A_{2g}$
(e) $[RhCl_6]^{3-}$	(diamagnetic)	(5) $^5T_{2g}$
(f) $[Mn(CN)_6]^{3-}$	(low spin)	(6) $^3A_{2g}$
(g) $[Ti(H_2O)_6]^{3+}$		(7) $^2T_{2g}$

7.6 *Intensities of electronic transitions.* (Read Figgis, Chapter 9). In each of the following pairs of transitions state which you would expect to be more intense and why.

(a) $^3A_{2g} \rightarrow {}^3T_{2g}$ in $[NiCl_6]^{4-}$ or $^3T_1 \rightarrow {}^3T_2$ in $[NiCl_4]^{2-}$

(b) $^4T_{2g} \rightarrow {}^4T_{1g}$ or $^4T_{2g} \rightarrow {}^4A_{2g}$ in $[Co(NH_3)_6]^{2+}$

(c) $^1A_{1g} \rightarrow {}^1T_{2g}$ in $[Co(NH_3)_6]^{3+}$ or $^1A_1 \rightarrow {}^1T_2$ in $[Co(en)_3]^{3+}$

(d) $^4A_2 \rightarrow {}^4E$ or $^4A_2 \rightarrow {}^2E$ in $[Cr(ox)_3]^{3-}$

(e) $^3A_2 \rightarrow {}^3E$ or $^3A_2 \rightarrow {}^3A_2$ in $[Ni(en)_3]^{2+}$

(f) $^1A_1$ (metal) $\rightarrow {}^1E$ (metal) or $^1A_1$ (metal) $\rightarrow {}^1E$ (π^a on ligand) in $[Fe(phen)_3]^{2+}$

(g) The most intense *d-d* band in $[CoCl_4]^{2-}$ or the most intense *d-d* band in $[MnCl_4]^{2-}$

7.7 *The crystal field parameter* Δ. (Read Figgis, Chapter 9 and Jørgensen, Chapter 7). Order the following pairs of complexes as to the expected magnitude of crystal field parameter Δ and give the basis of your choice:

(a) $[CoF_6]^{4-}$ versus $[CoF_6]^{3-}$

(b) $[CoCl_6]^{4-}$ versus $[CoCl_4]^{2-}$

(c) $[CoF_6]^{3-}$ versus $[CoCl_6]^{3-}$

(d) $[TiF_6]^{3-}$ versus $[VF_6]^{2-}$

(e) $[Fe(CN)_6]^{4-}$ versus $[Fe(CN)_6]^{3-}$

(f) $[V(H_2O)_6]^{2+}$ versus $[V(H_2O)_6]^{3+}$

(g) $Cr(CO)_6$ versus $W(CO)_6$

(h) $[Fe(CN)_6]^{4-}$ versus $[Os(CN)_6]^{4-}$

7.8 *The absorption spectrum of MnF_2.* (J. W. Stout, *J. Chem. Phys.*, 31, 709 (1959)). The electronic absorption spectrum of MnF_2 is shown in Figure 7.1. The crystal structure contains the Mn^{2+} ion surrounded by six fluoride ions in a nearly perfect octahedral array. Use the appropriate Tanabe-Sugano diagram and assign labels to the various bands.

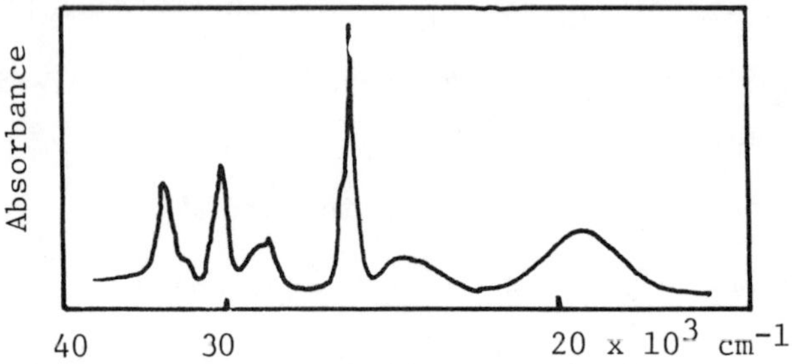

Figure 7.1

7.9 *Comparison of spectra of octahedral and tetrahedral complexes.* Compare the spectrum of the $[MnBr_4]^{2-}$ ion shown in Figure 7.2 with that of the $[MnF_6]^{4-}$ chromophore shown in Figure 7.1. Assign the spectral transitions.

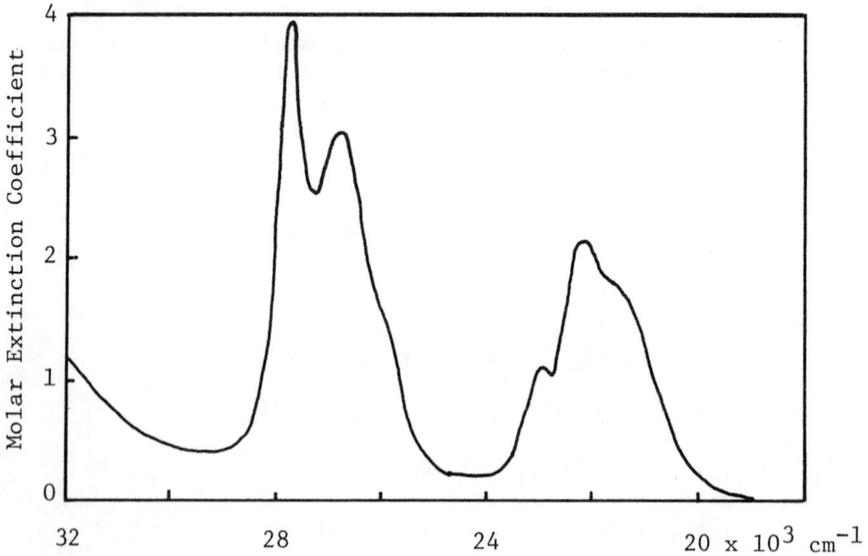

Figure 7.2

7.10 *Spectra of metal halides.* The absorption spectra of the divalent metal chlorides $TiCl_2$, VCl_2, $CrCl_2$, and $MnCl_2$ in molten $AlCl_3$ are shown in Figure 7.3. Assume the spectra arise from *d-d* transitions in the $[MCl_6]^{4-}$ ions, and decide which spectrum belongs to which ion.

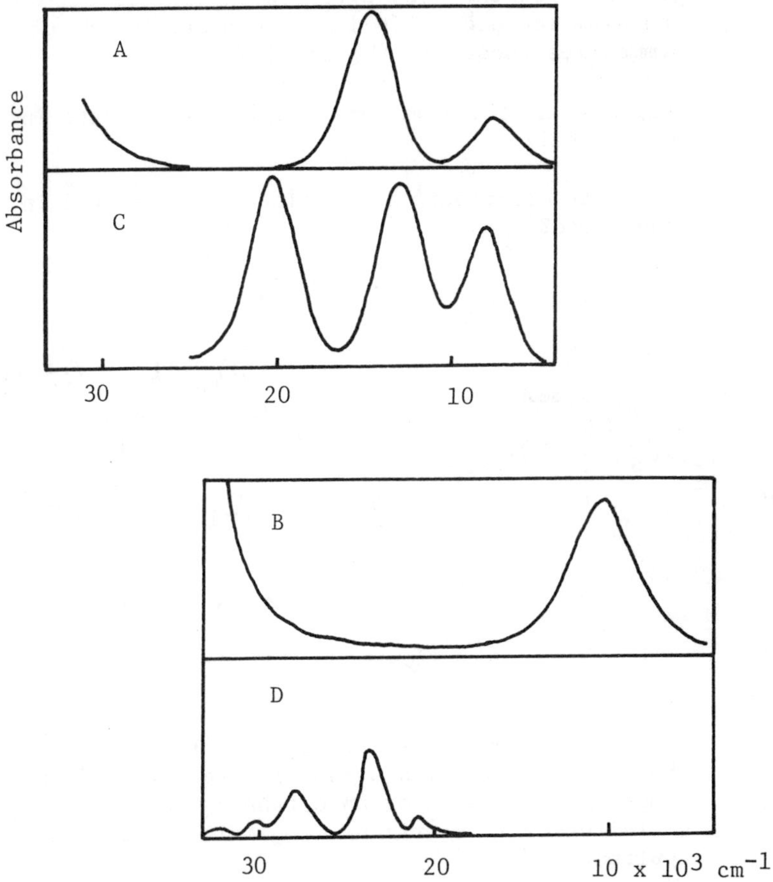

Figure 7.3

7.11 *Lower symmetry complexes.* (Read: Hush and Hobbs, p. 281ff). The molecular structure of the complex $VO(acac)_2$ is given in Figure 7.4.

(a) How many *d* electrons are there?

(b) What is the point group?

(c) In addition to using the rigorous symmetry of the molecule, one might try to analyse the spectrum assuming equivalence of all the *acac* oxygens and equality of all the "in plane" angles. What would the point group of this virtual symmetry be?

(d) Show how the octahedral energy levels of the vanadium ion are shifted by the axial fields of the symmetries chosen in (b) and (c).

(e) What is the symmetry of the ground state in each case?

(f) What electronic transitions are expected in each case?

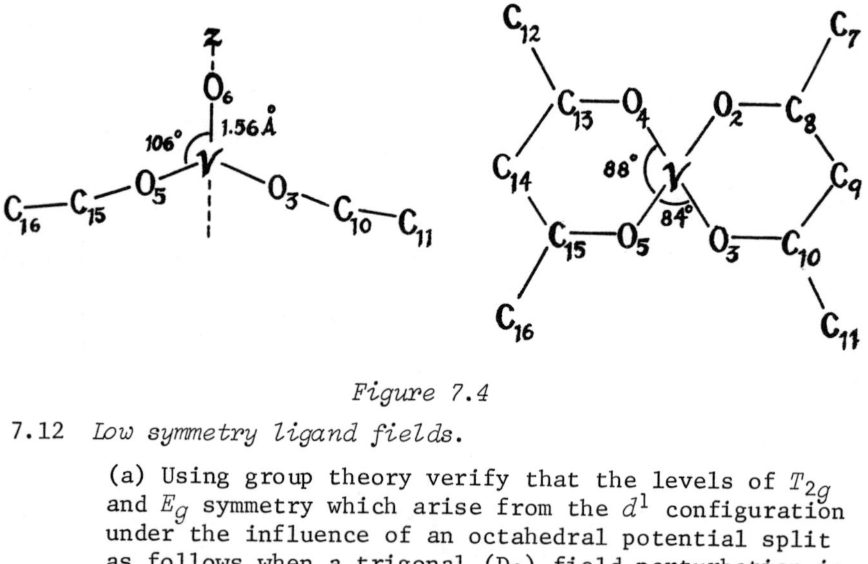

Figure 7.4

7.12 *Low symmetry ligand fields.*

(a) Using group theory verify that the levels of T_{2g} and E_g symmetry which arise from the d^1 configuration under the influence of an octahedral potential split as follows when a trigonal (D_3) field perturbation is applied:

$T_{2g} \rightarrow E + A_1$

$E_g \rightarrow E$

(b) Verify that these levels are further split by an additional rhombic distortion (lowering the symmetry to C_2) to 2B's and 3A's.

7.13 *Jahn-Teller distortions.* The Jahn-Teller theorem suggests that the structures of molecules with orbitally degenerate ground states are unstable with respect to structures with orbitally nondegenerate ground states.

(a) On the basis of this statement, which d^n ions in $ML_6{}^{q\pm}$ complexes with "octahedral" structures are expected to exhibit distortions?

(b) Which d^n ions in $ML_4{}^{q\pm}$ complexes with "tetrahedral" structures are expected to exhibit distortions?

(c) What simple distortions might be expected for $[Ti(H_2O)_6]^{3+}$, $[Fe(H_2O)_6]^{2+}$, and $[MnF_6]^{3-}$?

7.14 *Spectra of* $[NiX_4]^{2-}$ *ions.* The absorption spectra of $Cs_2(Zn,Ni)Cl_4$ and $Cs_2(Zn,Ni)Br_4$ are shown in Figure 7.5. Which spectrum belongs to which chromophore? Assign the various bands to the proper state transitions.

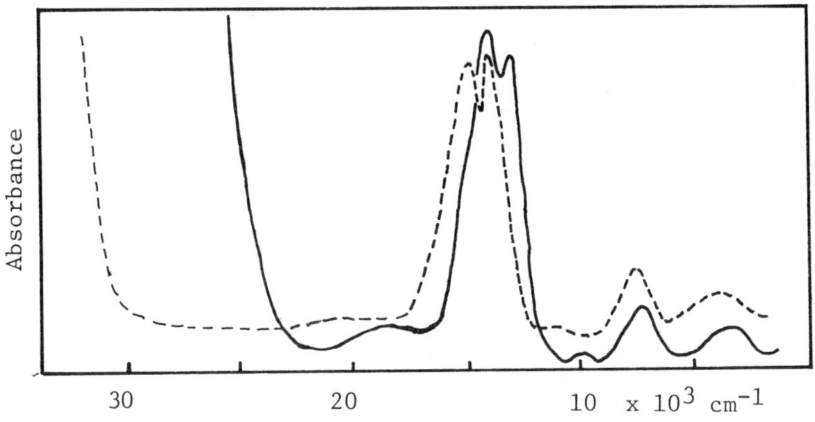

Figure 7.5

7.15 *Selection Rules.* (Read: Cotton, Chapters 5 and 8; Schonland, Chapter 10; Hochstrasser, Chapter 8). Determine which of the following transitions are allowed by an electric dipole mechanism:

(a) $B_1 \rightarrow B_2$ in C_{2v}

(b) $T_2 \rightarrow E$ in T_d

(c) $A_2 \rightarrow B_2$ in C_{4v}

(d) $A_1 \rightarrow A_2$ in D_3

7.16 *Selection rules.* (Read: Cotton, Chapter 8).

(a) For $[Cr(ox)_3]^{3-}$ determine the symmetries of the ground and excited electronic states by applying the group theoretical methods introduced in Chapter 4 to the appropriate Tanabe-Sugano diagram. Consider only the quartet states.

(b) The polarized absorption spectrum of $[Cr(ox)_3]^{3-}$ is shown in Figure 7.6. Give assignments for the transitions.

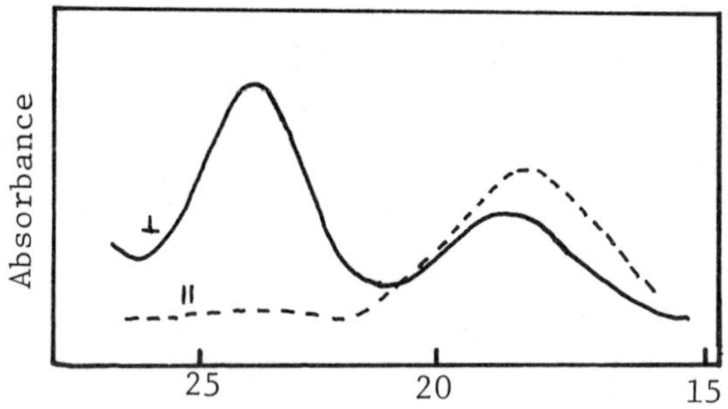

$cm^{-1} \times 10^3$

Figure 7.6 Absorption spectrum of $Cr(ox)_3{}^{3-}$.

7.17 *Molecular orbitals and electronic spectroscopy.* (Read: King, Chapter 10; Ballhausen and Gray, Chapter 6; and Herzberg, Chapter V.)

(a) A delocalized molecular orbital description of H_2O was constructed in Problem 5.34(b). From this diagram assign the lowest energy electronic transition in H_2O in terms of the orbitals involved and in terms of the symmetry of the ground and excited states. Assume the geometry of the excited state is the same as that of the ground state.

(b) From the delocalized molecular orbital description NH_3 in Problem 5.35, assign the lowest energy electronic transition assuming no change in symmetry. Is this transition allowed by the electric dipole mechanism?

7.18 *Spectroscopy and molecular orbital theory.* (Read: Ballhausen and Gray, Chapter 6). The one-electron energy level diagram for SO_2 is shown in Figure 7.7. Note that the configuration of ground state is $\ldots(1a_2)^2(3b_1)^2(4a_1)^2$.

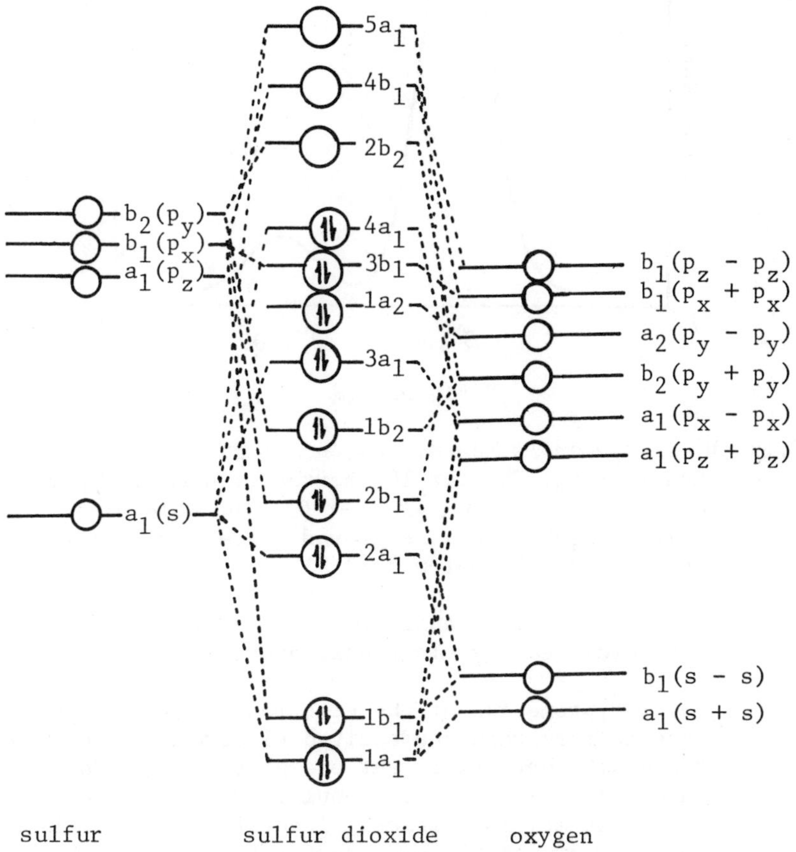

Figure 7.7

(a) What are the three lowest-energy transitions?

(b) Are all three transitions allowed?

(c) The spectrum of SO_2 is shown in Figure 7.8. Assign the bands in the spectrum in view of the experimental result that the band at 27,000 cm.$^{-1}$ is polarized perpendicular to the plane of the molecule.

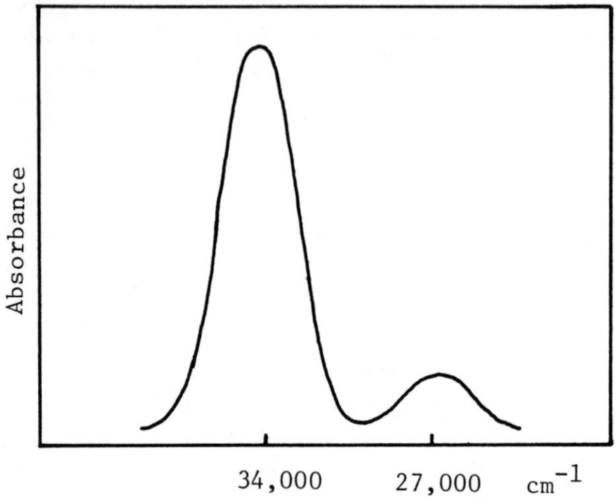

Figure 7.8

7.19 *Molecular orbitals and electronic spectroscopy.* (Read: King, Chapter 10; Herzberg, Chapter V; and Ballhausen and Gray, Chapter 6). The molecular orbital diagram for the ground state of CO_2 was constructed in Problem 5.32(b).

(a) What is the transition expected to give rise to the lowest energy electronic absorption in CO_2?

(b) The promotion of the electron as in (a) gives rise to more than one excited electronic state and these will have different energies owing to differences in inter-electron repulsions, as well as to different multiplicities. Identify these levels as to their symmetries and predict the assignment of the first strong absorption band in CO_2 which is observed at 1335Å (75,000 cm^{-1}).

(c) Two other CO_2 absorption bands are observed at lower energies (59,000 and 67,000 cm^{-1}) with extinction coefficients approximately an order of magnitude lower than the strong band at 75,000 cm^{-1}. These are thought to be the split components of another singlet-singlet transition forbidden in the linear molecule but allowed slightly owing to the lowering of the molecular symmetry to C_{2v} when an electron is excited to the anti-bonding level. Assign these two weak bands.

7.20 *Reduction of free ion term splittings.* (Read: Figgis, Chapter 7, and Jørgensen, Chapter 8.) The spectrum of the hexahydroxocobaltate(II) ion is given in Figure 7.9.

(a) Refer to the Tanabe-Sugano diagram for spin-free d^7 ions and suggest assignments for the spectral transitions.

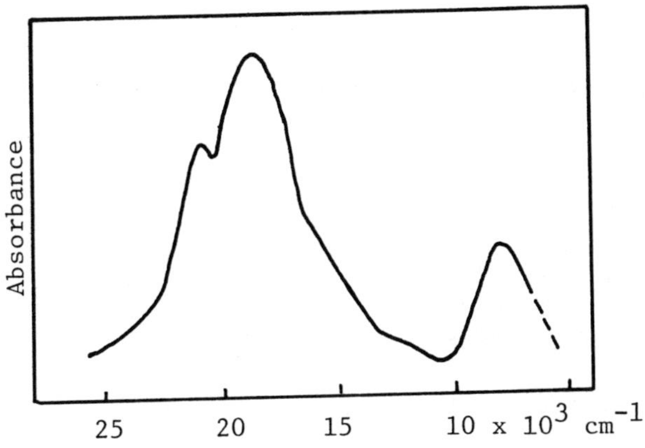

Figure 7.9 The reflectance spectrum of $Ba_2[Co(OH)_6]$.

(b) The energy levels for a spin-free d^7 ion are given by:

$$\begin{matrix} {}^4T_{1g}(F) \\ {}^4T_{1g}(P) \end{matrix} \begin{vmatrix} -6Dq - E & 4Dq \\ 4Dq & x - E \end{vmatrix} = 0$$

${}^4T_{2g}$: ${}^4T_{2g}$: $2Dq$; ${}^4A_{2g}$: $12Dq$.

where x is the ${}^4F$-${}^4P$ term splitting in the free ion. (See Chapter 3). In $Co(H_2O)_6^{2+}$ the three spin-allowed bands occur at 8350, 17850, and 20000 cm^{-1}. In crystal field theory the value which would be assumed for x would be taken from the spectrum of the gaseous Co^{2+} ion, but in ligand field theory the term splitting in the complex ion is taken as an adjustable parameter. Use the data for $[Co(H_2O)_6]^{2+}$ and evaluate Dq and x. (Note that $x = 15B = 15F_2 - 75F_4$.)

7.21 *Spectrum of* $[Cr(H_2O)_6]^{3+}$. The spin-allowed *d-d* bands of $[Cr(H_2O)_6]^{3+}$ occur at 17,400 cm^{-1}, 24,500 cm^{-1}, and 38,000 cm^{-1}. Using the assignment for these bands from the Tanabe-Sugano diagram, fit the spectral data to the parameters Dq and x in the expressions for the energies of the states:

$$E(^4A_{2g}) = -12Dq$$

$$E(^4T_{2g}) = -2Dq$$

$$E(^4T_{1g}) \begin{vmatrix} 6Dq - E & 4Dq \\ 4Dq & x - E \end{vmatrix} = 0$$

where x is the $^4P - {}^4F$ term separation in the complex.

7.22 *The nephelauxetic series.* (Read: Jørgensen, Chapter 8). The *d-d* bands for three octahedral chromium(III) complexes are listed in Table 7.1.

(a) Fit these data to the parameters Dq and $x = {}^4P - {}^4F$.

(b) The free ion $^4P - {}^4F$ term separation is 13,800 $cm.^{-1}$. Express the term splitting in the complexes as a fraction of the free ion value, that is, calculate

$$\beta = \frac{(^4P - {}^4F) \text{ in complex}}{(^4P - {}^4F) \text{ in free ion}}$$

(c) Order the ligands in terms of decreasing β.

(d) What is the significance of the series?

(e) How does it compare to the spectrochemical series?

Table 7.1

Spectral Data for $[CrL_6]^{q\pm}$

Chromophore	$^4A_{2g} \to {}^4T_{2g}$ cm^{-1}	$^4A_{2g} \to {}^4T_{1g}$ cm^{-1}
$[CrCl_6]^{3-}$	13,200	18,700
$[Cr(H_2O)_6]^{3+}$	17,400	24,500
$[Cr(NH_3)_6]^{3+}$	21,500	28,500

7.23 *The spectrum of* $[Ti(H_2O)_6]^{3+}$. The spectrum of the hexaaquotitanium(III) ion, $[Ti(H_2O)_6]^{3+}$, is shown in Figure 7.10. What explanation may be offered for the double-humped *d-d* spectrum?

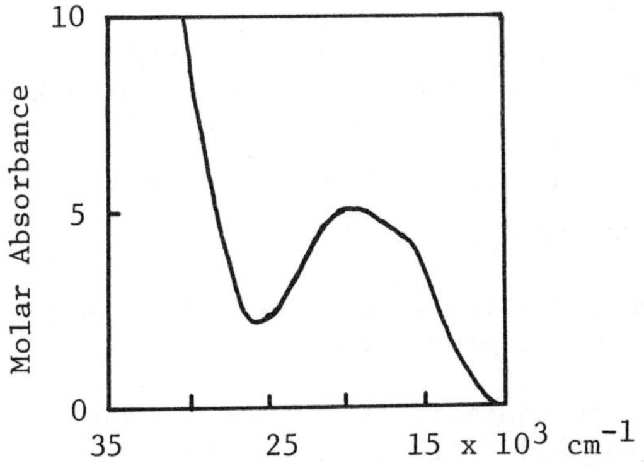

Figure 7.10 Absorption Spectrum of $[Ti(H_2O)_6]^{3+}$

7.24 *Comparison of* O_h *and* T_d *spectra.* The absorption spectra for $[Co(H_2O)_6]^{2+}$ and $[CoCl_4]^{2-}$ are shown in Figure 7.11. In view of the fact that $[Co(H_2O)_6]^{2+}$ is pink and $[CoCl_4]^{2-}$ is blue, and noting the extinction coefficients, which spectrum belongs to which ion?

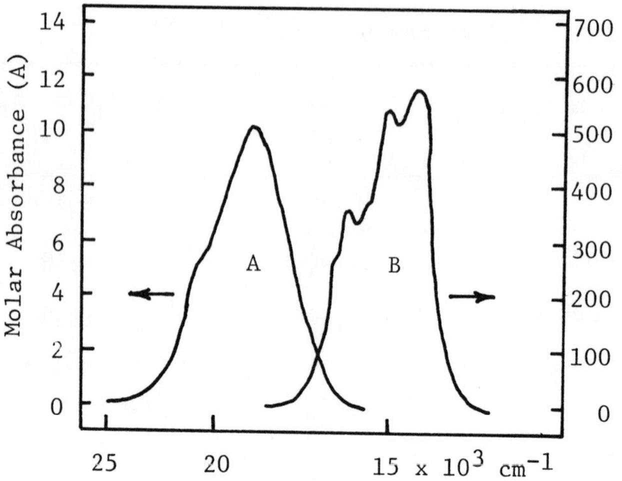

Figure 7.11 Absorption spectra of $[Co(H_2O)_6]^{2+}$ *and* $[CoCl_4]^{2-}$.

7.25 *Rule of average environment.* (Read: Jørgensen, Chapter 7.) The spectra of mixed complexes $[M(A_nB_{(6-n)})]$ may be interpreted in terms of a rule of average environment. The rule is:

$$Dq \text{ for } [M(A_nB_{(6-n)})] = \{n\,Dq_{(MA_6)} + (6-n)Dq_{(MB_6)}\}/6$$

The reduction in term splittings may be estimated by a similar procedure.

(a) An analysis of the absorption spectrum of $[V(H_2O)_6]^{3+}$ yields values for the parameters Dq and x of 1,840 and 9,300 cm.$^{-1}$, respectively. The values for Dq and x of $[VCl_6]^{3-}$ are 1,200 and 8,050 cm.$^{-1}$. The energies of the states of a d^2 ion are given by

$$E(^3T_{2g}) = 2Dq$$

$$E(^3A_{2g}) = 12Dq$$

$$E(^3T_{1g}) = \begin{vmatrix} -6Dq - E & 4Dq \\ 4Dq & x - E \end{vmatrix} = 0$$

where $x = {}^3P - {}^3F$.

Use the rule of average environment and calculate the energy of the ${}^3T_{1g} \rightarrow {}^3T_{1g}(F)$ transition for each compound in the series $[V(H_2O)_nCl_{6-n}]^{n-3}$.

(b) In view of the predicted transitions, which chromophores may be expected to present in the following compounds:

compound	observed band, cm^{-1}
$KVCl_4 \cdot 6H_2O$	22,990
$VCl_3 \cdot 6H_2O$	23,150
$VCl_3 \cdot 4H_2O$	22,880
$K_2VCl_5 \cdot H_2O$	19,300

7.26 *Selection rules.* Work out the electric dipole selection rules for transitions in molecules with symmetry C_{4v}.

7.27 *Spectroscopy and molecular orbital theory.* The stable crystalline form of $VOSO_4 \cdot 5H_2O$ contains the vanadium in a distorted octahedral site (exact symmetry C_{2v} ; approximate symmetry C_{4v}). A molecular orbital energy level diagram is given for the $O\text{-}V(O)_5$ chromophore in Figure 7.12. The energy level spacings are drawn approximately to scale. Use the selection rules derived for C_{4v} and the energy level diagram, and assign the bands in the polarized spectra in Figure 7.13. The ground state has the electronic configuration ....$(e\pi^b)^4 b_2$ (C. J. Ballhausen, B. F. Djurinskii, and K. J. Watson, *J. Am. Chem. Soc.*, 90, 3305 (1968)).

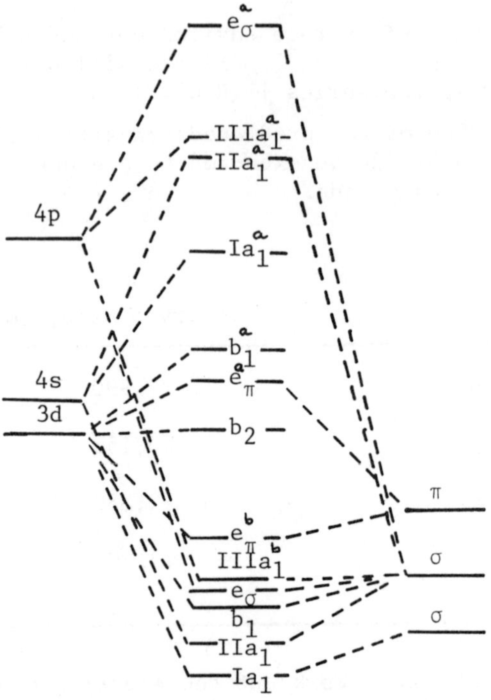

Figure 7.12

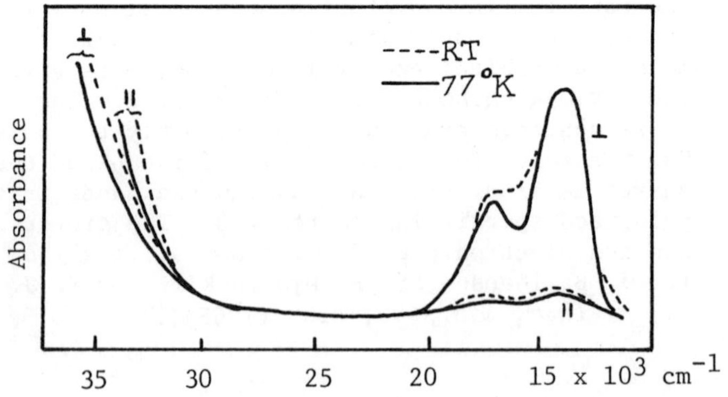

Figure 7.13

7.28 *Electronic structure and spectroscopy of the nitroprusside ion.* The molecular orbital energy level diagram for $[Fe(CN)_5NO]^{2-}$ is shown in Figure 7.14.

(a) Write the electronic configuration for the ground state.

(b) Is the predicted ground state configuration consistent with observed diamagnetism of the complex ion?

(c) Based on this ordering of the energy levels, what is the electronic configuration of $[Fe(CN)_5NO]^{3-}$?

(d) Suggest an experimental technique which may be used to confirm this hypothesis.

(e) Low energy transitions are possible from the metal-dominated molecular orbitals $6e$ and $2b_2$ to the $7e(\pi NO)$, $3b_1(x^2 - y^2)$, and $5a_1(z^2)$. Determine the symmetries of the excited states which arise from the excitation of electrons from $6e$ and from $2b_2$ to each of the low-lying empty orbitals listed above.

(f) Considering only the excited states in part (e), how many bands are expected in the absorption spectrum?

(g) When the electric vector is polarized perpendicular to the Fe-NO axis, the lowest energy band in the visible spectrum occurs at 20,080 cm^{-1}. In the parallel polarization, this band disappears, and a band appears at 25,380 cm^{-1}. Provide assignments for these electronic transitions.

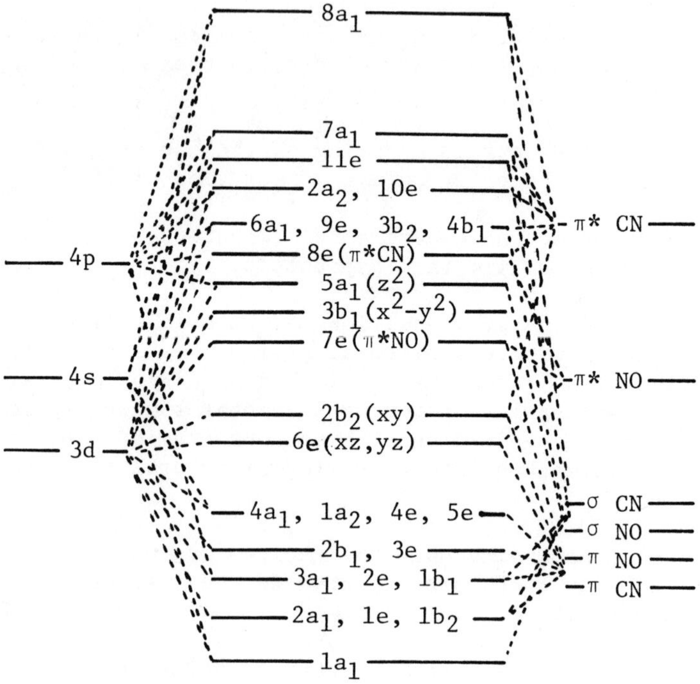

8a1
7a1
11e
2a2, 10e
6a1, 9e, 3b2, 4b1
8e(π*CN)
5a1(z2)
3b1(x2-y2)
7e(π*NO)
2b2(xy)
6e(xz,yz)
4a1, 1a2, 4e, 5e
2b1, 3e
3a1, 2e, 1b1
2a1, 1e, 1b2
1a1
4p
4s
3d
π* CN
π* NO
σ CN
σ NO
π NO
π CN

Figure 7.14

7.29 *Crystal field potential.* (Read: Ballhausen, Chapter 4.) The crystal field potential is expanded in terms of the normalized spherical harmonics. The formula is

$$V = \sum_{i} \sum_{\ell=0}^{n} \sum_{m} Y_{\ell}^{m_\ell}(\theta_i, \phi_i) R_n^{\ell}(r_i)$$

where the summation is carried out over the i ligands, $Y_\ell^{m_\ell}$ are the spherical harmonics which are conviently tabulated in Pauling and Wilson, and $R_n^\ell(r_i)$ are the radial functions. The second summation is truncated at $\ell = 4$ for d electrons and at $\ell = 6$ for f electrons. Much labor can be conserved if it is recognized that the crystal field potential must be invariant under all of the operations of the point group of the complex. All we have to do is to examine the symmetry properties of each term and reject those terms which do not transform as the totally symmetric irreducible representation.

(a) Use the rotation subgroup O, and determine the terms which contribute to the octahedral crystal field potential for a complex of a first row transition series metal ion.

(b) Determine which terms contribute to the crystal field potential for f electrons in a complex with D_3 symmetry.

(c) Trace the increasing complexity in crystal field calculations for titanium(III) complexes as the symmetry of the crystal field decreases from O_h to D_{4h}.

(d) In (a) above it was found that only certain $Y_\ell^{m_\ell}$ contributed to the octahedral crystal field potential. Choose the C_4 axis as the axis of quantization, that is, the axis about which the potential must be invariant, and verify that Y_4^0 and $Y_4^{\pm 4}$ are three terms in the octahedral potential. For this problem the Cartesian form of the spherical harmonics given in Appendix I of Chapter 4 in Ballhausen's book are most convenient.

(e) The octahedral crystal field potential, $V_0 = Y_4^0 + \alpha(Y_4^4 + Y_4^{-4})$ in terms of Cartesian coordinates is

$$V_0 = \frac{1}{8}\left(\frac{9}{4\pi}\right)^{1/2}\left(\frac{35z^4 - 30z^2r^2 + 3r^4}{r^4}\right)$$

$$+ \alpha\left(\frac{9}{4\pi}\right)^{1/2}\left(\frac{35}{128}\right)^{1/2}\left[\frac{(x + iy)^4}{r^4} + \frac{(x - iy)^4}{r^4}\right]$$

Solve for the factor α by carrying out a rotation of the potential about the C_3 axis and equating the resultant form of the potential to the initial form with which, by definition, it must be equal.

7.30 *Crystal field calculation.* (Read: Dunn, McClure, and Pearson, Chapter 1; and Figgis, Chapter 2.) The octahedral crystal field potential may also be written in Cartesian coordinates as

$$V_0 = \frac{35ze^2}{4a^5}\left(x^4 + y^4 + z^4 - \frac{3}{5}r^4\right).$$

According to perturbation theory the first order perturbation energy is given by

$$\int \psi^* V_0 \psi d\tau$$

where the ψ may be taken to be the hydrogenic free ion wave functions.

(a) Extract the ϕ dependent part from the potential and from the hydrogenic wave functions, and show that only the following integrals are non-zero:

$$\frac{1}{2\pi}\int_0^{2\pi} e^{im_\ell\phi}V_0(\phi)e^{-im'_\ell\phi}d\phi \equiv \frac{1}{2\pi}\langle e^{im_\ell\phi}|V_0(\phi)|e^{-im'_\ell\phi}\rangle$$

(For convenience the ℓ's have been dropped from the m_ℓ quantum numbers.)

(b) Evaluate the non-zero integrals in (a) above.

(c) In the perturbation calculation we are studying

integrals of the form

$$\int_0^\infty \int_0^\pi \int_0^{2\pi} \psi^*_{n,\ell,m_\ell}(r,\theta,\phi)[V_0(r,\theta,\phi)]\psi_{n,\ell,m_\ell}(r,\theta,\phi)r^2 \sin\theta\, dr d\theta d\phi$$

From (a) above, it is known that only seven of the integrals are non-zero. Recall that

$$\langle \Phi_0^* | V_0(\phi) | \Phi_0 \rangle = \frac{3}{4} r^4 \sin^4\theta$$

and that in the potential z^4 is θ-dependent. Verify that the integral

$$\int_0^\pi \Theta_2^{0*}[V_0(\theta,\phi)]\Theta_2^0 \sin\theta\, d\theta$$

is equal to $(\frac{5}{7})r^4$.

(d) The angular dependence of the crystal field potential has been evaluated in (b) and (c). The results are

$$\langle 0 | x^4 + y^4 + z^4 | 0 \rangle = (\frac{5}{7})r^4$$

$$\langle 1 | V_0 | 1 \rangle = \langle -1 | V_0 | -1 \rangle = (\frac{11}{21})r^4$$

$$\langle 2 | V_0 | 2 \rangle = \langle -2 | V_0 | -2 \rangle = (\frac{13}{21})r^4$$

$$\langle 2 | V_0 | -2 \rangle = \langle -2 | V_0 | 2 \rangle = (\frac{73}{105})r^4$$

Verify that the integral

$$\int_0^\infty R(r)[V_0]R(r)r^2 dr$$

is equal to $6Dq$ for m = 0, where

$$D = (\frac{35Ze^2}{4a^5})$$

and

$$q = \frac{2}{105}\int_0^\infty R^2_{n\ell} r^4 r^2 dr = \frac{2}{105}\overline{r^4}$$

(e) The integrals

$$\int\psi^*(r,\theta,\phi)\,|V_0(r,\theta,\phi)|\,\psi(r,\theta,\phi)dr$$

are

$$\langle\psi(1)|V_0|\psi(1)\rangle = \langle\psi(-1)|V_0|\psi(-1)\rangle = -4Dq$$

$$\langle\psi(2)|V_0|\psi(2)\rangle = \langle\psi(-2)|V_0|\psi(-2)\rangle = Dq$$

$$\langle\psi(-2)|V_0|\psi(2)\rangle = \langle\psi(2)|V_0|\psi(-2)\rangle = 5Dq$$

Set up the perturbation secular determinant and solve for the energy roots.

SOLUTIONS

7.1 (a) ^2D

(b) For O_h use the rotation group O and the formulas

$$\chi(E) = 2L + 1$$

$$\chi(C_{360/\alpha}) = \frac{\sin\,[(L + 1/2)\alpha]}{\sin\,\alpha/2}$$

(where $\chi(s)$ is the character element associated with the operation, s and α is the angle of rotation) to construct the reducible representation spanned by the D term. This gives the result:

O	E	$8C_3$	$6C_2$	$6C_4$	$3C_2$ $(=C_4{}^2)$
Γ_D	5	-1	1	-1	1

which reduces to $E + T_2$. The derivation of the above formula is given by Cotton, p. 191. In the group O_h, $E + T_2$ becomes $E_g + T_{2g}$. From the correlation table in Appendix 4.3 we find that $E_g + T_{2g}$ in O_h becomes $E + T_2$ in T_d. All the crystal field states are spin doublets since they arise from ^2D. Therefore the states are $^2T_{2g} + {}^2E_g$ in O_h and ^2T and ^2E in T_d.

(c) The three d-orbitals d_{xz}, d_{yz}, and d_{xy}, which form a basis for T_{2g}, are not directed toward ligands in O_h, and thus, T_{2g} is the ground state. The ground state in T_d by similar reasoning is E.

(d) For an O_h complex we expect $T_{2g} \rightarrow E_g$, and for a a T_d complex, $E \rightarrow T_2$.

7.2 (a) ^2D

(b) O_h : $E_g + T_{2g}$

T_d : $E + T_2$

(c) O_h : E_g
T_d : T_2

(d) O_h : $E_g \rightarrow T_{2g}$
T_d : $T_2 \rightarrow E$

7.3 (a) Refer to the answer to Problem 3.5. ^3F, ^3P, ^1G, ^1D, ^1S.

(b) Proceed as in the answer 7.1(b).

^3F gives $^3A_{2g} + {}^3T_{1g} + {}^3T_{2g}$

^3P gives $^3T_{1g}$

^1D gives $^1T_{2g} + {}^1E_g$

^1G gives $^1A_{1g} + {}^1E_g + {}^1T_{1g} + {}^1T_{2g}$

^1S gives $^1A_{1g}$

(c) The direct product of the symmetries of the two electrons in each configuration yields the symmetries of the strong field states arising from the configuration. The direct product $t_{2g} \times t_{2g}$ yields $A_{1g} + E_g + T_{1g} + T_{2g}$; $e_g \times t_{2g} = T_{1g} + T_{2g}$; and $e_g \times e_g = A_{1g} + A_{2g} + E_g$.

(d)

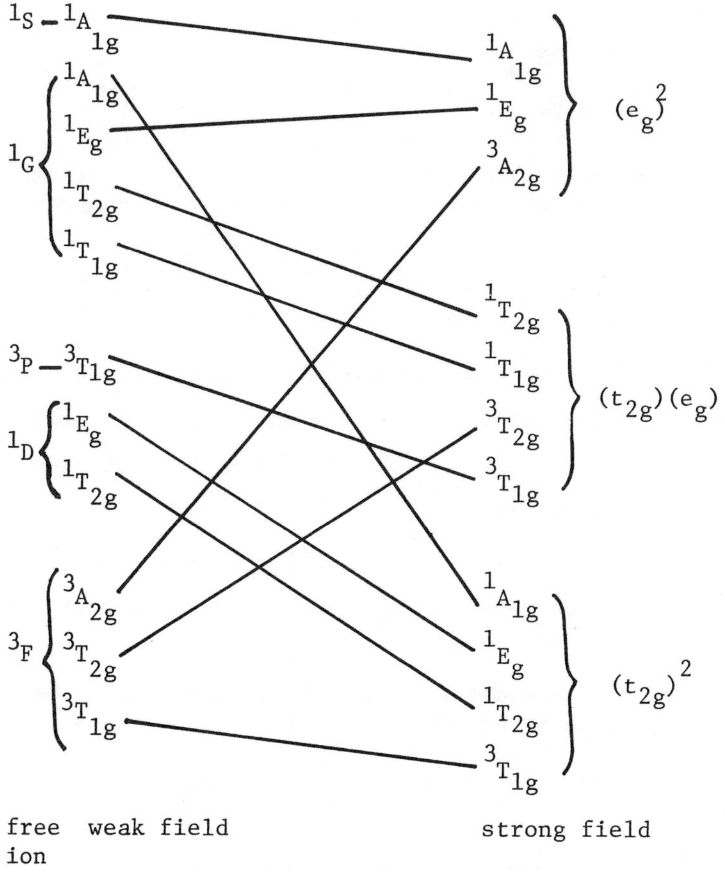

7.4 (a) Spin free (left side of diagram) : ${}^5E_g \rightarrow {}^5T_{2g}$

Spin paired (right side) : ${}^3T_{1g} \rightarrow {}^3E_g$

$\rightarrow {}^3T_{2g}$

$\rightarrow {}^3A_{1g}$

$\rightarrow {}^3A_{2g}$

7.5 (a) ${}^4A_{2g}$

(b) ${}^5T_{2g}$

(c) ${}^6A_{1g}$

(d) $^3A_{2g}$

(e) $^1A_{1g}$

(f) $^3T_{1g}$

(g) $^2T_{2g}$

7.6 (a) The intensity of the $[NiCl_4]^{2-}$ band should be greater since in the noncentric T_d symmetry the Laporte rule can be broken down via *d-p* (or other *g-u*) mixing. In O_h this is possible only through vibronic coupling.

(b) The transition to $^4T_{1g}$ should be more intense since it is formally a one-electron transition, whereas $^4T_{2g} \rightarrow {}^4A_{2g}$ is formally a two-electron transition.

(c) $^1A_1 \rightarrow {}^1E$ in $[Co(en)_3]^{3+}$ should be more intense since the D_3 point group is noncentric. (See part (a).)

(d) The quartet-doublet transition will be less intense because it is spin-forbidden.

(e) The transition to $^3A_2$ is symmetry forbidden and should be less intense.

(f) Metal to ligand charge transfer bands are more intense than *d-d* bands.

7.7 (a) $[CoF_6]^{3-} > [CoF_6]^{4-}$ For the same ligand, Δ increases with the oxidation state of the metal.

(b) $[CoCl_6]^{4-} > [CoCl_4]^{2-}$ $\Delta_T \simeq \frac{1}{2}\Delta_O$ for the same metal and ligand.

(c) $[CoF_6]^{3-} > [CoCl_6]^{3-}$

(e) $[Fe(CN)_6]^{3-} > [Fe(CN)_6]^{4-}$

(g) $W(CO)_6$ > $Cr(CO)_6$ All other factors being equal, Δ increases going down any group.

7.8 The Mn^{2+} ion has a d^5 electronic configuration. The assignments are in order of increasing energy.

$^6A_{1g} \rightarrow {}^4T_{1g}$

$^6A_{1g} \rightarrow {}^4T_{2g}$

$^6A_{1g} \rightarrow {}^4A_{1g}, {}^4E_g$

$^6A_{1g} \rightarrow {}^4T_{2g}$ (^4D)

$^6A_{1g} \rightarrow {}^4E_g$ (^4D)

$^6A_{1g} \rightarrow {}^4T_{1g}$ (^4P)

The low intensities indicate that charge transfer is not involved and that all the bands are associated with *d-d* transitions. The very sharp band at about 25,000 cm^{-1} is due to an excitation to the accidentally degenerate $^4A_{1g}$ and ^4E_g levels. As a general rule, sharp bands are associated with transitions between states whose energy differences are independent of the crystal field strength (that is, those which arise from the same strong-field configuration).

7.10 Spectrum *D* is assigned to the $[MnCl_6]^{4-}$ chromophore, based on its low intensity and the pattern and number of bands. (See Problem 7.9). Only for the configurations d^1, d^4, d^6, and d^9 should an octahedral MX_6 complex have a single band. Therefore, spectrum *B* must belong to $[CrCl_6]^{4-}$. There should be three spin-allowed bands in both the d^2 and d^3 cases. The fact that there are only two bands in spectrum *A* suggests that two of the excited states of the configuration giving rise to the spectrum have similar energies. Reference to the Tanabe-Sugano diagrams shows that such could result only for the d^2 ion, so spectrum *A* belongs to $[TiCl_6]^{4-}$ and spectrum *C* to $[VCl_6]^{4-}$.

7.11 (a) one

(b) C_{2v}

(c) C_{4v}

(d)

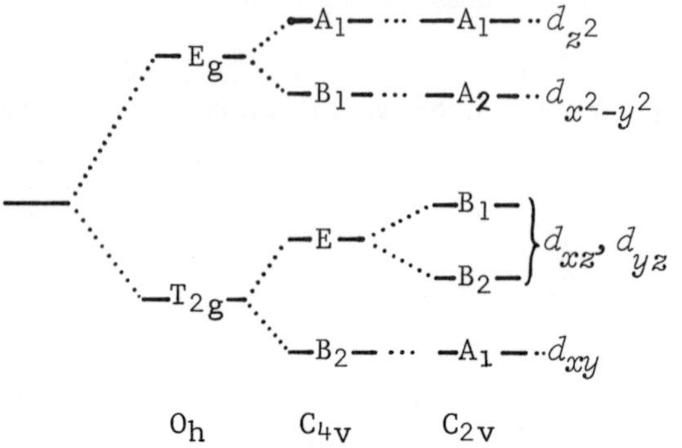

Note that the symmetry planes in C_{2v} correlate with the σ_d planes in C_{4v}, and the x and y axes have been taken along the bond directions.

(e) B_2 for C_{4v}; A_1 for C_{2v}

(f) 3 for C_{4v}; 4 for C_{2v}

The unambiguous assignment of this spectrum has not yet been made and other plausible energy level schemes have been proposed. See J. Selbin, *Coord. Chem. Rev.*, 1, 293 (1966), for a review of this work.

7.12 (a) The point group D_3 is a subgroup of O_h. To find the splitting of the O_h levels under the D_3 perturbation, we construct the representations spanned by T_{2g} and E_g and reduce them. The representations are made up of the characters of the operations which are present in both groups.

D_3	E	$2C_3$	$3C_2$
T_{2g}	3	0	1 $= E + A_1$
E_g	2	-1	0 $= E$

7.13 (a) d^1, d^2, d^4 (spin free), d^4 (spin paired), d^5 (spin paired), d^6 (spin free), d^7 (spin free and spin paired), and d^9.

(c) For $[Ti(H_2O)_6]^{3+}$ an axial compression would remove the degeneracy.

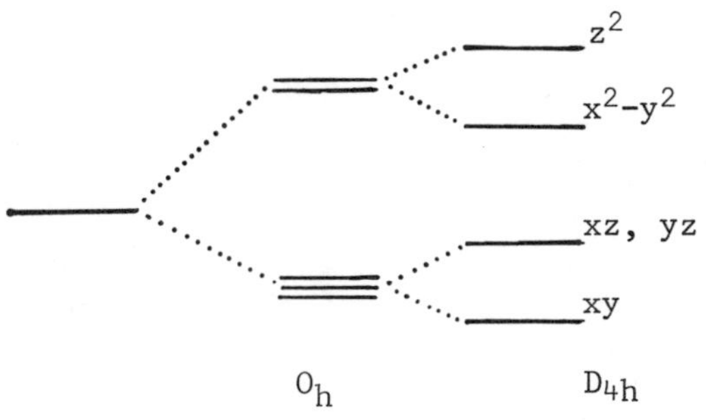

The same mechanism would work for $[Fe(H_2O)_6]^{2+}$, but either a compression or an elongation would remove the orbital degeneracy for the $d^4[MnF_6]^{3-}$ ion.

7.14 Since the bromide ion gives rise to a smaller crystal field splitting than the chloride ion and since the first two transitions are to states arising from the same free ion term as the ground state, we should expect to see the *d-d* bands in $[NiBr_4]^{2-}$ at lower energy than those in $[NiCl_4]^{2-}$. with this principle in mind, it is easy to see that spectrum *B* corresponds to $[NiBr_4]^{2-}$ and spectrum *A* to $[NiCl_4]^{2-}$.

There are five major bands in each spectrum. By reference to the Tanabe-Sugano diagram, we may make the following assignments for the $[NiBr_4]^{2-}$ spectrum:

Band position, cm^{-1}	Assignment
3,500	$^3T_1 \rightarrow {}^3T_2\ ({}^3F)$
7,500	$\rightarrow {}^3A_2\ ({}^3F)$
10,000	$\rightarrow {}^1T_2,\ {}^1E\ ({}^1D)$
14,000	$\rightarrow {}^3T_1\ ({}^3P)$
18,500	$\rightarrow$ components of ^1G

7.15 (a) In C_{2v} z transforms as A_1, x as B_1, and y as B_2. Therefore we must determine the products.

$B_2 \times A_1 \times B_1 = A_2$

$B_2 \times B_1 \times B_1 = B_2$

$B_2 \times B_2 \times B_1 = B_1$

Since A_1 does not appear in any of the products, the transition $B_2 \rightarrow B_1$ is not allowed by an electric dipole mechanism.

(b) In T_d (x, y, z) form a basis for T_2. We determine the character of the product $T_2 \times T_2 \times E$ and decompose it.

T_d	E	$8C_3$	$3C_2$	$6S_4$	$6\sigma_d$
$T_2 \times T_2 \times E$	18	0	2	0	0

$= A_1 + A_2 + 2E + 2T_1 + 2T_2$

Since the direct product contains the totally symmetric representation, A_1, the $T_2 \rightarrow E$ transition is orbitally allowed.

7.16 (a) The molecular symmetry is D_3. From the d^3 Tanabe-Sugano diagram, the octahedral states are found and then their transformation properties in D_3 symmetry are determined by decomposing the characters of the elements common to O_h and D_3 (E, C_3, and C_2).

O_h	E	C_3	C_2'
A_{2g}	1	1	-1
T_{1g}	3	0	-1
T_{2g}	3	0	1

D_3	E	$2C_3$	$3C_2$
A_1	1	1	1
A_2	1	1	-1
E	2	-1	0

In this way the energy level diagram below is derived. (The order of the *E*-*A* splittings shown is arbitrary; it cannot be derived from simple group theory.)

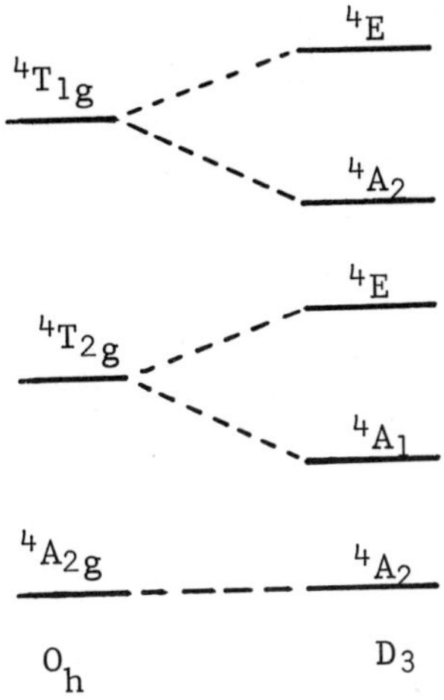

(b) The z, or parallel (||), component of the dipole moment operator transforms as A_2 and the x,y components (⊥) as E in D_3. The direct products are

$$E \times E \times A_2 = A_1 + A_2 + E$$

$$E \times A_2 \times A_2 = E$$

$$A_2 \times E \times A_2 = E$$

$$A_2 \times A_2 \times A_2 = A_2$$

$$A_1 \times E \times A_2 = E$$

$$A_1 \times A_2 \times A_2 = A_1$$

The selection rules are

$^4A_2 \rightarrow {}^4A_1$	not allowed ⊥ C_3
$^4A_2 \rightarrow {}^4A_1$	allowed || C_3
$^4A_2 \rightarrow {}^4E$	not allowed || C_3
$^4A_2 \rightarrow {}^4E$	allowed ⊥ C_3

$^4A_2 \rightarrow {^4A_2}$ not allowed $\perp$ C_3

$^4A_2 \rightarrow {^4A_2}$ not allowed $||$ C_3

The only transition allowed when the electric dipole is polarized parallel to the C_3 axis is $^4A_2 \rightarrow {^4A_1}$; this accounts for the band at 17,400 cm $^{-1}$. With the electric dipole polarized perpendicular to the C_3 axis, the transition $^4A_1 \rightarrow {^4E}$ is allowed.

Thus, both E components are seen in the $\perp$ spectrum.

7.17 (a) The lowest energy electronic transition should involve transfer of an electron from the b_1 level to the $4a_1$ level. In terms of state symmetries this would be a $^1A_1 \rightarrow {^3B_1}$ transition. However, the triplet, lower in energy than the singlet by Hund's rule, would probably not be directly observed, and a more reasonable assignment would be $^1A_1 \rightarrow {^1B_1}$. The lowest energy band in water occurs at 1800-1500 cm $^{-1}$.

7.18 (a) $\ldots(1a_2)^2(3b_1)^2(4a_1)^2$; $^1A_1 \rightarrow$

$$(1a_2)^2(3b_1)^2(4a_1)^1(2b_2)^1;\ {^1B_2}$$

$$(1a_2)^2(3b_1)^1(4a_1)^2(2b_2)^1;\ {^1A_2}$$

$$(1a_2)^1(3b_1)^2(4a_1)^2(2b_2)^1;\ {^1B_1}$$

(b) The components of the dipole moment operator transform as:

μ_x as B_1

μ_y as B_2

μ_z as A_1

Therefore, $^1A_1 \rightarrow {^1B_2}$ is allowed with μ_y

$^1A_1 \rightarrow {^1B_1}$ is allowed with μ_x

$^1A_1 \rightarrow {^1A_2}$ is not allowed

7.19 (a) The lowest energy absorption must involve the transfer of an electron from the filled π_g level to the π_u level, that is, from a configuration ... π_g^4 to a configuration ... $\pi_g^3\ \pi_u^1$.

(b) The states arising from the configuration ... $\pi_g^3\ \pi_u^1$ can be derived via group theory. The configuration has the same symmetry as the $\pi_g^1\ \pi_u^1$ "hole" configuration and thus the states will be given by the direct product in $D_{\infty h}$ of $\pi_g \times \pi_u$. The character elements of this reducible representation are

$D_{\infty h}$	E	$2C_\infty^\phi$	$\infty\sigma_v$	i	$2S_\infty^\phi$	∞C_2
$\chi\pi_g \times \pi_u$	4	$4\cos^2\Phi$	0	−4	$-4\cos^2\Phi$	0

This can be decomposed by standard techniques to give $\Sigma_u^+ + \Sigma_u^- + \Delta_u$. Actually, six states arise since the spin state can be either S = 0 or S = 1 for each of the three orbital symmetries. The spacial symmetry of either set will be the same. The spacial selection rules can be worked out by taking the direct products

$\Gamma_i \times \Gamma_{\mu_{xyz}} \times \Gamma_{f_j}$, where Γ_i is Σ_g^+, the irreducible representation of the ground state, $\Gamma_{\mu_{xyz}}$ are each of the representations spanned by the components of the electric dipole operator, and Γ_{f_j} are each of the excited state representations. In $D_{\infty h}$, μ_z forms a basis for Σ_u^+ and μ_{xy}, a basis for Π_u. Since the ground state is totally symmetric, it will have no effect on the product.

$$\Sigma_u^+ \times \Sigma_u^+ \rightarrow \Sigma_g^+$$

$$\Sigma_u^+ \times \Sigma_u^- \rightarrow \Sigma_g^-$$

$$\Sigma_u^+ \times \Delta_u \rightarrow \Delta_g$$

$\Pi_u \times \Sigma_u^+ \rightarrow \Pi_g$

$\Pi_u \times \Sigma_u^- \rightarrow \Pi_g$

$\Pi_u \times \Delta_u \rightarrow \Pi_g + \Phi_g$ (Φ_g is the next even 2-dimensional representation after Δ_g in $D_{\infty h}$)

Therefore, only the transition ${}^1\Sigma_g^+ \rightarrow {}^1\Sigma_u^+$ is allowed and this must be the assignment of the first strong absorption band. (The transition ${}^1\Sigma_g^+ \rightarrow {}^3\Sigma_u^+$ is spin-forbidden and probably not observable.)

(c)

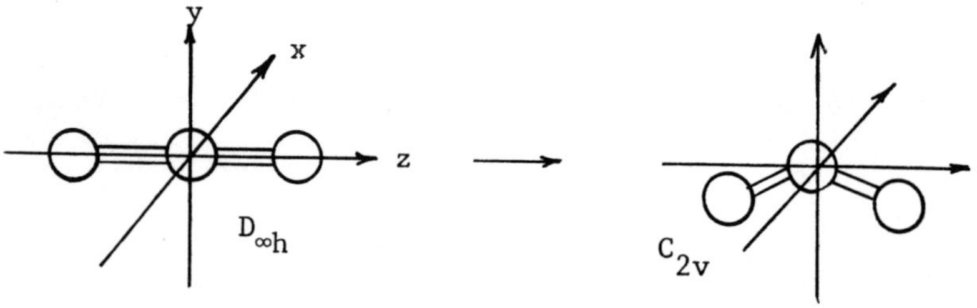

	$D_{\infty h}$	E	∞C_2	$i \times C_2 = \sigma_h(xy)$	$\infty\sigma_v(yz)$
	C_{2v}	E	$C_2(y)$	$\sigma_v(xy)$	$\sigma_v'(yz)$
$D_{\infty h}$	$\chi\Delta_u$	2	0	-2	0
C_{2v}	χA_2	1	0	-1	-1
	χB_2	1	0	-1	1

From the above it can be seen that Δ_u correlates to $A_2 + B_2$ in C_{2v}. If we assume that the Δ_u state is at lower energy than the Σ_u^+ or Σ_u^- states (as actually indicated by detailed calculations), then the two weak bands may be assigned to the ${}^1\Sigma_g^+ \rightarrow {}^1A_2$ and ${}^1\Sigma_g^+ \rightarrow {}^1B_2$ transitions.

7.20 (a)

	Band maxima, cm^{-1}	Spectral assignments
$[Co(OH)_6]^{4-}$	8,200	$^4T_{1g} \rightarrow {}^4T_{2g}$
	11,700	$\rightarrow {}^2E_g$
	15,000 sh(?)	$\rightarrow {}^4A_{2g}$
	19,200	$\rightarrow {}^4T_{1g}$
	21,200	$\rightarrow {}^2T_{2g}, {}^2T_{1g}$

(b) The energy separation between $^4T_{2g}$ and $^4A_{2g}$ corresponds to $10Dq$ and is 9500 cm^{-1}. Since the energy of the $^4T_{2g}$ is $2Dq$ = 1900 cm^{-1} and $E(^4A_{2g}) - E(^4T_{1g})$ = 8,350 cm^{-1}, the energy of $^4T_{1g}(F)$ is 1900 - 8350 = -6450 cm^{-1}.

Substituting this root into the determinant for configuration interaction yields

$$\begin{vmatrix} -6(950) + 6450 & 4(950) \\ 4(950) & x + 6450 \end{vmatrix} = 0$$

from which x = 12,800 cm^{-1}; B = 854 cm^{-1}. (Compare this with a B of 1080 cm^{-1} for the free ion.)

7.22 (a) Expressions for the energy levels in an octahedral complex of a d^3 ion are given in Problem 7.21.

By procedures used above we find the following:

	Dq, cm^{-1}	$E(^4A_{2g})$, cm^{-1}	$E(^4T_{1g})$, cm^{-1}	x, cm^{-1}
$[CrCl_6]^{3-}$	1320	-15,840	2860	8370
$[Cr(NH_3)_6]^{3+}$	1740	-25,800	2700	9950
$[Cr(H_2O)_6]^{3+}$	2150	-20,880	3600	10,640

(b) $$\beta_{Cl^-} = \frac{8370\ cm^{-1}}{13,800\ cm^{-1}} = 0.606$$

$$\beta_{NH_3} = \frac{9950\ \text{cm}^{-1}}{13{,}800\ \text{cm}^{-1}} = 0.721$$

$$\beta_{H_2O} = \frac{10{,}640\ \text{cm}^{-1}}{13{,}800\ \text{cm}^{-1}} = 0.771$$

(c) $Cl^- > NH_3 > H_2O$

(d) The relative decrease in the term splitting is a measure of the covalency of the metal-ligand bond; the greater the decrease the more covalent the bond.

(e) There is no correlation with the spectrochemical series.

7.23 The Jahn-Teller theorem would require that the symmetries of both the ground and excited states in a d^1 ion be lower than octahedral. Thus, the double-humped band might be due to transitions as illustrated in the figure below:

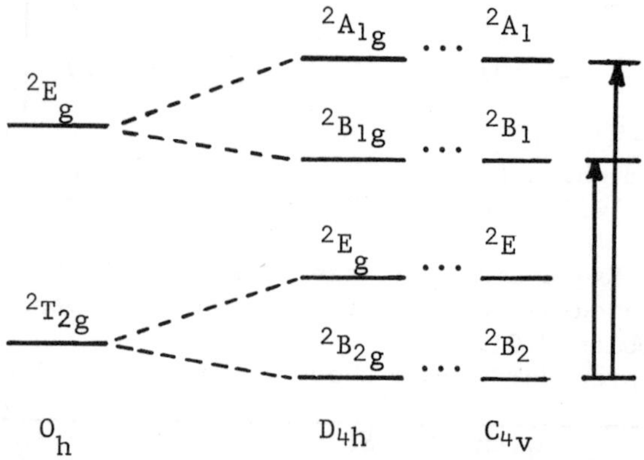

However, the interpretation is complicated by vibronic interactions and spin-orbit coupling.

7.25 (a)

Species	Dq, cm $^{-1}$	x, cm $^{-1}$	$^3T_{1g}(P)$ - $^3T_{1g}(F)$	
$[V(H_2O)_6]^{3+}$	1840	9300	25,200	obs.
$[V(H_2O)_5Cl]^{2+}$	1733	9092	23,900	calc.
$[V(H_2O)_4Cl_2]^+$	1627	8833	22,800	calc.

Species	Dq, cm^{-1}	x, cm^{-1}	$^3T_{1g}(P)$ - $^3T_{1g}(F)$	
$[V(H_2O)_3Cl_3]$	1520	8675	21,600	calc.
$[V(H_2O)_2Cl_4]^-$	1413	8467	20,300	calc.
$[V(H_2O)Cl_5]^{2-}$	1307	8258	19,200	calc.
$[VCl_6]^{3-}$	1200	8050	18,000	obs.

7.28 See P. T. Manoharan and H. B. Gray, *J. Am. Chem. Soc.*, 87, 3340 (1965).

7.29 (a) The first term in the expansion with $\ell = 0$ is spherically symmetrical and consequently transforms as A_{1g}. This term does not split the degeneracy of the *d*-orbitals. The second term, with $\ell = 1$, is given by a linear combination of $Y_1^{m_\ell}$. The representation

	E	$8C_3$	$3C_2$	$6C_4$	$6C_2$
$\Gamma_{Y_1^{m_\ell}}$	3	0	-1	1	-1

$= T_1$ is derived by the same techniques as used in Problem 7.1.

Since the combination of $Y_1^{m_\ell}$ transforms as T_{1g}, the second term does not contribute to the potential. In fact all terms with an odd number for ℓ can be ignored because the $\Theta_\ell^{m_\ell}$ functions are odd functions and thus the integrals vanish.

The third term, with $\ell = 2$, gives the representation

	E	$8C_3$	$3C_2$	$6C_4$	$6C_2$
$\Gamma_{Y_2^{m_\ell}}$	5	-1	1	-1	1

$= E + T_2$

Again, no contribution is made to the potential.

The term with $\ell = 4$ yields the representation

	E	$8C_3$	$3C_2$	$6C_4$	$6C_2$
$\Gamma_{Y_4^{m_\ell}}$	9	0	1	1	1

$= A_1 + E + T_1 + T_2$. The inclusion of A_1 in this product signals a contribution to the potential.

Since terms with ℓ greater than 4 exert no influence, the octahedral crystal field potential will be composed of a linear combination of $Y_4^{m_\ell}$ terms plus the spherically symmetric term.

(b) We need to investigate the terms $Y_2^{m_\ell}$, $Y_4^{m_\ell}$, and $Y_6^{m_\ell}$ under the group D_3.

D_3	E	$2C_3$	$3C_2$	
$\Gamma_{Y_2^{m_\ell}}$	5	-1	1	$= A_1 + 2E$
$\Gamma_{Y_4^{m_\ell}}$	9	0	1	$= 2A_1 + A_2 + 3E$
$\Gamma_{Y_6^{m_\ell}}$	13	1	-1	$= 2A_1 + 3A_2 + 4E$

Since all span representations which contain A_1, all will contribute terms to the potential.

(c) From (a) above, we know that the octahedral crystal field potential contains Y_0^0 and some combination of $Y_4^{m_\ell}$. To find the additional terms which constitute the tetragonal (D_{4h}) perturbation we need only investigate the transformation properties of the sets of even spherical harmonics with $\ell \leqslant 4$ under the operations of the rotation group D_4.

D_4	E	$2C_4$	$C_2 (=C_4^2)$	$2C_2'$	$2C_2''$
$\Gamma_{Y_2^{m_\ell}}$	5	-1	1	1	1
			$= A_1 + B_1 + B_2 + E$		
$\Gamma_{Y_4^{m_\ell}}$	9	1	1	1	1
			$= 2A_1 + A_2 + B_1 + B_2 + 2E$		

Thus, in addition to extra $Y_4^{m_\ell}$ terms, $Y_2^{m_\ell}$ terms enter into the potential.

(d) We must investigate the symmetry properties of each of the nine spherical harmonics $Y_4^{m_\ell}$. Only those for which $Y_4^{m_\ell} = C_4 Y_4^{m_\ell}$ contribute. Space will not permit us to carry out each operation; however, an example follows: Omitting constant terms,

$$C_4 Y_4^4 = C_4\left[\frac{(x + iy)^4}{r^4}\right] = \frac{[y + i(-x)]^4}{r^4}$$

Now multiply y by (i)(-i), that is by one.

$$= \frac{[(i)(-i)y - ix]^4}{r^4}$$

$$= \frac{(-i)^4 (x + iy)^4}{r^4}$$

$$= \frac{(x + iy)^4}{r^4} = Y_4^4$$

(e) Under the operation C_3,

$$x \to y$$

$$y \to z$$

$$z \to x$$

Therefore,

$$C_3 V_0 = \frac{1}{8}\left(\frac{9}{4\pi}\right)^{1/2}\left(\frac{35x^4 - 30x^2 r^2 + 3r^4}{r^4}\right)$$

$$+ \left(\frac{9}{4\pi}\right)^{1/2}\left(\frac{35}{128}\right)^{1/2}\left[\frac{(y + iz)^4}{r^4} + \frac{(y - iz)^4}{r^4}\right]$$

Now setting $V_0 = C_3 V_0$ and collecting terms in z^4, we get

$$\frac{5}{8} z^4 = 2\alpha\left(\frac{35}{128}\right)^{1/2} z^4$$

(Recall that

$$r^2 = x^2 + y^2 + z^2.)$$

Thus, $30z^2 r^2$ yields a term $30z^4$. Therefore,

$$\alpha = (\frac{5}{14})^{1/2}.$$

So we may write

$$V_0 = Y_4{}^0 + (\frac{5}{14})^{1/2}(Y_4{}^4 + Y_4{}^{-4})$$

7.30 (a) The ϕ-dependent part of the potential is $x^4 + y^4$ since the r^4 term is spherically symmetrical and z is independent of ϕ. Recalling that $x = r\ sin\ \theta\ cos\ \phi$ and $y = r\ sin\ \theta\ sin\ \phi$, we can write

$$x^4 + y^4 = r^4\ sin^4\theta\ (sin^4\phi + cos^4\phi).$$

We wish to evaluate integrals of the type

$$\langle e^{im\phi} | sin^4\phi + cos^4\phi | e^{-im'\phi} \rangle$$

which upon substitution for the operator will give

$$\langle e^{im\phi} | \left\{\frac{e^{i\phi} - e^{-i\phi}}{2i}\right\}^4 + \left\{\frac{e^{i\phi} + e^{-i\phi}}{2}\right\}^4 | e^{-im\phi} \rangle$$

Expanding the operator we get

$$(e^{i\phi} - e^{-i\phi})^4 = e^{4i\phi} - 4e^{2i\phi} + 6 - 4e^{-2i\phi} + e^{-4i\phi}$$

and

$$(e^{i\phi} + e^{-i\phi})^4 = e^{4i\phi} + 4e^{2i\phi} + 6 + 4e^{-2i\phi} + e^{-4i\phi}$$

which upon collecting terms yields

$$\langle e^{im\phi} | \frac{1}{8}\ (e^{4i\phi} + e^{-4i\phi} + 6 | e^{-im'\phi} \rangle$$

where m and m' take the values 0, ±1, ±2. Now examine a typical integral.

$$\int_0^{2\pi} e^{im\phi}\ e^{iq\phi}\ e^{-im'\phi}\ d\phi = \int_0^{2\pi} e^{i(m + q - m')\phi}\ d\phi$$

or letting m + q - m' = r

$$\int_0^{2\pi} e^{ir\phi}\ d\phi$$

Upon integration this yields

$$\frac{1}{r} e^{ir\phi}\Big|_0^{2\pi} = \frac{1}{r}\,(\cos r\phi + i\sin r\phi)\Big|_0^{2\pi}$$

which will vanish unless m + q - m' = r = 0. Noting that the values of q (from the operator) are 0 or ±4, we can see that only those matrix elements indicated by <> will be non-zero in the secular equation:

$$\begin{array}{c|ccccc|l}
m_\ell & +2 & +1 & 0 & -1 & -2 & \\
\hline
+2 & \langle\rangle & 0 & 0 & 0 & \langle\rangle & \\
+1 & 0 & \langle\rangle & 0 & 0 & 0 & \\
0 & 0 & 0 & \langle\rangle & 0 & 0 & = 0 \\
-1 & 0 & 0 & 0 & \langle\rangle & 0 & \\
-2 & \langle\rangle & 0 & 0 & 0 & \langle\rangle &
\end{array}$$

(b) We were asked to evaluate such integrals as

$$\frac{1}{2\pi}\ \langle e^{2i\phi}|V_0(\phi)|e^{-2i\phi}\rangle$$

where

$$\begin{aligned}
V_0(\phi) &= x^4 + y^4 \\
&= r^4 \sin^4\theta\ (\sin^4\phi + \cos^4\phi) \\
&= \frac{1}{8}\, r^4 \sin^4\theta\ (e^{4i\phi} + e^{-4i\phi} + 6)
\end{aligned}$$

Rewriting the integral,

$$\frac{1}{16\pi}\, r^4 \sin^4\theta\ \langle e^{2i\phi}|e^{4i\phi} + e^{-4i\phi} + 6|e^{-2i\phi}\rangle$$

$$= \frac{3}{8\pi}\, r^4 \sin^4\theta\ \langle e^{2i\phi}|e^{-2i\phi}\rangle$$

$$= \frac{3}{4}\, r^4 \sin^4\theta$$

(Recall that the limits of integration are 0 and 2π.) This same value obtains for all m = m' integrals. For m = 2, m' = -2, and m = -2, m' = 2 it is easy to show that

$$\langle\Phi(m_\ell)|V(\phi)|\Phi(m_\ell')\rangle = \frac{1}{8} r^4 \sin^4\theta$$

(c) The operator is now $(3/4)(r^4 \sin^4\theta) + r^4 \cos^4\theta$ since $z^4 = r^4 \cos^4\theta$. From Pauling and Wilson, page 134, we see

$$\Theta_2^0 = \left(\frac{5}{8}\right)^{1/2} (3 \cos^2\theta - 1).$$

The integral is

$$\frac{5}{8}\int_0^\pi (3 \cos^2\theta - 1)\left[\frac{3}{4} r^4 \sin^4\theta + r^4 \cos^4\theta\right](3 \cos^2\theta - 1)\sin\theta \, d\theta$$

$$= \frac{5}{8}\int_0^\pi r^4\left(\frac{27}{4} \sin^5\theta \cos^4\theta - \frac{9}{2}\cos^2\theta \sin^5\theta + \frac{3}{4}\sin^5\theta\right.$$

$$\left. + 9 \sin\theta \cos^8\theta - 6 \sin\theta \cos^6\theta + \sin\theta \cos^4\theta\right)d\theta$$

From the integral tables we find

$$\int_0^{\pi/2} \sin^{n-1}(x)\cos^{m-1}(x)\,dx = \frac{1}{2} B\left(\frac{n}{2}, \frac{m}{2}\right)$$

where

$$B(n, m) = \frac{\Gamma_{(m)}\Gamma_{(n)}}{\Gamma_{(m+n)}}$$

$$\Gamma_{(n)} = (n-1)!$$

$$\Gamma_{(n+\frac{1}{2})} = \frac{1 \cdot 3 \cdot 7 \cdots (2n - 1)}{2^n}\sqrt{\pi}$$

Changing the limits on the integral we have

$$\frac{5}{4} r^4 \int_0^{\pi/2} (\cdots) \, d\theta$$

For the first term $\sin^5\theta\cos^4\theta$, $n = 6$ and $m = 5$. Therefore we want

$$\frac{1}{2}B\left(\frac{n}{2},\frac{m}{2}\right) = \frac{1}{2}B\left(3, 2 + \frac{1}{2}\right)$$

$$= \frac{1}{2}\frac{\Gamma_{(3)}\Gamma_{(2 + \frac{1}{2})}}{\Gamma_{(5 + \frac{1}{2})}} = \frac{8}{(5 \cdot 7 \cdot 9)}$$

The result is

$$\frac{5}{4}r^4\left[\frac{27}{4}\cdot\frac{8}{5\cdot 7\cdot 9} - \frac{9}{2}\cdot\frac{8}{3\cdot 5\cdot 7} + \frac{3}{4}\cdot\frac{8}{3\cdot 5} + \frac{9}{9} - \frac{6}{7} + \frac{1}{5}\right] = \frac{5}{7}r^4$$

(d) The integral is

$$<R(r)\ \left|D(x^4 + y^4 + z^4 - \frac{3}{5}r^4)\right|\ R(r)>$$

$$= \frac{5}{7}\,D\int_0^\infty R(r)^2 r^4 r^2 dr - \frac{3}{5}\,D\int_0^\infty R(r)^2 r^4 r^2 dr$$

$$= \frac{4}{35}\,D\int_0^\infty R^2 r^4 r^2 dr$$

$$= 6Dq$$

(e) The secular determinant is

m_ℓ	2	1	0	-1	-2
+2	$Dq-E$	0	0	0	$5Dq$
+1	0	$-4Dq-E$	0	0	0
0	0	0	$6Dq-E$	0	0
-1	0	0	0	$-4Dq-E$	0
-2	$5Dq$	0	0	0	$Dq-E$

The roots from the diagonal are

$$E = -4Dq$$
$$E = -4Dq$$
$$E = 6Dq$$

From the two dimensional determinant

$$\begin{vmatrix} Dq - E & 5Dq \\ 5Dq & Dq - E \end{vmatrix} = 0$$

there results the quadratic equation

$$E^2 - 2DqE + Dq^2 = 25Dq^2$$

which has the solutions

$$E = 6Dq$$
$$E = -4Dq.$$

This yields the familar splitting of five *d*-orbitals into a triply degenerate set at $-4Dq$ and a doubly degenerate set at $+6Dq$.

Appendix 7.1

Tanabe-Sugano Diagrams

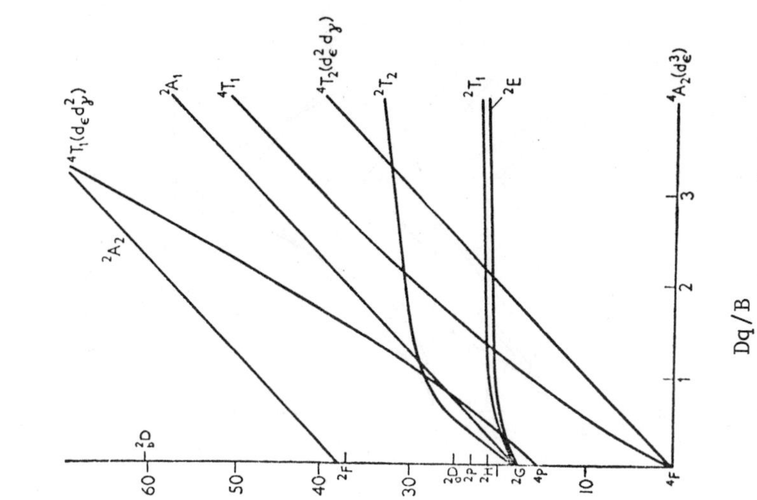

Energy diagram for d^3. C/B = 4.50

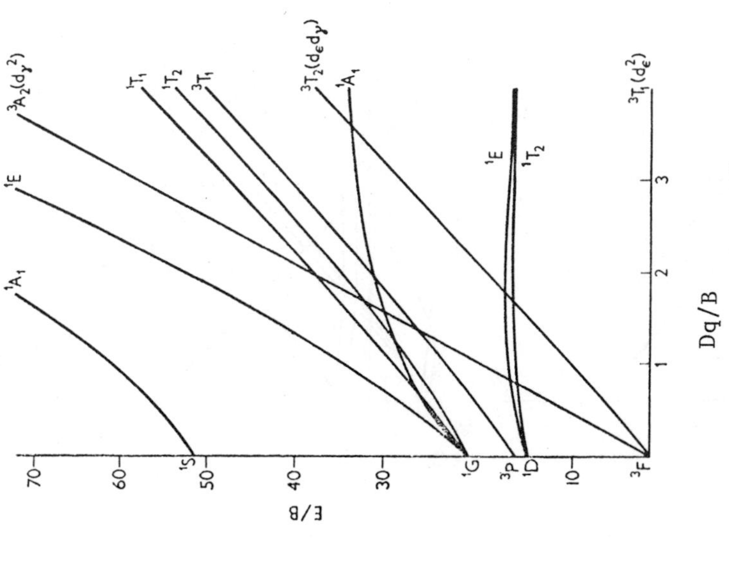

Energy diagram for d^2. C/B = 4.42

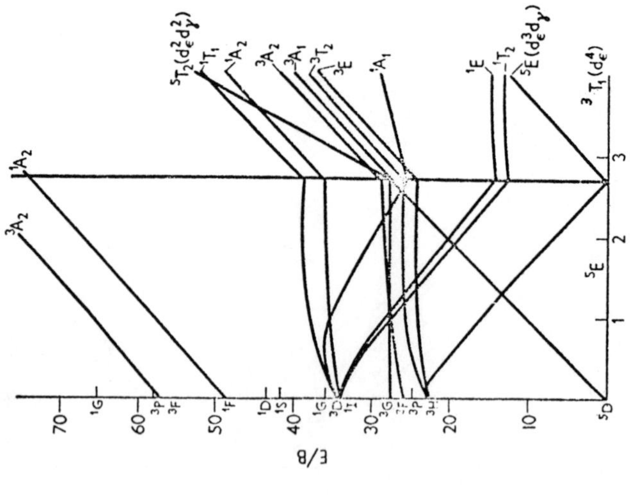

Energy diagram for d^4. C/B = 4.6

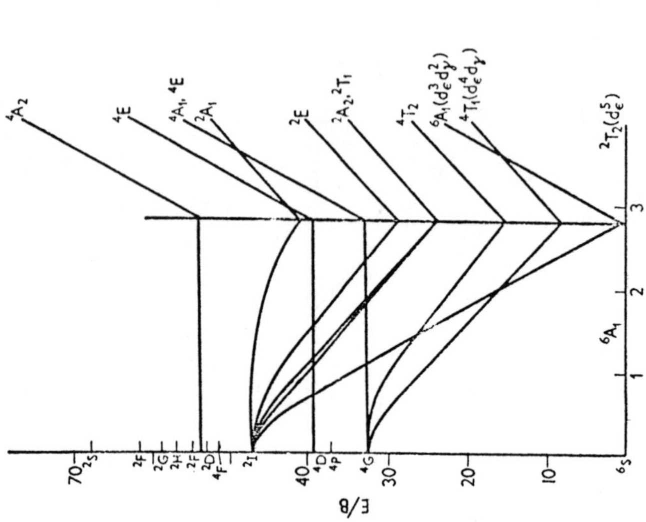

Energy diagram for d^5. C/B = 4.48

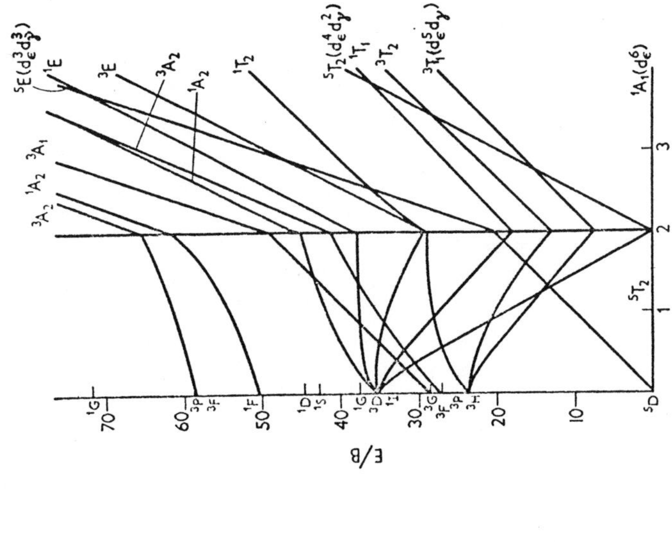

Energy diagram for d^6.

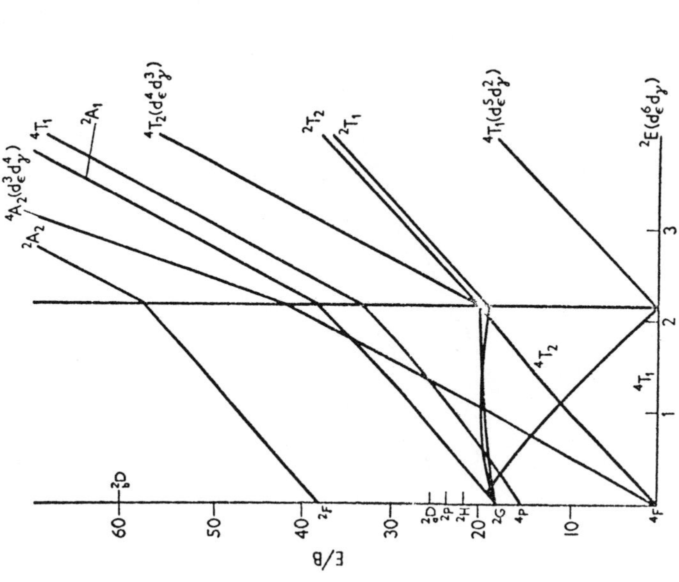

Energy diagram for d^7. C/B = 4.63

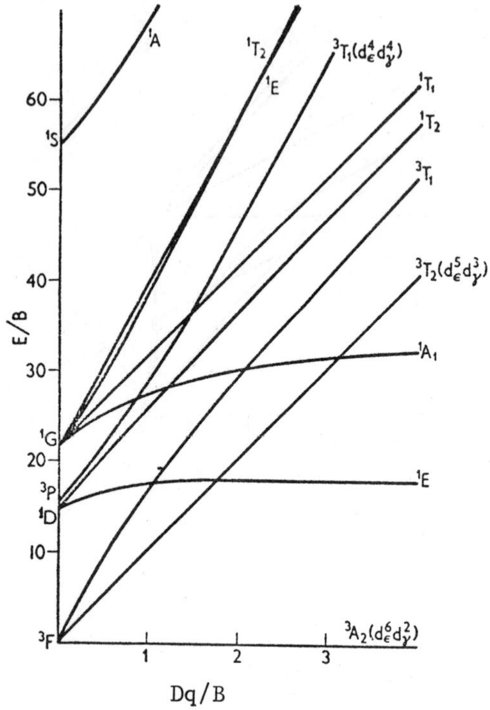

Energy diagram for d^8. C/B = 4.71

Chapter 8

MAGNETIC SUSCEPTIBILITY

The behavior of materials in magnetic fields yields extremely valuable chemical information. Experiments which are possible cover a wide spectrum of complexity and sophistication. Among the most valuable are static magnetic susceptibility determinations, electron spin resonance spectroscopy and nuclear magnetic resonance spectroscopy. Here we shall present some problems which demonstrate the utility of static magnetic susceptibility information in determining the stereochemistries, bonding, and electronic structures of metal complexes.

Important reference books and review articles include:

1. J. H. Van Vleck, *The Theory of Electric and Magnetic Susceptibilities,* Oxford University Press, 1932. (This classical book is available in a paper cover.)

2. J. B. Goodenough, *Magnetism and the Chemical Bond,* Interscience Publishers, New York, 1963.

3. E. König, *Magnetic Properties of Coordination and Organometallic Transition Metal Compounds,* Landolt-Börnstein Series, Group II, Vol. 2, Springer-Verlag, New York, 1966.

4. B. N. Figgis and J. Lewis, "The Magnetic Properties of Transition Metal Complexes," in *Progress in Inorganic Chemistry,* F. A. Cotton, ed., Interscience

Publishers, Inc., New York, Vol. 6, pp. 37 ff.

5. R. S. Drago, *Physical Methods in Inorganic Chemistry,* Reinhold Publishing Corp., New York, 1965.

6. A. Carrington and A. D. McLachlan, *Introduction to Magnetic Resonance,* Harper and Row, Publishers, New York, 1967.

7. B. N. Figgis, *Introduction to Ligand Fields,* Interscience Publishers, New York, 1966.

8. J. A. McMillan, *Electron Paramagnetism,* Reinhold Book Corporation, New York, 1968.

9. A. Earnshaw, *Introduction to Magnetochemistry,* Academic Press, New York, 1968.

PROBLEMS

8.1 *Spin-only moments.* (Read: Figgis and Lewis, or Figgis, Chapter 10.) For each of the ions listed below, write the electronic configuration for both spin-paired and spin-free (where defined) octahedral complexes, and calculate the spin-only magnetic moment of each:

Ti^{3+}, V^{3+}, Cr^{3+}, Mn^{3+}, Mn^{2+}, Fe^{2+}, Co^{3+}, Co^{2+},

Ni^{2+}, Cu^{2+}. Compare the calculated values with experimentally determined moments for a representative complex for each ion.

8.2 *Spin-only susceptibilities.* (Read: Figgis and Lewis). The spin-only formula for magnetic susceptibility is

$$\chi_M = \frac{Ng^2\beta^2}{3kT}S(S + 1)$$

Calculate the room temperature (298°K) susceptibilities for the various spin states exhibited by metal ions of the first transition series.

8.3 *Spin-only susceptibilities.* (Read: Carlin, *J. Chem. Ed.*, 43, 521 (1966).) Show that the expression derived for the magnetic susceptibility for a one electron system,

$$\chi = \frac{Ng^2\beta^2}{4kT}$$

is a special case of the spin-only formula

$$\chi = \frac{Ng^2\beta^2}{3kT}S(S + 1)$$

8.4 *Temperature independent paramagnetism.* (Read: Figgis and Lewis).

(a) What is the effect of temperature independent paramagnetism on the temperature dependence of χ_M and μ_{eff}? Explain.

(b) With respect to the spin-only values for μ_{eff}, how large are TIP moments? Where will they be most important?

8.5 *Oribital contribution to moments.* With relation to the spin-only moment, predict the moments (μ_{eff}) of the following substances:

(a) $[Ti(H_2O)_6]^{3+}$ (b) $[Co(en)_3](NO_3)_3$

(c) $[Mn(H_2O)_6]^{2+}$ (d) $[NiCl_4]^{2-}$

8.6 *Spin pairing energy versus ligand field stabilization energy.* (Read: Cotton and Wilkinson, Chapter 26). Octahedrally coordinated transition metal ions with electronic configurations d^4 through d^7 may either be high-spin (spin-free) or low-spin (spin-paired) depending on the strength of the ligand field.

(a) Calculate the energies of the ground states for both high-spin and low-spin d^4, d^5, d^6, and d^7 in terms of Dq and P. (P is the pairing energy, the energy required to pair one electron in the t_{2g} level, and $10Dq$ is the ligand field splitting, the e_g - t_{2g} energy difference.)

(b) Calculate the relationship between $10Dq$ and P when both high-spin and low-spin configurations have the same energy.

8.7 *The Van Vleck equation.* (Read: Ballhausen, Chapter 6; Carlin, *J. Chem. Ed.*, <u>43</u>, 521 (1966)). In the standard derivation of the Van Vleck equation for magnetic susceptibility the energy, E_n, of level n in a magnetic field H is given by

$$E_n = E_n^{0} + HE_n^{(1)} + H^2E_n^{(2)} + \ldots$$

where E_n^{0} is the zero-field energy of the level, the term in H is the first-order Zeeman term, and the term in H^2 is the second-order Zeeman term. Since the magnetic moment of level n is given by

$$\mu_n = \frac{-\partial E_n}{\partial H},$$

then the total moment of the system of n levels is given by

$$M = \frac{N\sum_n[\mu_n \; exp(-E_n/kT)]}{\sum_n[exp(-E_n/kT)]}$$

Using the above information and the following definitions:

$$\chi_m = M/H$$

$$exp(-E_n/kT) \simeq [1 - HE_n^{(1)}/kT] \; exp(-E_n^{\circ}/kT)$$

$$\sum_n[- E_n^{(1)} \; exp(-E_n^{\circ}/kT)] = 0,$$

derive the Van Vleck equation:

$$\chi_m = \frac{N\sum_n[(E_n^{(1)})^2/kT - 2E_n^{(2)}] \; exp(-E_n^{\circ}/kT)}{\sum_n[exp(-E_n^{\circ}/kT)]}$$

8.8 *Magnetic susceptibility of a* d^1 *ion in an octahedral field.* (Read: Ballhausen, Chapter 6). In this problem you are asked to derive an expression for the magnetic susceptibility of a d^1 ion in an octahedral crystal field. For the purposes of the problem we will only consider contributions to the magnetism from the $^2T_{2g}$ electronic state. There are three steps in the derivation. First the effect of spin-orbit coupling on the $^2T_{2g}$ state is determined, then the first and second order Zeeman terms are evaluated, and finally the results are substituted into the Van Vleck equation. The wave functions including spin are

$$\Psi_1 = (1/\sqrt{2})(d_2 - d_{-2})\alpha$$

$$\Psi_2 = (1/\sqrt{2})(d_2 - d_{-2})\beta$$

$$\Psi_3 = d_1\alpha$$

$$\Psi_4 = d_1\beta$$

$$\Psi_5 = d_{-1}\alpha$$

$$\Psi_6 = d_{-1}\beta$$

Step 1. The spin-orbit interaction operator is

$$\lambda L\cdot S = \lambda\{\tfrac{1}{2}(L_x + iL_y)(S_x - iS_y) + \tfrac{1}{2}(L_x - iL_y)(S_x + iS_y) + L_zS_z\}$$

where $(L_x \pm iL_y)$ and $(S_x \pm iS_y)$ are the raising and lowering operators. (See Problem 3.14). Calculate the matrix elements $\langle\Psi_1|\lambda L\cdot S|\Psi_1\rangle$ and $\langle\Psi_1|\lambda L\cdot S|\Psi_6\rangle$. The remaining non-zero elements are

$$\langle\Psi_2|\lambda L\cdot S|\Psi_3\rangle = \langle\Psi_3|\lambda L\cdot S|\Psi_2\rangle = \lambda/\sqrt{2}$$

$$\langle\Psi_3|\lambda L\cdot S|\Psi_3\rangle = \langle\Psi_6|\lambda L\cdot S|\Psi_6\rangle = \lambda/2$$

$$\langle\Psi_4|\lambda L\cdot S|\Psi_4\rangle = \langle\Psi_5|\lambda L\cdot S|\Psi_5\rangle = -\lambda/2$$

$$\langle\Psi_6|\lambda L\cdot S|\Psi_1\rangle = -\lambda/\sqrt{2}$$

Set up the secular determinant and determine the energy levels. Substitute the energies thus determined into the secular equations and generate the wave functions, ϕ_i, corrected for spin-orbit coupling.

Step 2. The operator for the interaction of the system with the external magnetic field is

$$(L_z + 2S_z)\beta H$$

since in O_h all directions are equivalent. Calculate the first and second order Zeeman terms, and demonstrate that the final energy level diagram is that shown in Figure 8.1, if the only non-zero matrix elements are

$$\langle\phi_1|L_z + 2S_z|\phi_1\rangle\beta H$$

$$\langle\phi_2|L_z + 2S_z|\phi_2\rangle\beta H$$

$$\langle\phi_1|L_z + 2S_z|\phi_3\rangle\beta H$$

$$\langle\phi_2|L_z + 2S_z|\phi_4\rangle\beta H$$

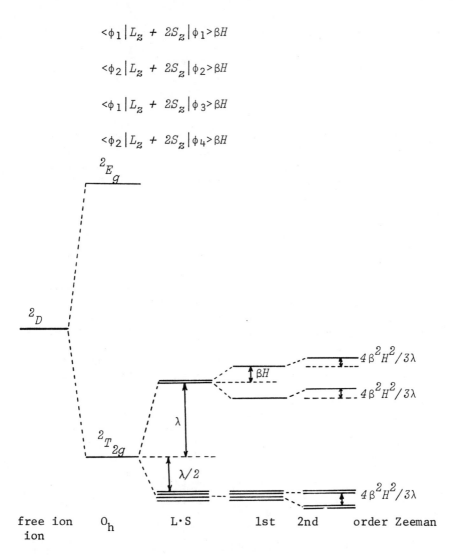

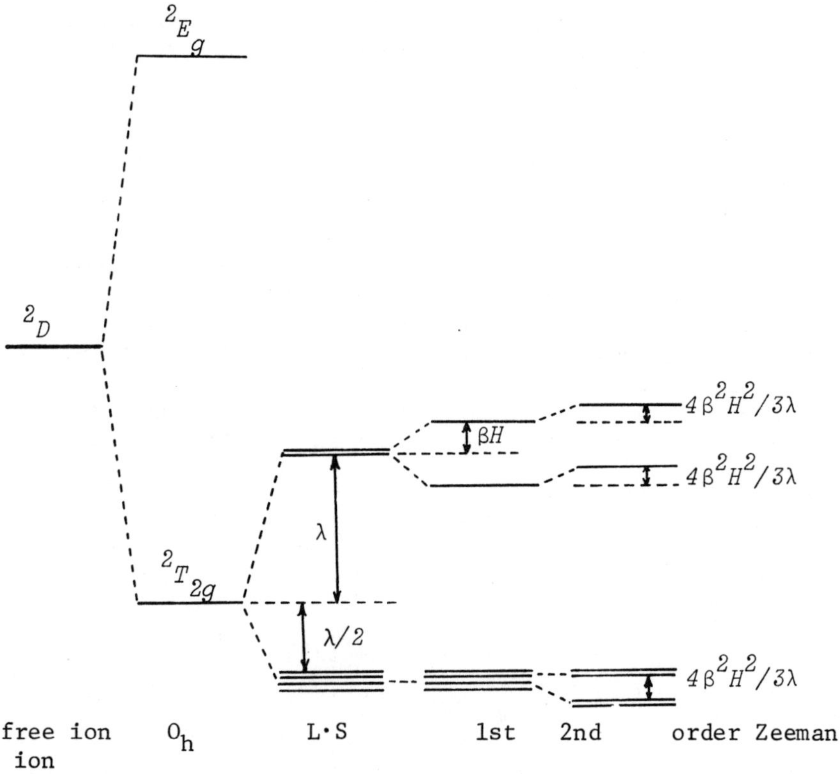

Figure 8.1

Step 3. Substitute the various quantities into the Van Vleck equation, and show that the following expression results

$$\chi_m = \frac{N\beta^2}{3kT} \cdot \frac{8 + (\frac{3\lambda}{kT} - 8)exp(\frac{-3\lambda}{2kT})}{(\frac{\lambda}{kT})[2 + exp(\frac{-3\lambda}{2kT})]}$$

8.9 *Double groups*. (Read: Cotton, Chapter 8). In order to determine the representation spanned by an orbital wave function characterized by the angular momentum quantum number ℓ, the characters for the members of the rotation subgroup are determined by

$$\chi(\alpha) = \frac{sin(\ell + \frac{1}{2})\alpha}{sin(\frac{\alpha}{2})}$$

where α is the angle of rotation. It is also desirable in many applications to determine the representations spanned by the wave functions characterized by the spin angular momentum quantum number s, or by the total angular momentum quantum number J. A difficulty arises when J (or s) takes on half integer values as can be seen by substitution of $(\alpha + 2\pi)$:

$$\chi(\alpha + 2\pi) = \frac{sin[(J + \frac{1}{2})(\alpha + 2\pi)]}{sin(\frac{\alpha + 2\pi}{2})}$$

$$= \frac{sin[(J + \frac{1}{2})(\alpha + 2\pi)]}{sin[\frac{\alpha}{2} + \pi]}$$

$$= \frac{sin(J + \frac{1}{2})\alpha}{-sin(\frac{\alpha}{2})}$$

$$= -\chi(\alpha)$$

Since rotation by 2π is normally the identity operation $\chi(\alpha + 2\pi)$ might be expected to equal $\chi(\alpha)$. In order to circumvent the difficulty, rotation by 2π is treated as a distinct symmetry operation, R, but not as the identity operation. Groups which are formed by the addition of the new element R are called *double groups*. The character $\chi(\alpha = 0)$ is $(2J + 1)$, and $\chi(\alpha = 2\pi)$ is $(2J + 1)$ if J is an integer (for which there is no need for double groups) or $-(2J + 1)$ if J is a half integer. Write out the members of the double group D_4' and determine the class structure.

8.10 *Spin-orbit coupling in* d^1 *octahedral complex.* Use the double group O' and determine the representations spanned by the states which arise from spin-orbit coupling in a complex of a d^1 ion in an octahedral field. (Hint: Construct the representation of the direct product of the representations spanned by the orbital wave functions and the spin wave functions.)

8.11 *Magnetic susceptibility equation for a* d^2 *ion.* The energy level diagram for a d^2 ion in an octahedral field is shown in Figure 8.2. Derive an expression for the magnetic susceptibility.

8.12 *Magnetic susceptibility for a chromium(III) ion in an octahedral field.* A chromium(III) ion in an octahedral crystal field has a $^4A_{2g}$ ground state. Derive an equation for the temperature dependence of the magnetic susceptibility for $[CrL_6]^{q\pm}$ complexes.

8.13 *Effect of an axial distortion on the magnetic susceptibility.* An axial distortion of a chromium(III) complex $[CrL_6]^{q\pm}$ may partially resolve the four-fold degeneracy of the $^4A_{2g}$ ground state such that the $M_s = \pm\frac{1}{2}$ levels are separated by a quantity δ, the zero field splitting, from the $M_s = \pm\frac{3}{2}$ levels. Assume that the $M_s = \pm\frac{1}{2}$ levels lie lowest in energy and set up an equation for the parallel component of the magnetic susceptibility, *i.e.* with H_z parallel to the principal molecular axis.

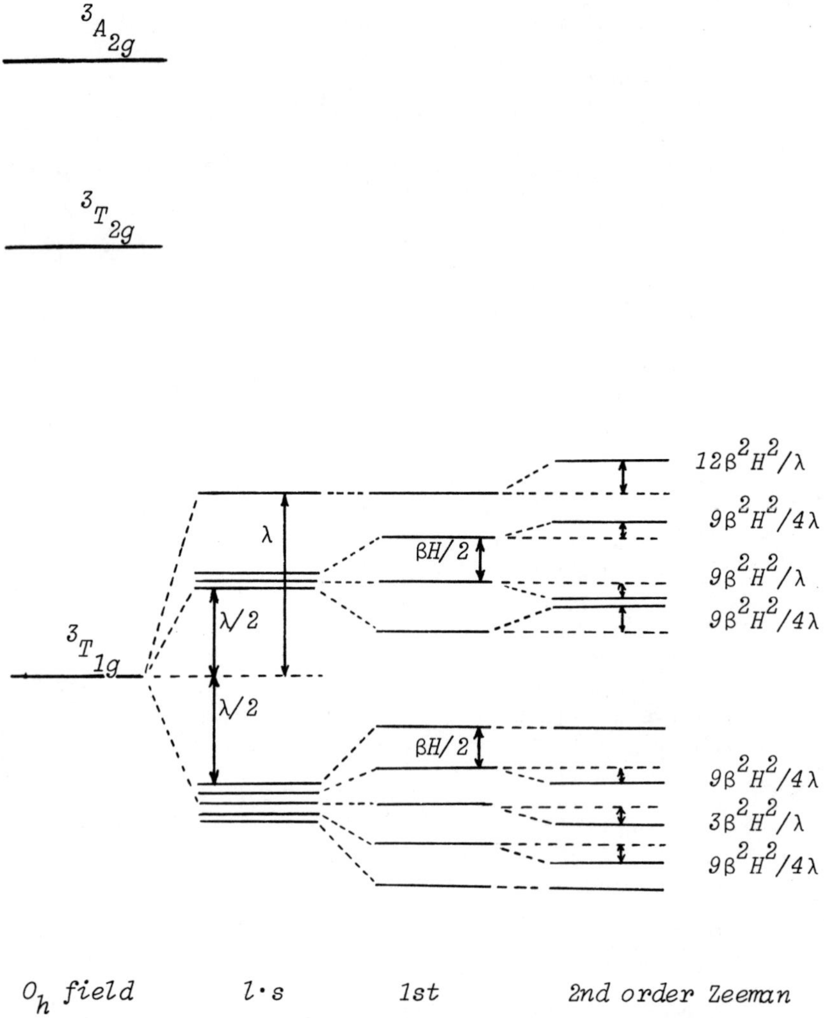

Figure 8.2

8.14 *Magnetic susceptibility of tetragonal nickel(II) complexes.* A nickel(II) ion in a cubic crystal field with a small tetragonal component has the following energy level diagram:

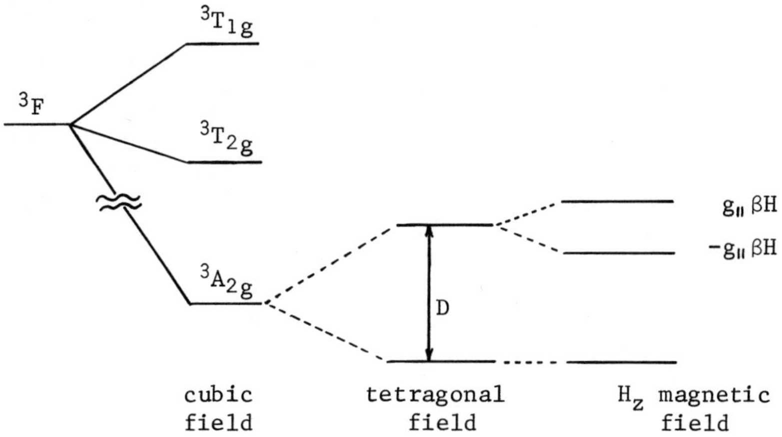

Calculate the parallel component of the magnetic susceptibility at room temperature if $g_{\parallel} = 2.25$ and $D = -2.24\ cm^{-1}$.
[Hint: For D<<kT, $\exp(\frac{-D}{kT}) \simeq (1 - \frac{D}{kT})$]

8.15 *Moments of cobalt(II) complexes.*

(a) Calculate the spin-only magnetic moments for cobalt(II) in: (i) a weak octahedral field, (ii) a strong octahedral field, and (iii) a weak tetrahedral field.

(b) In octahedral cobalt(II) complexes $\mu_{eff^{300}}$ values fall in the following ranges: 4.8--5.6 and 1.73--2.0 B. M. for spin-free and spin-paired complexes, respectively. However, there are known a group of cobalt(II) complexes which have $\mu_{eff^{300}}$ values in the range of 2.3 to 4.3 B. M. How might these "anomlous" moments be explained?

8.16 *Spin pairing and exchange effects.* (Read: Figgis, Chapter 10). What is the difference between spin pairing and antiferromagnetic exchange?

8.17 *Moments in dimeric copper complexes.* Copper acetate monohydrate has a dimeric structure both in the solid state and in certain solvents. The Cu-Cu internuclear distance in the crystalline material is 2.64 A. Explain why the magnetic susceptibility per copper ion increases from nearly zero at 77°K to a maximum of about 800 x 10^{-1} c.g.s.u. at about 260°K.

8.18 *The Curie-Weiss Law.* (Read: Figgis and Lewis). The behavior of the magnetic properties of many transition metal complexes may be approximated by the Curie-Weiss Law. From the data given below, indicate which compounds obey the Curie-Weiss Law.

Compound	Molar Susceptibility 10^{-6} c.g.s. units	T,°K
(a) $Cu(C_2H_5OCH_2CO_2)_2 \cdot 2H_2O$	1511	298
	2307	196
	5803	77
(b) $Cu(C_6H_5CO_2)_2 \cdot C_4H_9OH$	889	299
	902	271
	871	233
	818	203
	190	82
(c) $[Cu(C_6H_5N_2O)H]NO_3$	1404	301
	2208	196
	5505	77

8.19 *Moments in dodecahedral ligand fields.* The compound $K_4Mo(CN)_8 \cdot 2H_2O$ is known to have a dodecahedral structure. Account for the diamagnetism of the compound in terms of a *d*-orbital splitting diagram.

8.20 *Moments of lanthanide ions.* (Read: Van Vleck). For each of the ions listed below, look up the ground state term, and calculate the spin-only magnetic moment and the moment predicted for those instances in which J is a good quantum number. Compare the calculated values with experimentally determined moments for magnetically dilute complexes. Explain the divergencies.

La^{3+}, Ce^{3+}, Pr^{4+}, Pr^{3+}, Nd^{3+}, Pm^{3+}, Sm^{3+}, Eu^{3+},

Sm^{2+}, Gd^{2+}, Eu^{2+}, Tb^{3+}, Dy^{3+}, Ho^{3+}, Er^{3+}, Tu^{3+}.

SOLUTIONS

8.1

	spin-free			spin-paired		
Ion	config.	mag. moment		config.	mag. moment	
		calc.	obs. range		calc.	obs. range
Ti^{3+}	$(t_{2g})^1$	1.73	1.7 - 1.8	------	-----	----------
V^{3+}	$(t_{2g})^2$	2.83	2.7 - 2.9	------	-----	----------
Mn^{3+}	$(t_{2g})^3(e_g)^1$	4.90	4.75 - 4.9	$(t_{2g})^4$	2.83	3.2 - 3.3
Co^{3+}	$(t_{2g})^4(e_g)^2$	4.90	5.3 - 5.7	$(t_{2g})^6$	0.0	----------
Co^{2+}	$(t_{2g})^5(e_g)^2$	3.88	4.3 - 5.2	$(t_{2g})^6(e_g)^1$	1.73	1.8 - 2.0
Cu^{2+}	$(t_{2g})^6(e_g)^3$	1.73	1.7 - 2.2	----------	----	----------

8.2 For $S = \frac{1}{2}$, we get upon substitution for the quantities in the susceptibility expression:

$$\chi_M = \frac{(6.023 \times 10^{23})\ (2.0)^2\ (0.927 \times 10^{-20})^2}{3(1.38 \times 10^{-16})\ (298)} \{\tfrac{1}{2}(\tfrac{1}{2} + 1)\}$$

$$= (1.67 \times 10^{-3})(\tfrac{3}{4})$$

$$= 1,250 \times 10^{-6} \text{ c.g.s. units}$$

$$S = 1;\ \chi_M = 3,340 \times 10^{-6} \text{ c.g.s. units}$$

$$S = \tfrac{3}{2};\ \chi_M = 6,260 \times 10^{-6} \text{ c.g.s. units}$$

8.3 We need only substitute into the spin-only equation the spin quantum number $S = \frac{1}{2}$. Therefore

$$\chi = \frac{Ng^2\beta^2}{3kT}\{(\tfrac{1}{2})(\tfrac{1}{2} + 1)\}$$

$$= \frac{Ng^2\beta^2}{4kT}$$

8.4 (a) $\chi_M = \frac{N\beta^2}{3kT}\mu_{eff}^2$

considering only the first order Zeeman effect.

However, the TIP term (from the second order Zeeman effect) gives

$$\chi_M = \frac{N\beta^2}{3kT}\mu_{eff}^2 + N\alpha$$

That is, the TIP term ($N\alpha$) is a temperature independent term in χ_M. Because of the relationship between χ_M and μ_{eff} the latter becomes temperature dependent as a result of temperature independent paramagnetism. TIP will cause μ_{eff} to decrease with temperature.

(b) TIP terms are usually at least one order of magnitude smaller than the first order effects. They will be most important in species having no spin moments.

8.6 We will give a detailed calculation for d^4 and the results for d^5, d^6.

(a) For d^4, the electronic configurations are

d^4: e_g

t_{2g}

high spin   low spin

	high spin	low spin
d^4	$E_{g.s.} = 3(-4Dq) + 6Dq = -6Dq$	$E_{g.s.} = 4(-4Dq) + P = -16Dq + P$
d^5	$= 0$	$= -20Dq + 2P$
d^6	$= -4Dq + P$	$= -24Dq + 3P$

(b) For d^4 : $E(\text{high spin}) = E(\text{low spin})$

$$-6Dq = -16Dq + P$$

or, $$10Dq = P$$

This relationship holds for all configurations. That is, if $10Dq < P$, then spin-free complexes are expected, but if $10Dq > P$, then spin-paired complexes are expected.

8.8 Step 1. The matrix element $<\Psi_1|\lambda L\cdot S|\Psi_1>$:

$$< \frac{1}{\sqrt{2}}(d_2 - d_{-2})\alpha|\lambda\{\frac{1}{2}(L_x + iL_y)(S_x - iS_y)$$

$$+ \frac{1}{2}(L_x - iL_y)(S_x + iS_y) + L_zS_z\}| \frac{1}{\sqrt{2}}(d_2 - d_{-2})\alpha>$$

$$= \frac{1}{2}\lambda\{<(d_2 - d_{-2})\alpha|\frac{1}{2}(L_x + iL_y)(S_x - iS_y)|(d_2 - d_{-2})\alpha >$$

$$+ 0 + 0\}$$

$$= \frac{1}{4}\lambda\{<(d_2 - d_{-2})|L_x + iL_y|(d_2 - d_{-2})> <\alpha|S_x - iS_y|\alpha>\}$$

$$= \frac{1}{4}\lambda\{<(d_2 - d_{-2})|L_x + iL_y|(d_2 - d_{-2})> <\alpha|\beta>\}$$

$$= 0$$

The matrix element $<\Psi_1|\lambda L\cdot S|\Psi_6>$:

$$<\frac{1}{\sqrt{2}}(d_2 - d_{-2})\alpha|\lambda L\cdot S|d_{-1}\beta>$$

$$= \lambda<\frac{1}{\sqrt{2}}(d_2 - d_{-2})\alpha|\frac{1}{2}(L_x - iL_y)(S_x + iS_y)|d_{-1}\beta>$$

$$= \lambda<\frac{1}{\sqrt{2}}(d_2 - d_{-2})|\frac{1}{2}(L_x - iL_y|d_{-1}> <\alpha|\alpha>$$

$$= \lambda<\frac{1}{\sqrt{2}}(d_2 - d_{-2})|\frac{1}{2}\sqrt{(2 + 1 + 1)(2 - 1)}|d_{-2}>$$

$$= \frac{\lambda}{\sqrt{2}}\{<d_2|d_{-2}> - <d_{-2}|d_{-2}>\}$$

$$= -\frac{\lambda}{\sqrt{2}}$$

The secular determinant is:

$$\begin{array}{c|cccccc|}
 & \Psi_1 & \Psi_2 & \Psi_3 & \Psi_4 & \Psi_5 & \Psi_6 \\
\Psi_1 & -E & 0 & 0 & 0 & 0 & \frac{-\lambda}{\sqrt{2}} \\
\Psi_2 & 0 & -E & \frac{\lambda}{\sqrt{2}} & 0 & 0 & 0 \\
\Psi_3 & 0 & \frac{\lambda}{\sqrt{2}} & (\frac{\lambda}{2})-E & 0 & 0 & 0 \\
\Psi_4 & 0 & 0 & 0 & (\frac{-\lambda}{2})-E & 0 & 0 \\
\Psi_5 & 0 & 0 & 0 & 0 & (\frac{-\lambda}{2})-E & 0 \\
\Psi_6 & \frac{-\lambda}{\sqrt{2}} & 0 & 0 & 0 & 0 & (\frac{+\lambda}{2})-E
\end{array} = 0$$

Which simplifies to

$$\text{A:} \quad \begin{vmatrix} -E & \frac{-\lambda}{\sqrt{2}} \\ \frac{-\lambda}{\sqrt{2}} & (\frac{+\lambda}{2})-E \end{vmatrix} = 0$$

$$\text{B:} \quad \begin{vmatrix} -E & \frac{\lambda}{\sqrt{2}} \\ \frac{\lambda}{\sqrt{2}} & (\frac{\lambda}{2})-E \end{vmatrix} = 0$$

$$\text{C:} \quad \left| (\frac{-\lambda}{2}) - E \right| = 0$$

$$\text{D:} \quad \left| (\frac{-\lambda}{2}) - E \right| = 0$$

These give rise to the solutions

$E = \frac{1}{2}\lambda$ once each from C and D and once each from A and B

$E = +\lambda$ once each from A and B

Substituting the energies into the secular equations yields:

A: $-EC_1 - \frac{\lambda}{\sqrt{2}} C_2 = 0$

$$\frac{-\lambda}{\sqrt{2}} C_1 + [(\frac{\lambda}{2}) - E]C_2 = 0$$

For $E = +\lambda$, if $C_1 = 1$, then $C_2 = -\sqrt{2}$ and the normalized function is $\phi_1 = \frac{1}{\sqrt{3}}(\Psi_1 - \sqrt{2}\Psi_6)$.

For $E = \frac{-\lambda}{2}$, if $C_2 = 1$, then $C_1 = \sqrt{2}$ and the function is $\phi_3 = \frac{1}{\sqrt{3}}(\sqrt{2}\Psi_1 - \Psi_6)$.

B: $\phi_2 = \frac{1}{\sqrt{3}}(\Psi_2 + \sqrt{2}\Psi_3)$

$$\phi_4 = \frac{1}{\sqrt{3}}(\Psi_3 - \sqrt{2}\Psi_2)$$

C: $\phi_5 = \Psi_4$

D: $\phi_6 = \Psi_5$

Step 2. Now for the Zeeman effect. We are told that only $\langle\phi_1|H|\phi_1\rangle$, $\langle\phi_2|H|\phi_2\rangle$, $\langle\phi_1|H|\phi_3\rangle$ and $\langle\phi_2|H|\phi_4\rangle$ are non-zero. The first order correction to the energy for level 1 is

$$E_1^{(1)} = \langle\phi_1|H|\phi_1\rangle$$

$$= \frac{1}{\sqrt{3}}(\langle\Psi_1 - \sqrt{2}\Psi_6)|L_z + 2S_z| \frac{1}{\sqrt{3}}(\Psi_1 - \sqrt{2}\Psi_6)\rangle\ \beta H$$

$$= \beta H$$

and for level 2, $E_2^{(1)} = -\beta H$. The second order correction

$$E_1^{(2)} = \frac{\langle\phi_1|H|\phi_3\rangle\langle\phi_3|H|\phi_1\rangle}{E_1 - E_3}$$

$$= \frac{(\sqrt{2}\beta H)(\sqrt{2}\beta H)}{\lambda - (\frac{-\lambda}{\sqrt{2}})} \qquad = \frac{4\beta^2 H^2}{3\lambda}$$

Also, $E_2^{(2)} = \frac{4\beta^2H^2}{3\lambda}$; $E_3^{(2)} = \frac{-4\beta^2H^2}{3\lambda}$; and $E_4^{(2)} = \frac{-4\beta^2H^2}{3\lambda}$.

The energy level diagram has now been verified.

Step 3. Substitution into the Van Vleck equation is straight forward:

$$\chi = \frac{N\{2(\frac{0^2}{kT} + \frac{8\beta^2}{3\lambda})exp(\frac{\lambda}{2kT}) + 2(\frac{0^2}{kT} - 2\cdot 0)exp(\frac{\lambda}{2kT})}{4exp(\frac{\lambda}{2kT}) + 2exp(\frac{-\lambda}{kT})}$$

$$+ \frac{(\frac{\beta^2}{kT} - \frac{8\beta^2}{3\lambda})exp(\frac{-\lambda}{kT}) + (\frac{\beta^2}{kT} - \frac{8\beta^2}{3\lambda})exp(\frac{-\lambda}{kT})\}}{4exp(\frac{\lambda}{2kT}) + 2exp(\frac{-\lambda}{kT})}$$

$$= \frac{N\{\frac{8\beta^2}{3\lambda}exp(\frac{\lambda}{2kT}) + (\frac{\beta^2}{kT} - \frac{8\beta^2}{3\lambda})exp(\frac{-\lambda}{kT})\}}{2exp(\frac{\lambda}{2kT}) + exp(\frac{-\lambda}{kT})}$$

$$= \frac{N\beta^2}{3kT}\cdot\left\{\frac{8 + (\frac{3\lambda}{kT} - 8)exp(\frac{-3\lambda}{2kT})}{(\frac{\lambda}{kT})[2 + exp(\frac{-\lambda}{kT})]}\right\}$$

8.9 The members of the group are E, R, C_4, $C_4{}^3$, C_2, $2C_2{}'$, $2C_2{}''$, C_4R, $C_4{}^3R$, C_2R, $2C_2{}'R$, and $2C_2{}''R$. The class structure is

$$E \quad R \quad \begin{Bmatrix} C_4 \\ C_4{}^3R \end{Bmatrix} \begin{Bmatrix} C_4{}^3 \\ C_4R \end{Bmatrix} \begin{Bmatrix} C_2 \\ C_2R \end{Bmatrix} \begin{Bmatrix} 2C_2{}' \\ 2C_2{}'R \end{Bmatrix} \begin{Bmatrix} 2C_2{}'' \\ 2C_2{}''R \end{Bmatrix}$$

8.10 For $S = \frac{1}{2}$, in O', $\Gamma_{spin} = \Gamma_6$. Thus, from ${}^2T_{2g}$ we have $\Gamma_6 \times \Gamma_5 = \Gamma_7 + \Gamma_8$, and for ${}^2E_g$ we have $\Gamma_6 \times \Gamma_3 = \Gamma_8$. Spin-orbit coupling, therefore splits the ${}^2T_{2g}$ state into states with two-fold and four-fold degeneracies.

8.11

$$\chi_m = \frac{N\beta^2}{kT} \cdot \left\{ \frac{\frac{5\lambda}{2kT} + 15 + (\frac{\lambda}{2kT} + 9)exp(\frac{-\lambda}{kT}) - 24exp(\frac{-3\lambda}{2kT})}{(\frac{\lambda}{kT})\{5 + 3exp(\frac{-\lambda}{kT}) + exp(\frac{-3\lambda}{2kT})\}} \right\}$$

8.12 The ground state is orbitally nondegenerate and experiences only first order Zeeman splitting in a magnetic field, viz.,

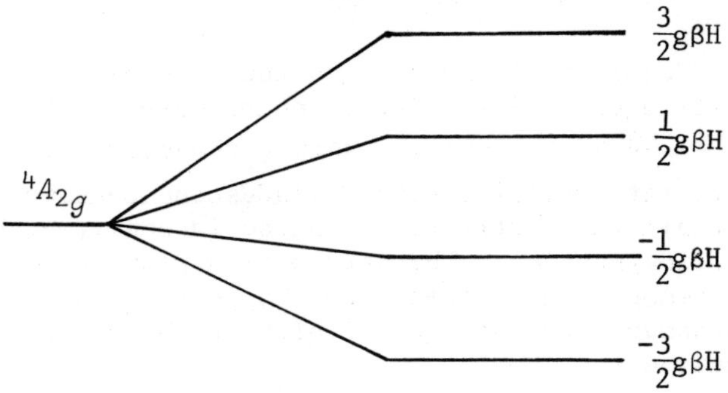

Thus,

$$\chi = \frac{N\{\frac{9g^2\beta^2}{4kT} + \frac{g^2\beta^2}{4kT} + \frac{g^2\beta^2}{4kT} + \frac{9g^2\beta^2}{4kT}\}exp(\frac{-E_0}{kT})}{4\ exp\ (\frac{-E_0}{kT})}$$

$$= \frac{5Ng^2\beta^2}{4kT}$$

Note that this is the spin-only formula for $S = \frac{3}{2}$.

8.13 The energy level diagram is

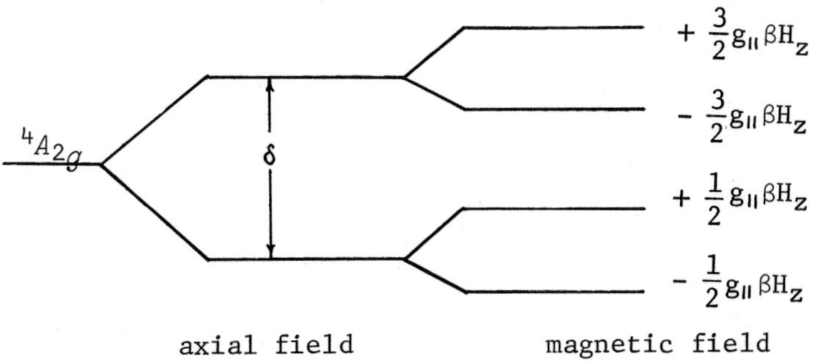

From Van Vleck's equation:

$$\chi = \frac{N\{\frac{g_{\parallel}{}^2\beta^2}{2kT} + \frac{9g_{\parallel}{}^2\beta^2}{2kT}\, exp(\frac{-\delta}{kT})\}}{2 + 2\ exp\ (\frac{-\delta}{kT})}$$

$$= \frac{Ng_{\parallel}{}^2\beta^2}{4kT} \cdot \frac{\{1 + 9\ exp\ (\frac{-\delta}{kT})\}}{\{1 + exp\ (\frac{-\delta}{kT})\}}$$

8.17 The d^9 copper(II) ion has one unpaired electron, and if magnetically dilute, should have $\chi_M \approx 1{,}500 \times 10^{-6}$ c.g.s. units. However, the short internuclear distance leads one to suspect a metal-metal interaction, which is confirmed by the diamagnetism. The increase in paramagnetism with temperature arises from the population of a paramagnetic state (a triplet) at higher temperatures.

8.18 The Curie-Weiss law shows that the magnetic susceptibility of a complex depends inversely on the temperature T, such that

$$\chi_M = \frac{C}{(T - \Theta)}$$ where C and Θ are constants.

Thus $$\frac{C}{\chi_M} = T - \Theta$$

A plot of $1/\chi_M$ vs. T should give a straight line with the intercept on the T axis equal to θ.

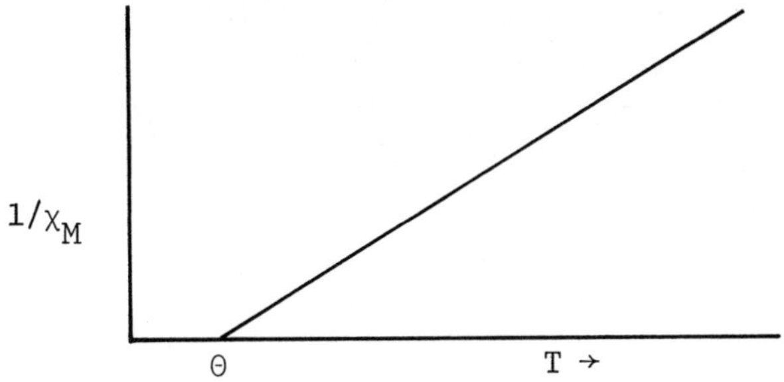

The constant C is termed the Curie constant and is equal to $\frac{N\mu^2}{3k}$ where N is Avogadro's number, k is Boltzmann's constant and μ is the magnetic moment.

Thus

$$\chi_M = \frac{(\frac{N\mu^2}{3k})}{T - \theta}$$

$$\text{or } \mu = (\frac{3k}{N})^{1/2}[\chi_M(T - \theta)]^{1/2}$$

$$= 2.83[\chi_M(T - \theta)]^{1/2}$$

The term θ is the Weiss constant; it is difficult to place any precise interpretation on it. For many paramagnetic complexes θ is approximately zero, and is generally 20°. (Thus, neglecting θ in magnetic moment calculations rarely affects the value of μ_{eff}). When is much greater than 50° then the interactions present in the system are generally so large that the magnetic behavior does not obey the Cuie-Weiss law anyway.

To find which compounds in the examples obey the Curie-Weiss law, it is necessary to plot the reciprocal of the magnetic susceptibility against the temperature. Only example (b) does not obey the Curie-Weiss law.

8.20 The results tabulated in Table 8.1 show that the moments of rare earth complexes can be predicted with good agreement by $\mu_{eff} = g[J(J + 1)]^{1/2}$ for all ions except Sm^{3+}, Eu^{3+}, Sm^{2+}. For these three ions excited J states are close enough to the ground state to be populated at room temperature.

Table 8.1.

Magnetic Data for Rare Earth Ions

Ion	Spectroscopic State	L	S	J	Effective Magnetic Moment, B.M.		
					spin only	$g[J(J+1)]^{1/2}$	Experimental
La^{3+}	$^1S_0$	0	0	0	0	0	diamagnetic
Ce^{3+} Pr^{4+}	$^2F_{5/2}$	3	$\frac{1}{2}$	$\frac{5}{2}$	1.73	2.54	$\begin{Bmatrix} 2.50 \\ 2.62 \end{Bmatrix}$
Pr^{3+}	$^3H_4$	5	1	4	2.83	3.58	3.61
Nd^{3+}	$^4I_{9/2}$	6	$\frac{3}{2}$	$\frac{9}{2}$	3.88	3.62	3.80
Pm^{3+}	$^5I_4$	6	2	4	4.91	2.68	——
Sm^{3+}	$^6H_{5/2}$	5	$\frac{5}{2}$	$\frac{5}{2}$	5.92	0.84	1.51
Eu^{3+} Sm^{2+}	$^7F_0$	3	3	0	6.93	0.00	3.63
Gd^{3+} Eu^{2+}	$^8S_{7/2}$	0	$\frac{7}{2}$	$\frac{7}{2}$	7.94	7.94	7.79 7.90

CHAPTER 9

MAGNETIC RESONANCE

The development of magnetic resonance spectroscopy has permitted many advances in structural and dynamical inorganic chemistry. This field is divided into two areas, those being electron paramagnetic resonance (EPR) and nuclear magnetic resonance (NMR). Although the principle ideas underlying NMR and EPR are similar, the differences in the magnitudes of the interactions involved require different experimental techniques.

Data from magnetic resonance experiments can be utilized in a wide range of problems from the simple determination of the number of chemically different atoms in molecules to a detailed description of the electronic distribution in metal-ligand bonds. In this chapter we present some problems which illustrate how magnetic resonance spectroscopy can be used to answer questions concerning the structure and bonding of inorganic compounds.

There are many useful reference works dealing with NMR and EPR. Some valuable ones include:

1. A. Carrington and A. D. McLachlan, *Introduction to Magnetic Resonance*, Harper & Row, Publishers, New York, 1967.

2. G. E. Pake, *Paramagnetic Resonance*, W. A. Benjamin, Inc., New York, 1962.

3. B. R. McGarvey, "Electron Spin Resonance of Transition Metal Complexes," in *Transition Metal Chemistry*, edited by R. L. Carlin, Vol. 3, p. 89 (1967).

4. R. S. Drago, *Physical Methods in Inorganic Chemistry*, Reinhold Publishing Co., New York, 1965, Chapter 4.

5. J. D. Roberts, *An Introduction to the Analysis of Spin-Spin Splitting in High-Resolution Nuclear Magnetic Resonance Spectra*, W. A. Benjamin, Inc., New York, 1962.

6. J. A. Pople, W. G. Schneider, and H. J. Bernstein, *High-Resolution Nuclear Magnetic Resonance*, McGraw-Hill Book Co., Inc., New York, 1959.

PROBLEMS

9.1 *Nuclei useful in* NMR *spectroscopy*. (Read: Drago, Chapter 4, and Pople, Schneider and Bernstein, Appendix A.)

(a) In evaluating the prospects of using a particular nucleus for *NMR* experiments several characteristics of the nucleus must be considered. What are these?

(b) Evaluate the following nuclei as to their potential usefullness in *NMR*:

(i) ^{19}F, (ii) ^{115}Sn, (iii) ^{15}N, (iv) ^{29}Si, (v) ^{11}B.

9.2 *Nuclear resonance condition*. (Read: Carrington and McLachlan, Chapter 1). For a static magnetic field of 14,100 gauss, what frequency of the oscillating perpendicular magnetic field will cause resonance absorption for the following nuclei: (a) ^{2}H, (b) ^{14}N, (c) ^{13}C, (d) ^{19}F, (e) ^{31}P, (f) ^{17}O, (g) ^{16}O, (h) ^{1}H.

9.3 *Comparison of* EPR *and* NMR *energies*. (Read: Carrington and McLachlan). In comparing the *EPR* experiment to the *NMR* experiment, it is useful to consider the origin and magnitude of the splittings of electronic spin states on the one hand and nuclear spin states on the other, each under the influence of the same magnetic field. Assuming no complicating factors, the energy splittings are given as

NMR : $h\nu_n = g_n \beta_n H_0$

and

EPR : $h\nu_e = g_e \beta_e H_0$

where the g's are the nuclear and electronic Landé factors and the β's are the nuclear and electronic (Bohr) magnetons. The β's are related to the magnetic moments as follows:

$$\beta_e = \frac{e_e \hbar}{2m_e c} = -9,273 \times 10^{-24}$$; e_e and m_e are the charge and mass of electron, respectively

$$\beta_n = \frac{e_p \hbar}{2m_p c} = 5.0493 \times 10^{-24};$$ e_p and m_p are the charge and mass of proton, respectively

$$\mu_e = g_e \beta_e S = -9,273 \times 10^{-24} \text{ erg/gauss}$$

$$\mu_n = g_n \beta_n I = +14.1 \times 10^{-24} \text{ erg/gauss}$$

For a field of 14,000 gauss, calculate the energy separations for an electronic system with $S = \frac{1}{2}$ and for a nuclear system with $I = \frac{1}{2}$:

(a) In cm^{-1}

(b) In Mc

(c) What are the corresponding wave lengths?

(d) Repeat the calculations for a field of 10,000 gauss.

9.4 EPR *resolution*. If in a certain system a resolution of two *EPR* lines one gauss apart is possible, what is the resolution in wave numbers?

Assume $g = 2.0$, $\nu = 9{,}000$ Mc , and $\beta = 0.927 \times 10^{-20}$ erg/gauss.

9.5 *Phosphorus-31* NMR. The phosphorus-31 resonance spectra of H_3PO_3 and H_3PO_2 consist of a doublet and a triplet, respectively. On the basis of this information assign structures to the two molecules.

9.6 *Fluorine-19* NMR. The ^{19}F high resolution spectrum of $[SnF_5 \cdot C_2H_5OH]$ at $-28°C$ consists of a doublet at 160 ppm and a quintet at 167 ppm (relative to the external standard $CFCl_3$). Propose a structure.

9.7 *Phosphorus-31* NMR. Phenylphosphorus dibromide, $C_6H_5PBr_2$, reacts with Br_2 and NH_4Br in *sym*-tetrabromoethane to yield two forms of the cyclic tribromotriphenyltriphosphonitrile. One trimer melts at 152-53°C and the other at 194-95°C. In benzene solution the higher melting compound exhibits only one phosphorus-31 resonance peak while the lower melting compound exhibits two peaks with an intensity ratio of 2:1. Assign structures to the two isomeric forms.

9.8 *Fluorine-19* NMR. Sketch the expected fluorine-19 resonance spectrum of ClF_3.

9.9 *Fluorine-19* NMR.

(a) The ^{19}F nuclear magnetic resonance spectrum of SF_4 at -98°C is shown in Figure 9.1. Is this spectrum consistent with the structure predicted by valence-shell-electron-repulsion theory (full hybridization theory)?

Figure 9.1

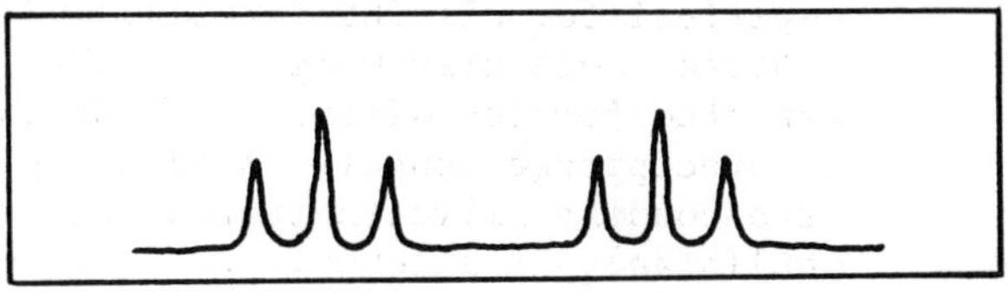

(b) When SF_4 is warmed the two sets of triplets merge to a single peak. What is the explanation for this change?

9.10 *Boron-11* NMR.

(a) Look up the boron-11 chemical shifts for the series BF_3, BCl_3, BBr_3, and BI_3.

(b) Provide an explanation for the chemical shifts observed for the series of boron halides.

(c) Is your explanation for part (b) consistent with the chemical shifts observed for the tetrahaloborate ions?

9.11 *Temperature effects.* The *NMR* spectrum of the compound

$$PtCl_2[CH_3\ddot{S}(CH_2)_2\ddot{S}CH_3]$$

may be expected to consist of a triplet for the methyl proton resonance — a doublet due to spin-spin coupling with the ^{195}Pt nucleus (34% abundance, $I = 1/2$), symmetrical about a single line from the methyl protons in molecules containing platinum isotopes of zero spin. However, in deuterated dimethylsulfoxide, two such triplets with $J_{Pt-H} = 48.0 \pm 0.5\ sec^{-1}$ appeared in the methyl region. Upon warming the solution to 95°C the triplets converged into a single triplet, and upon cooling, the original spectrum reappeared. Suggest an explanation.

9.12 NMR *as a structural tool*. Rhodium trichloride trihydrate reacts with hexafluoroacetylacetone in anhydrous ethanol to yield $RhCl(hfac)_2 \cdot 3H_2O$. The hydrate can be decomposed to yield a material with empirical formula $RhCl(hfac)_2$ which exhibits an apparent molecular weight in CCl_4 very near that expected for the dimer. In addition, no cloudiness or precipitate appears immediately upon the addition of alcoholic silver nitrate to a solution of $RhCl(hfac)_2$ in alcohol. Finally the *NMR* spectrum in CCl_4 shows one proton absorbance at -6.54 ppm, and two ^{19}F bands at -5.42 and -5.27 ppm. (The monomeric $Rh(hfac)_3$ has one proton band at -6.59 ppm). Propose a structure for $[RhCl(hfac)_2]_2$ which is consistent with these experimental data and assign the two fluorine-19 resonances.

9.13 *Virtual coupling*. The *NMR* spectrum of free dimethylphenylphosphine shows that the resonance of the methyl protons appears as a symmetrical doublet. When two molecules of the phosphine are present in positions *trans* to each other in a metal complex, the same resonance appears not as a doublet but as a well-defined 1:2:1 triplet. When the phosphines are present in *cis* positions the doublet formation is retained. Explain these observations and predict the stereochemistry of the following complexes from the given data:

	Compound	Methyl proton resonance
(a)	$IrCl_3(PMe_2Ph)_3$	1:2:1 triplet + symmetrical doublet. Intensity ratio triplet: doublet = 2:1.
(b)	yellow-$RuCl_2(CO)(PMe_2Ph)_3$	1:2:1 triplet + symmetrical doublet. Intensity ratio triplet: doublet = 2:1. Dipole moment = 3.90 D.
(c)	white-$RuCl_2(CO)(PMe_2Ph)_3$	1:3:3:1 quartet + sym. doublet. Intensity ratio quartet:doublet = 2:1. Dipole moment = 7.4 D.
(d)	$[IrCl_4(PMe_2Ph)_2]^-$	1:2:1 triplet

9.14 NMR *as a structural tool.* The amino acids L-alanine, L-leucine and L-proline form tris-complexes with the cobalt(III) ion. (a) Represent the amino acid as N—O and draw structural models for the isomers expected in any one amino acid-cobalt(III) system. (b) Both a red and a violet form of $(-)$-$[Co(L\text{-alanine})_3]$ were isolated. The *NMR* spectra were recorded and the methyl resonances are shown in Figure 9.2. Analyze the *NMR* spectral features and determine which isomer is *cis* and which is *trans*.

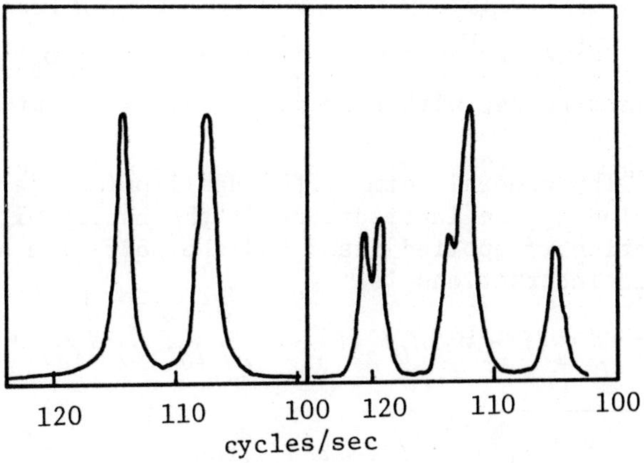

Figure 9.2.

9.15 NMR *and* IR *as structural tools*. Explain why PF_5 exhibits an infrared spectrum consistent with a trigonal bipyramidal structure but gives rise to only a single fluorine resonance (split into a doublet by P-F coupling).

9.16 *Sodium-23* NMR. (Read: Carrington and McLachlan, Chapter 1) A sample containing sodium-23 nuclei is placed in a homogeneous magnetic field with strength of 14,100 gauss. The gyromagnetic ratio, γ_N, for ^{23}Na is 7,081 radians sec^{-1} gauss^{-1}. Assume quadrupole interactions are negligible.

(a) Draw an energy level diagram for the ^{23}Na nucleus showing the Zeeman splittings, and designate each energy level by the appropriate quantum number.

(b) Which transitions may be induced by an oscillating electromagnetic field whose magnetic vector is perpendicular to the static magnetic field?

(c) Give the frequencies of the allowed transitions.

(d) Calculate the ratios of the populations of the $M_I = \frac{3}{2}$ and $M_I = \frac{1}{2}$ energy levels at thermal equilibrium at 300°K.

9.17 *Spin-Spin Splitting in* NMR. (Read: Roberts, Chapter 2). This problem is concerned with the spin-spin interaction between two nuclei.

(a) Write the spin wave functions for two nonequivalent nuclei with spin $\frac{1}{2}$.

(b) The general form of the Hamiltonian operator including the interactions of the nuclei with the stationary applied field and the pair-wise spin-spin interactions is

$$H = \sum_i \gamma_i(1 - \sigma_i)H_0 I_z(i) + \sum_{i<j} J_{ij}\vec{I}_{(i)}\cdot\vec{I}_{(j)}$$

$$= \sum_i \nu_i I_z(i) + (1/4) \sum_{i<j} J_{ij}(2p_{ij} - 1)$$

Here, $\nu_i = \gamma_i(1 - \sigma_i)H_0$ and is in cycles per second if γ_i is in cycles per second per gauss, I_z is the spin angular momentum operator, J_{ij} is the spin-spin coupling constant, and p_{ij} is a permutation operator which interchanges the index numbers of specified pairs of nuclei in the wave functions. Write the explicit Hamiltonian operator to be used in the calculation of the Zeeman energies and the spin-spin interactions.

(c) Calculate the energy levels for two nonequivalent magnetic nuclei undergoing spin-spin coupling.

(d) Indicate the possible transitions on an energy level diagram. (Let $\gamma_1 = \gamma_2$, and assume the chemical shift is large with respect to J_{12}). The probability of a nuclear spin transition is given by $P_{i\to j} = [\langle\Phi_j|I_+|\Phi_i\rangle]^2$ if $I(z)_i < I(z)_j$. (For $I(z)_j > I(z)_i$ use I_-. $I_\pm$ are nuclear spin raising and lowering operators.) Calculate the probabilities of nuclear spin transitions between the energy levels determined in part (c).

(e) Write the spin-wave functions for two equivalent nuclei undergoing spin-spin coupling, calculate the energy level diagram, and indicate the allowed transitions.

(f) Compare the spectrum calculated in part (d) with that calculated in part (e).

(g) What important experimental result will permit one to distinguish between lines in *NMR* spectra due to spin-spin coupling and lines due to chemical shifts.

(h) Repeat each step in (e) for two nonequivalent nuclei for which the spin-spin coupling constant is comparable in magnitude to the chemical shift.

9.18 *Spin-spin coupling in* EPR. Consider the tris-(2,2'-bipyridine)copper(II) ion. The rigidity of the bipyridine ligands would suggest that all the nitrogens might be equivalent, but this symmetry, D_3, is thought to be unlikely for copper(II). Rhombic symmetry is more probable. Predict the hyperfine structure of the *EPR* spectrum of $[Cu(bipy)_3]^{2+}$:

(a) with coupling of the electron spin to the copper nucleus only.

(b) with strong coupling to the copper nucleus and weak coupling to six equivalent nitrogens. (Assume the copper coupling constant is much larger than the nitrogen coupling constant.)

(c) with strong coupling to the copper nucleus and weak coupling to four equivalent nitrogens.

9.19 *Spin-spin coupling in* EPR. Interpret the *EPR* spectrum of $Fe(CN)_5NO^{3-}$ shown below.

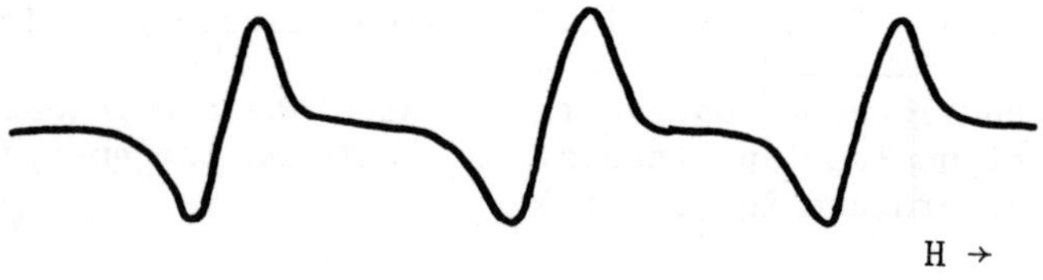

9.20 *Spin-spin coupling in* EPR. The *EPR* spectrum of $Mo(CN)_8{}^{3-}$ in solution consists of one line, but that of a ^{13}C-enriched sample exhibits nine lines. Explain.

9.21 *Hyperfine coupling in Cu(phen)Cl₂*. In the compound $Cu(phen)Cl_2$ diluted about one percent in the isomorphous $Zn(phen)Cl_2$ crystal, the copper ions are in a distorted tetrahedral environment. The *EPR* spectrum of the single crystal with the field directly perpendicular to the Cl-Cu-Cl plane shows large coupling of the single unpaired electron with the copper nucleus and additional splitting of these peaks by coupling with the two nitrogens. Assume the interaction of the electron moment with the copper nuclear moment is much greater than that with the nitrogen nuclear moments and sketch the expected spectrum.

9.22 *Correlation of* EPR *spectra with electronic spectra.* (Read: McGarvey or Ballhausen, Chapter 6). The site symmetry of Cu^{2+} in $BaCuSi_4O_{10}$ is D_{4h}. From the energies of the transitions as assigned below, predict the values of $g_{\parallel}$ and $g_{\perp}$. Assume ξ = 830 cm^{-1}.

$E(^2B_{1g} \rightarrow {}^2B_{2g})$ = 12,900 cm

$E(^2B_{1g} \rightarrow {}^2E_g)$ = 15,800

9.23 EPR *spectra of bis(acetylacetonato)copper(II) in solution and in the vitreous state.* The *EPR* spectrum of bis(acetylacetonato)copper(II), $Cu(acac)_2$, in 60% toluene-40% chloroform is shown in Figure 9.3(a), and the *EPR* spectrum of $Cu(acac)_2$ in a glass of tolune-chloroform is shown in Figure 9.3(b). Provide a detailed explanation for these spectra which includes the identification of the hyperfine coupling constants and the g-tensor anisotropy.

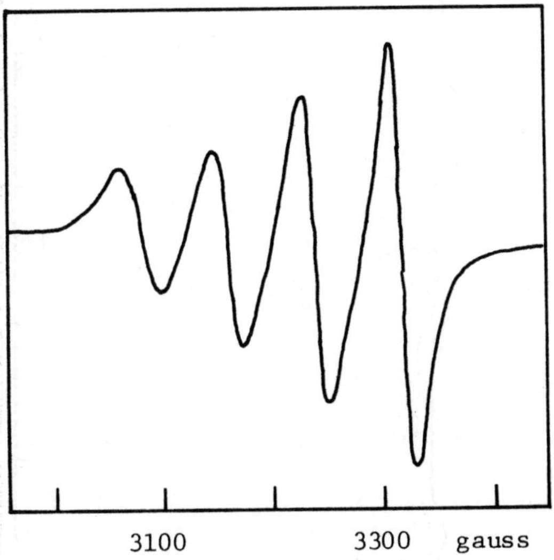

Figure 9.3 (a)

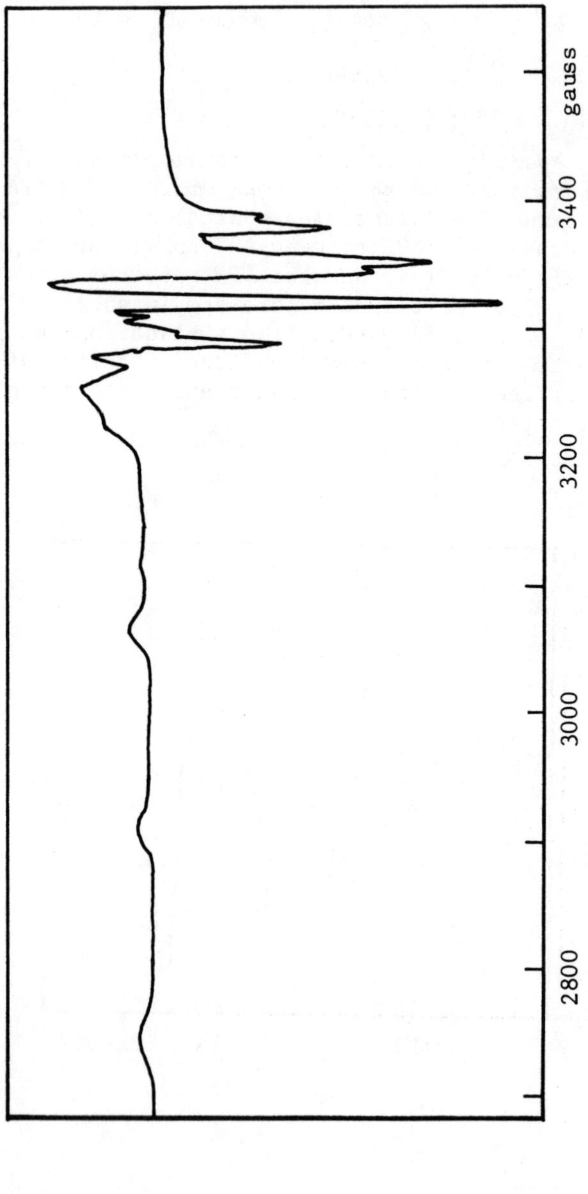

Figure 9.3 (b)

9.24 *Analysis of* EPR *spectra.* The *EPR* spectrum of bis(salicylaldiminato)copper(II) - (copper-63)

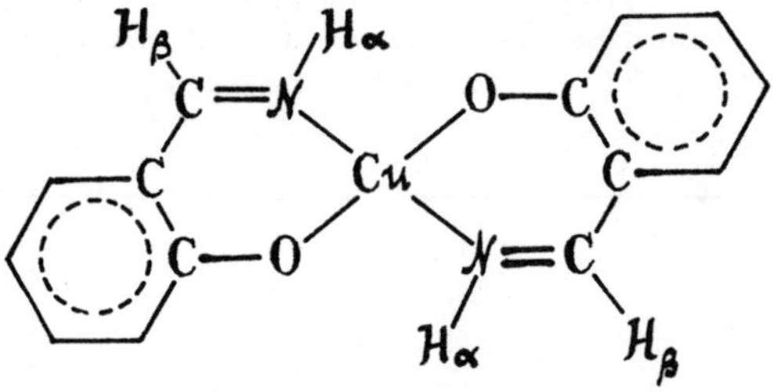

consists of four main groups of lines as shown in Figure 9.4 (a). Each of the four peaks is split into eleven equally spaced components separated by 5.5 gauss with the intensity ratio 1:2:3:4:5:6:5:4:3:2:1. Upon substitution of the hydrogen atoms bonded to nitrogen (the H_α atoms) by deuterium, no change was produced in the spectrum, but when the H_β atoms were replaced by methyl groups, the spectrum in Figure 9.4(b) resulted. The splitting of the hyperfine components in (b) is 11 gauss. Provide a consistent explanation of these experimental data.

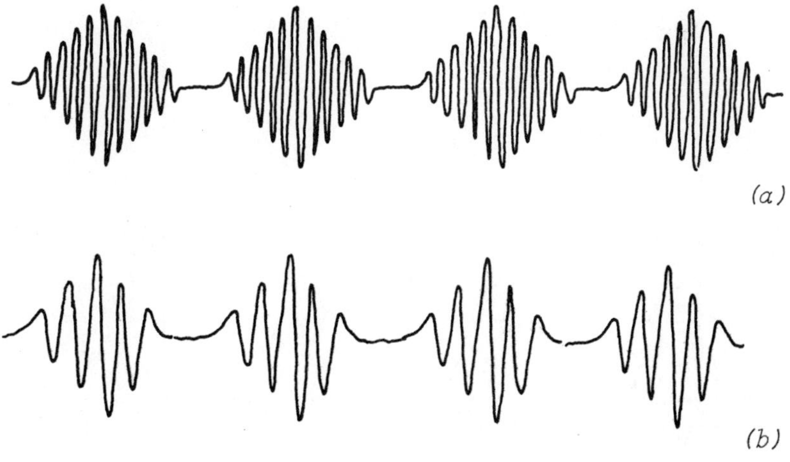

Figure 9.4

9.25 *Display of* EPR *spectra.* *EPR* spectra are usually recorded as the first derivative of the absorption curve as a function of field strength. In Figure 9.5, the *EPR* absorption spectrum for a frozen solution of $[Ag(phen)_2]S_2O_8$ is shown. Sketch the shape of the first derivative of this line, and indicate the magnetic field value to be used in the calculation of the g-values.

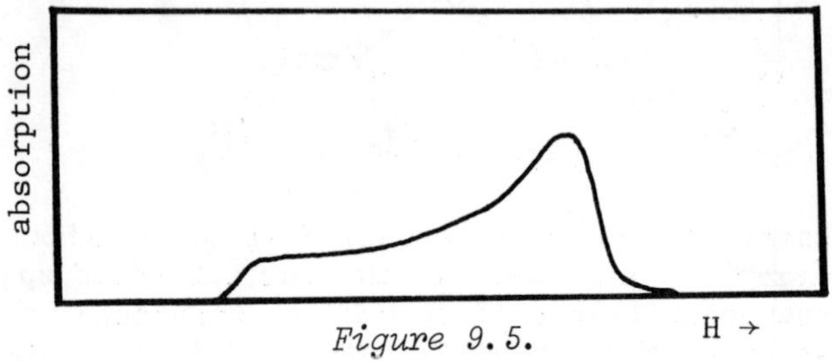

Figure 9.5.

9.26 *Selection rules for* EPR *transitions.* (Read: Carrington and McLachlan, Chapter 2). The zero-field splitting and Zeeman splitting for a chromium complex with a quartet ground state is shown in Figure 9.6. How many lines are expected in the *EPR* spectrum? To which transitions do they correspond?

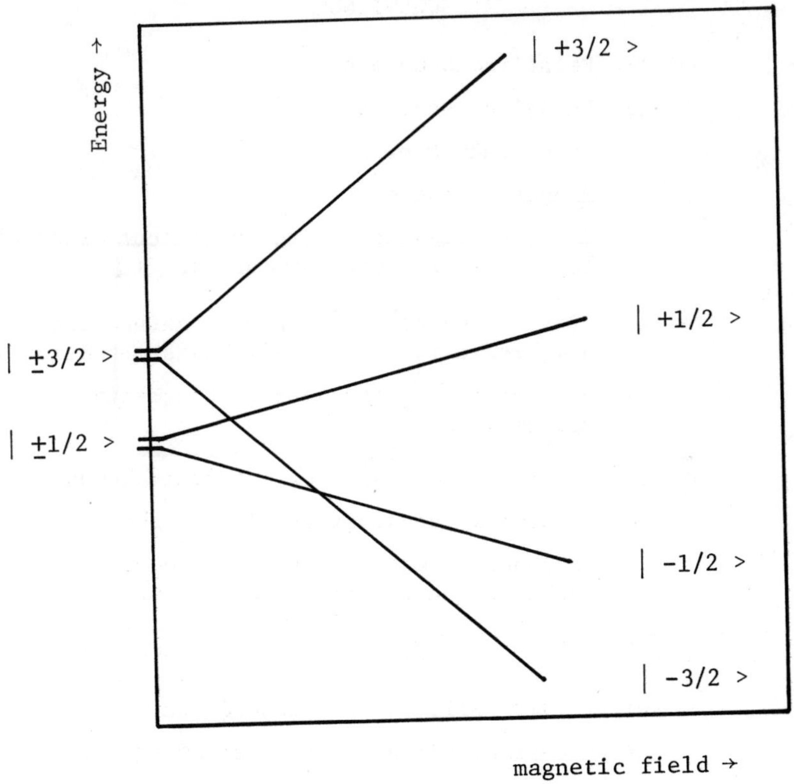

Figure 9.6.

SOLUTIONS

9.1 (a) 1. Relative abundance

2. Relative sensitivity

3. Spin quantum number

4. Magnetic moment

5. Range of resonance frequencies corresponding to available range of magnetic field.

(b) i) ^{19}F: Excellent; 100 percent abundance, high relative sensitivity, high magnetic moment, $I = \frac{1}{2}$ (therefore, no quadrupole moment).

ii) ^{115}Sn: Poor; very low natural abundance; reasonable sensitivity, $I = \frac{1}{2}$, but not abundant enough; better use one of the other Sn nuclei, ^{117}Sn or ^{119}Sn.

iii) ^{15}N: Poor; very low natural abundance.

iv) ^{29}Si: Fair; low abundance, but $I = \frac{1}{2}$, and other Si nuclei do not interfere.

v) ^{11}B: Excellent from point of view of abundance and sensitivity and moment; however $I = \frac{3}{2}$; therefore, $eQ \neq 0$.

9.2 (a) 9.21, (b) 4.34, (c) 15.10, (d) 56.48, (e) 24.31, (f) 8.14, (g) no resonance possible, (h) 60.04 megacycles sec^{-1}.

9.3 First calculate the g factors:

$$g_e = \frac{-9,273 \times 10^{-24}\ erg/gauss}{-9,273 \times 10^{-24}\ erg/gauss \times \frac{1}{2}} = 2.00$$

$$g_n = \frac{14.1 \times 10^{-24} \; erg/gauss}{5.0493 \times 10^{-24} \; erg/gauss \times \frac{1}{2}} = 5.58$$

For the *NMR* experiment $E = h\nu_n = g_n \beta H$.

(a) $E = 5.58 \times 5.0493 \times 10^{-24} \; erg/gauss \times 1.4 \times 10^{4} \; gauss$

$= 39.5 \times 10^{-20} \; ergs \times 5.0348 \times 10^{15} \; cm^{-1}/erg$

$= 1.98 \times 10^{-3} \; cm^{-1}$

(b) $1.98 \times 10^{-3} \; cm^{-1} \times 2.998 \times 10^{10} \; cm \; sec^{-1}$

$= 6.0 \times 10^{7} \; sec^{-1}$

$= 60 \times 10^{6} \; sec^{-1}$

$= 60 \; Mc$

(c) $\lambda = 500 \; cm$

For the *EPR* experiment $E = h\nu_e = g_e \beta H$.

(a) $E = 2.00 \times -9,273 \times 10^{-24} \; erg/gauss \times 1.4 \times 10^{4} \; gauss$

$= 2.60 \times 10^{-16} \; erg$

$2.60 \times 10^{-16} \; erg \times 5.0348 \times 10^{15} \; cm^{-1}/erg = 1.30 \; cm^{-1}$

(b) $1.30 \; cm^{-1} \times 2.998 \times 10^{10} \; cm \; sec^{-1} = 3.90 \times 10^{10} \; sec^{-1}$

$= 39,000 \times 10^{6} \; sec^{-1}$

$= 39,000 \; Mc$

(c) $\lambda = 0.758 \; cm$

9.5 The only atom that can split the phosphorus resonance is hydrogen, and it is likely that only hydrogen atoms bonded to phosphorus will be effective. For H_3PO_3, from the spin-spin splitting equation the number of peaks

$n = 2\sum S_H + 1 = 2$. The total spin of the interacting hydrogens, $\sum S_H = \frac{(2 - 1)}{2} = \frac{1}{2}$.

Therefore only one hydrogen is bound to phosphorus and the structure is

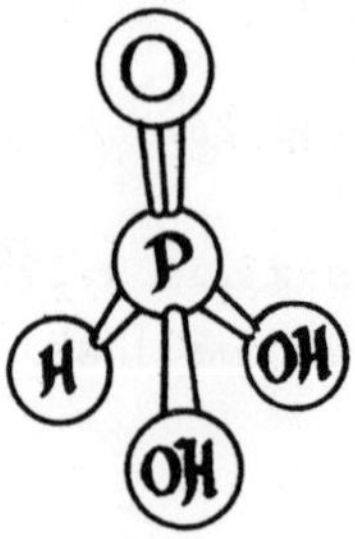

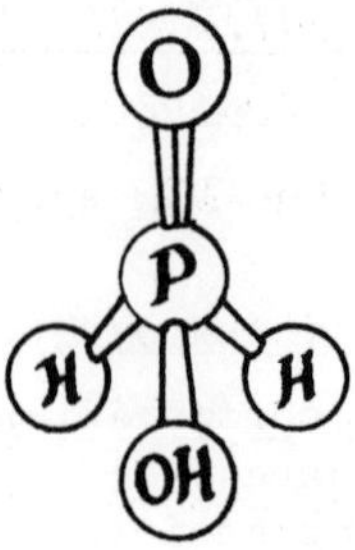

For H_3PO_2, $\sum S_H = \frac{(3 - 1)}{2} = 1$. This indicates two hydrogens bonded to phosphorus as in the structure shown above.

9.7 In the *cis* isomer (I) all phosphorus atoms are equivalent and should give rise to one resonance, while the *trans* isomer (II) should exhibit two phosphorus resonances with an intensity ratio of 2:1.

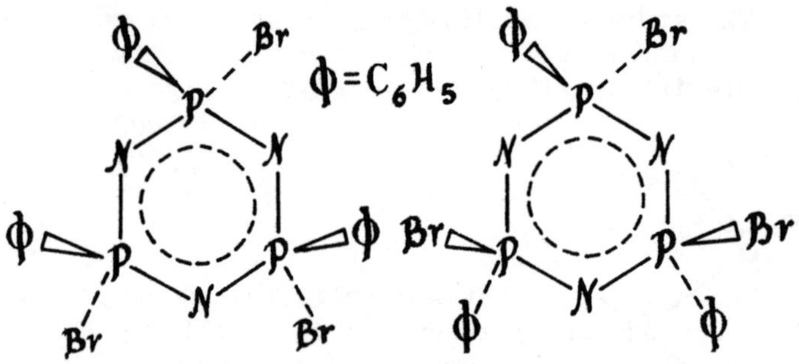

(I) (II)

m.p. = 194-95°C m.p. = 152-53°C

9.9 (a) VSER predicts the structure

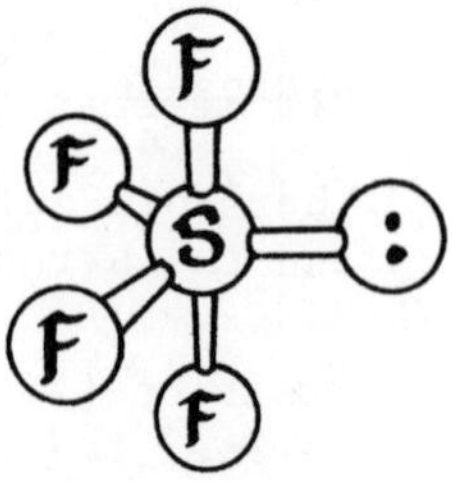

with two pairs of equivalent fluorines. For each pair we expect a triplet, since the number of spin-spin splitting components is

$$2\sum S_B + 1 = 2(\frac{1}{2} + \frac{1}{2}) + 1 = 3 \text{ lines}$$

The intensities check also. The ratio is given by the coefficients of the polynomial $(1 + r)^{2\Sigma S_B}$. We have $(1 + r)^2 = 1 + 2r + r^2$ for the 1:2:1 observed intensity ratios.

(b) Upon heating, the fluorine atoms begin to exchange and as a result the triplets broaden and move toward each other. It is estimated that complete exchange occurs at -20 ± 20°. Reference: F. A. Cotton and J. W. George, *J. Chem. Phys.*, 28, 994 (1958).

9.10 (a) BF_3, -11.6; BCl_3, -45.6; BBr_3, -40.0; BI_3, 6.7 ppm. The chemical shifts given here are relative to $BF_3 \cdot Et_2O$.

(c) Chemical shifts for BX_4^- are BF_4^-, 1.8; BCl_4^-, -6.6; BBr_4^-, 23.9 ppm.

9.11 The stereochemically-active lone pair of electrons on sulfur permits the formation of isomers which are in equilibrium.

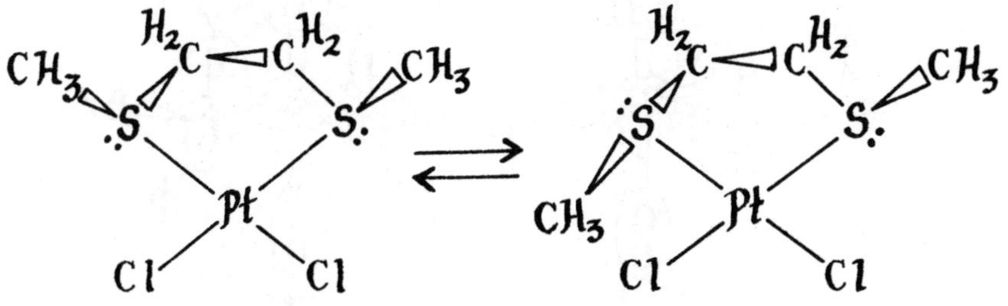

Reference: Abel, Bush, Hopton, and Jenkins, *Chem. Comm.*, 58 (1966).

9.13 The resonance of the methyl protons in free $[PMe_2Ph]$ appears as a symmetrical doublet due to spin-spin interaction with the phosphorus nucleus (^{31}P, spin = $\frac{1}{2}$). When the phosphines are in *trans* positions in a complex, strong coupling occurs between the phosphorus atoms, and the methyl protons couple equally with both phosphorus nuclei, giving rise to a 1:2:1 triplet. This is called *virtual* coupling. For the *cis* complexes the observation of the symmetrical doublet, as in the free phosphine, indicates that the phosphorus nuclei in *cis* positions do not couple strongly.

(a) $IrCl_3(PMe_2Ph)_3$

The observation of a triplet and doublet in intensity ratio 2:1 indicates that 2 phosphine ligands are in mutual *trans* positions.

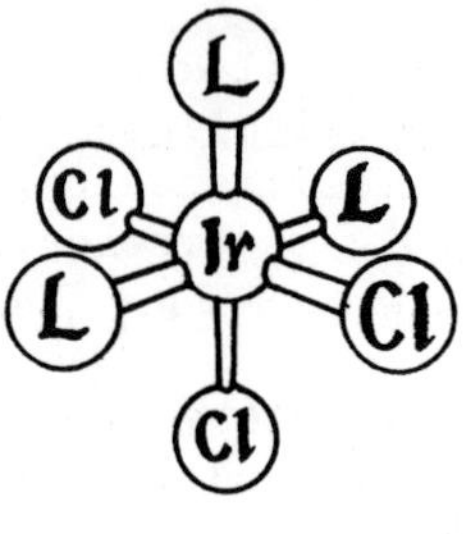

(b) $RuCl_2(CO)(PMe_2Ph)_3$ (yellow isomer)

The ratio of the intensities again indicates that 2 phosphines are in mutually *trans* positions. However, there are two structural isomer possibilities here, one with the third phosphine ligand *trans* to Cl and the other with it *trans* to CO.

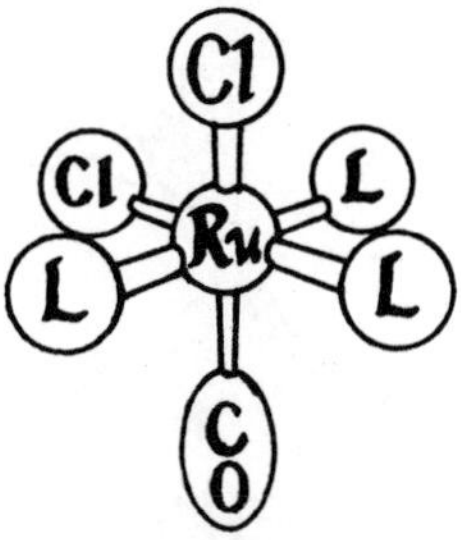

I

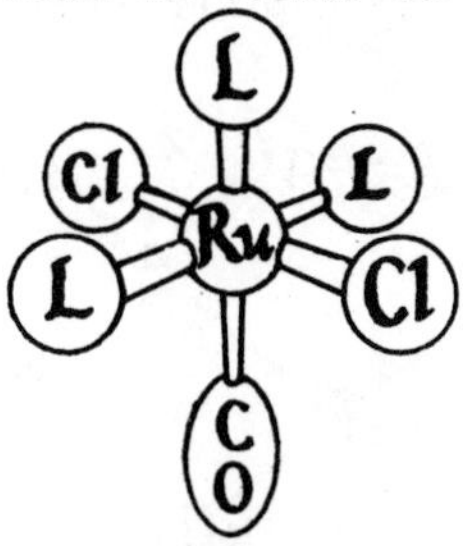

II

The relatively low dipole moment of the yellow isomer (3.9D) suggests that the third phospine ligand is situated *trans* to the carbonyl group (i.e., structure II).

(c) $Ru(CO)Cl_2(PMe_2Ph)_3$ (white isomer)

The intensity ratios again indicate that 2 phosphines are mutually *trans* and the relatively high dipole moment (7.4D) suggests that the third ligand is situated *trans* to Cl (i.e., structure I). The observation of a quartet rather than a triplet must arise from restricted rotation about the Ru-P bond causing the 2 methyl groups on the same phosphorus atom to be non-equivalent.

9.16 (a)

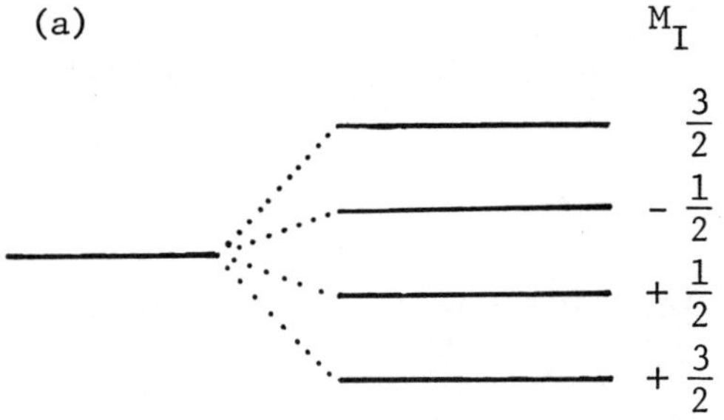

(b) $\Delta M_I = \pm 1$; Thus $-\frac{3}{2} \leftrightarrow -\frac{1}{2}$; $-\frac{1}{2} \leftrightarrow +\frac{1}{2}$; $+\frac{1}{2} \leftrightarrow +\frac{3}{2}$ are allowed.

(c) $\omega_0 = \gamma H_0$; $\nu_0 = (\frac{\gamma}{2\pi})H_0 = \frac{(7.081 \times 10^3)(14.1 \times 10^3)}{6.28}$

$$= 15.9 \times 10^6 Hz$$

(d) $\frac{N_{1/2}}{N_{3/2}} = exp\frac{(-15.9 \times 10^6)(6.625 \times 10^{-27})}{(1.38)(3.)(10^{-14})}$

$\simeq 1 - (2.54 \times 10^{-6})$

$\simeq 0.99999746$

9.17 (a) $\Phi_1 = \alpha\alpha$; $I_z = 1$

$\Phi_2 = \alpha\beta$; $I_z = 0$

$\Phi_3 = \beta\alpha$; $I_z = 0$

$\Phi_4 = \beta\beta$; $I_z = -1$

(b) $H = \gamma_1(1 - \sigma_1)H_0 I_z(1) + \gamma_2(1 - \sigma_2)H_0 I_z(2) + (\frac{1}{4})J_{12}(2p_{12} - 1)$

(c) The energies are

$$\langle\alpha\alpha|\gamma_1(1-\sigma_1)H_0I_z(1)+\gamma_2(1-\sigma_2)H_0I_z(2)$$
$$+\left(\frac{1}{4}\right)J_{12}(2p_{12}-1)|\alpha\alpha\rangle = \frac{\gamma}{2}(2-\sigma_1-\sigma_2)H_0+\frac{J_{12}}{4}$$

where it is assumed $\gamma_1 = \gamma_2 = \gamma$.

$$\langle\alpha\beta|Operator|\alpha\beta\rangle = \left(\frac{\gamma}{2}\right)(\sigma_2-\sigma_1)H_0-\frac{J_{12}}{4}$$

$$\langle\beta\alpha|Operator|\beta\alpha\rangle = \left(\frac{\gamma}{2}\right)(\sigma_1-\sigma_2)H_0-\frac{J_{12}}{4}$$

$$\langle\beta\beta|Operator|\beta\beta\rangle = -\frac{\gamma}{2}(2-\sigma_1-\sigma_2)H_0+\frac{J_{12}}{4}$$

To construct the energy level diagram, assume $\sigma_2 > \sigma_1$.

	E	
1	———	$\frac{\gamma}{2}(2-\sigma_1-\sigma_2)H_0+\frac{J_{12}}{4}$
2	———	$\frac{\gamma}{2}(\sigma_2-\sigma_1)-\frac{J_{12}}{4}$
3	———	$\frac{\gamma}{2}(\sigma_1-\sigma_2)-\frac{J_{12}}{4}$
4	———	$-\frac{\gamma}{2}(2-\sigma_1-\sigma_2)H_0+\frac{J_{12}}{4}$

(d) $P(\alpha\beta\rightarrow\alpha\alpha) = [\langle\alpha\alpha|I_+|\alpha\beta\rangle]^2$
$= [\langle\alpha\alpha|\alpha\alpha\rangle]^2$
$= 1$

$P(\beta\alpha\rightarrow\alpha\alpha) = [\langle\alpha\alpha|I_+|\beta\alpha\rangle]^2 = 1$

$P(\beta\beta\rightarrow\alpha\beta) = [\langle\alpha\beta|I_+|\beta\beta\rangle]^2$

$= [\langle\alpha\beta|\alpha\beta\rangle + \langle\alpha\beta|\beta\alpha\rangle]^2$

$= 1$

$P(\beta\beta\rightarrow\beta\alpha) = [\langle\beta\alpha|I_+|\beta\beta\rangle]^2 = 1$

These transitions are equally probable. However, transitions of the sort $\beta\beta\leftrightarrow\alpha\alpha$ are not allowed, *viz.*

$$P = [\langle\alpha\alpha|I_+|\beta\beta\rangle]^2 = [\langle\alpha\alpha|\alpha\beta\rangle + \langle\alpha\alpha|\beta\alpha\rangle]^2 = 0$$

The energies of the transitions are

$\alpha\beta\leftrightarrow\alpha\alpha$; $\gamma(1-\sigma_2)H_0+\frac{J_{12}}{2}$

$$\beta\alpha \leftrightarrow \alpha\alpha; \quad \gamma(1 - \sigma_1)H_0 + \frac{J_{12}}{2}$$

$$\beta\beta \leftrightarrow \alpha\beta; \quad \gamma(1 - \sigma_1)H_0 - \frac{J_{12}}{2}$$

$$\beta\beta \leftrightarrow \beta\alpha; \quad \gamma(1 - \sigma_2)H_0 - \frac{J_{12}}{2}$$

(e) The wave functions are

$$\Phi_1 = \alpha\alpha$$

$$\Phi_2 = \frac{(\alpha\beta + \beta\alpha)}{\sqrt{2}}$$

$$\Phi_3 = \frac{(\alpha\beta - \beta\alpha)}{\sqrt{2}}$$

$$\Phi_4 = \beta\beta$$

The Hamiltonian operator is

$$H = \gamma(1 - \sigma)H_0 I_z(1) + \gamma(1 - \sigma)H_0 I_z(2) + (\tfrac{1}{4})J_{12}(2p_{12} - 1)$$

The energies of the states are

$$\langle 1|Op|1\rangle = \gamma(1 - \sigma)H_0 + \frac{J_{12}}{4}$$

$$\langle 2|Op|2\rangle = \frac{J_{12}}{4}$$

$$\langle 3|Op|3\rangle = \frac{-3J_{12}}{4}$$

$$\langle 4|Op|4\rangle = -\gamma(1 - \sigma)H_0 + \frac{J_{12}}{4}$$

Arranged on an energy level diagram we have

E (increasing upward)

——— $\gamma(1 - \sigma)H_0 + \frac{J_{12}}{4}$

——— $+\frac{J_{12}}{4}$

——— $\frac{-3J_{12}}{4}$

——— $-\gamma(1 - \sigma)H_0 + \frac{J_{12}}{4}$

The transition probabilities are

$$P[\beta\beta \rightarrow (\alpha\beta + \beta\alpha)] = [< \frac{1}{\sqrt{2}}(\alpha\beta + \beta\alpha)|I_+|\beta\beta>]^2$$

$$= [\frac{1}{\sqrt{2}}\{<\alpha\beta|\alpha\beta> + <\alpha\beta|\beta\alpha> + <\beta\alpha|\alpha\beta> + <\beta\alpha|\beta\alpha>\}]^2$$

$$= [\frac{1}{\sqrt{2}}\{1 + 0 - 0 - 1\}]^2$$

$$= 0$$

$$P[(\alpha\beta + \beta\alpha) \rightarrow \alpha\alpha] = [<\alpha\alpha|I_+|\frac{(\alpha\beta + \beta\alpha)}{\sqrt{2}}>]^2 = 2$$

$$P[(\alpha\beta - \beta\alpha) \rightarrow \alpha\alpha] = 0$$

The energies of the transitions are

$\beta\beta \leftrightarrow (\alpha\beta + \beta\alpha)$; $\gamma(1 - \sigma)H_0$

$(\alpha\beta + \beta\alpha) \leftrightarrow \alpha\alpha$; $\gamma(1 - \sigma)H_0$

(f) The spectrum of the two nonequivalent interacting nuclei will show four lines of equal intensity. Sets of two lines will be symmetrically displaced about the chemical shifts of the two nonequivalent atoms. This spin problem is referred to in *NMR* literature as an *AX* case. The spectrum of the two equivalent nuclei will consist of only one line. This is an A_2 case.

(g) The spin-spin splittings are not dependent on the frequency of the oscillating electromagnetic field while the chemical shifts are.

9.18 (a) $I_{Cu} = \frac{3}{2}$

The number of hyperfine lines is n·2·I+1, where n is the number of equivalent nuclei, in this case one. Therefore, the number of lines = 1·2·3/2 +1 = 4.

(b) The number of lines = $\prod_i (n \cdot 2 \cdot I + 1)$ where the product is over the *i* sets of different nuclei.

$I(^{14}N) = 1$; therefore, the number of lines = $(1 \cdot 2 \cdot \frac{3}{2} + 1)(6 \cdot 2 \cdot 1 + 1) = 52$. Because of the relative sizes of the coupling constants this might be observed as 13 small peaks superimposed on each of the four larger copper hyperfine lines.

9.19 The unpaired electron interacts with the NO nitrogen nucleus ($I = 1$) to give three lines with the same intensity ($1 \cdot 2 \cdot I + 1 = 3$).

9.20 In $Mo(CN)_8{}^{3-}$, the Mo(IV) ion has a d^1 configuration and only one line is expected, but for the ^{13}C-enriched sample the number of hyperfine lines is $n \cdot 2 \cdot I + 1$, or nine, since ^{13}C has I equal to 1/2. By covalent bonding, unpaired electron density is transmitted to the carbon atom.

9.21 $I_{Cu} = \frac{3}{2}$. Therefore $(2 \cdot \frac{3}{2} + 1) = 4$ major peaks.
$I_N = 1$. Therefore $2(2 \cdot 1) + 1 = 5$.
that is, each of the 4 is split into 5 lines.

9.23 See H. R. Gersmann and J. D. Swalen, *J. Chem. Phys.*, 36, 3221 (1962).

9.24 First, there are four major groups of lines; these are due to ^{63}Cu ($I = 3/2$) hyperfine coupling. Since ^{14}N and ^{1}H have nuclear moments, hyperfine coupling to give the components of each major group could arise only if the unpaired electron is delocalized onto the ligand. Since substitution of ^{2}H ($I = 1$) for hydrogen atoms in the α-position had no effect, then this hydrogen must not participate in hyperfine coupling. Replacement of the hydrogen atoms in the β-position reduced the number of lines from eleven to five. This latter number is expected for two equivalent nitrogen atoms.

$$h.f.s. = (n \cdot 2 \cdot I + 1) = (2 \cdot 2 \cdot 1 + 1) = 5$$

Note that the eleven gauss splitting in the nitrogen hyperfine structure is exactly twice that observed in the hydrogen compound. Consequently, there is a considerable amount of overlapping to account for the number of lines. For two equivalent hydrogen atoms and two equivalent nitrogen atoms, the expected hyperfine structure would consist of 15 lines, that is,

$$(2 \cdot 1 \cdot 2 + 1)(2 \cdot 1/2 \cdot 2 + 1) = 15$$

The hyperfine splitting in one of the four major peaks is constructed in Figure 9.7.

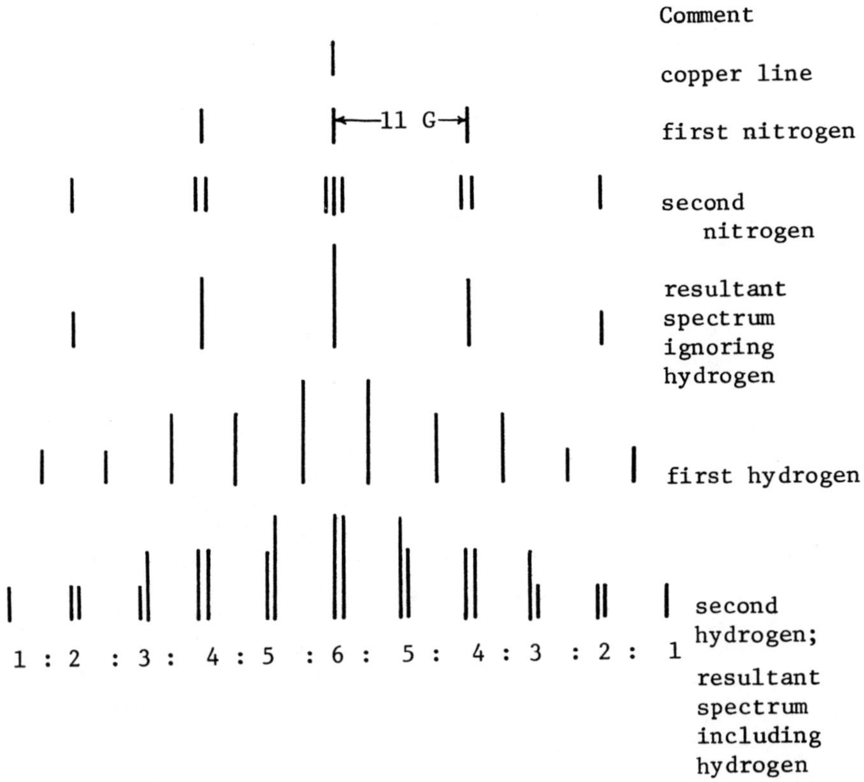

Figure 9.7

9.25

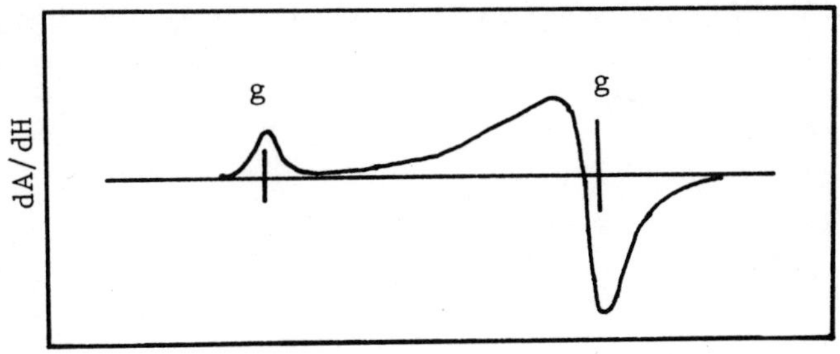

9.26 In solution or along one of the principle axes, three lines are expected. The ^{53}Cr nucleus has $I = \frac{3}{2}$ and $g_n = -.3163$ but occurs in only 9.55% natural abundance and hyperfine lines are not usually seen. If the magnitude of the zero-field splitting is less than the energy of the incident radiation, then the low field transition corresponds to $+\frac{1}{2} \leftrightarrow +\frac{3}{2}$, the mid-field transition is $-\frac{1}{2} \leftrightarrow +\frac{1}{2}$, and $-\frac{3}{2} \leftrightarrow -\frac{1}{2}$ occurs at high field. If the zero-field splitting is larger than $h\nu$, then $-\frac{1}{2} \leftrightarrow +\frac{1}{2}$ may be the only observable transition with $-\frac{3}{2} \leftrightarrow -\frac{1}{2}$ occuring at very high fields.

(There are some complications here, and for additional discussion the student is referred to McGarvey's review article, page 129.)

Chapter 10

BASIC CRYSTALLOGRAPHY

This chapter deals with the elements of symmetry which apply to crystal structure analysis. The orientation is toward the needs of chemists to understand and use the results reported by crystallographers. The problems also stand as an introduction to the study of x-ray crystallography, although they stop short of covering more than the basic principles and theory of x-ray diffraction. Several exercises on the use of the International Tables for X-ray Crystallography are included and access to a copy of Volume I of this series is required. A prior mastery of the problems on point group symmetry in Chapter 4 is essential.

Texts which cover the material in this chapter include:

1. Donald E. Sands, *Introduction to Crystallography,* W. A. Benjamin, Inc., New York, 1969.
2. Martin J. Buerger, *Elementary Crystallography: An Introduction to the Fundamental Geometrical Features of Crystals*. Rev. printing, J. Wiley & Sons, Inc., New York 1963.
3. Charles W. Bunn, *Chemical Crystallography: An Introduction to Optical and X-ray Methods*, 2d ed., The Clarendon Press, Oxford, 1961.
4. George H. Stout and Lyle H. Jensen, *X-ray Structure Determination, A Practical Guide*, The McMillan Co., New York, 1968.

PROBLEMS

10.1 *Crystal systems.* Classify the following unit cells into their proper crystal systems:

(a) a = 6.39Å, b = 7.04Å, c = 10.13Å, $\alpha = \beta = \gamma =$ 90°.

(b) a = b = c = 9.43Å, $\alpha = \beta = \gamma$ = 82°.

(c) a = b = 7.36Å, c = 12.45Å, $\alpha \neq \beta \neq \gamma$.

(d) a = b = 10.73Å, c = 14.30Å, $\alpha = \beta$ = 90°, γ = 120°.

(e) a = 4.47Å, b = 7.64Å, c = 11.32Å, $\alpha = \gamma$ = 90°, β = 108°.

10.2 *Atomic positions from fractional coordinates.*

(a) The crystals of α-$Ni(H_2O)_6SO_4$ are tetragonal; a = 6.790Å, c = 18.305Å. The nickel atom is surrounded by six oxygens (in two sets of three, related by symmetry.) The fractional coordinates are given below for the nickel and three oxygens. Calculate the atomic positions of these four atoms in Ångstroms from the origin.

	x	y	z
Ni	0.2101	0.2101	0.000
O(1)	0.1714	-0.0489	0.0518
O(2)	0.4720	0.2449	0.0564
O(3)	0.3564	0.0641	-0.0852

(b) The crystals of tetrasulfur tetranitride are monoclinic; a = 8.75Å, b = 7.16Å, c = 8.65Å, β = 92.5°. The fractional coordinates of the atoms of one molecule are given below. Calculate the atomic positions in Ångstroms.

	$x \times 10^4$	$y \times 10^4$	$z \times 10^4$
S(1)	106	9260	2996
S(2)	1500	7088	943
S(3)	-1488	7951	652
S(4)	-353	5535	2787
N(1)	70	7587	-238
N(2)	172	7309	3923
N(3)	-1761	6075	1597
N(4)	1663	8851	2083

10.3 *Interatomic distances.* The distance between any two points in a general three-dimensional coordinate system is given by

$$\ell = [(x_1-x_2)^2a^2 + (y_1-y_2)^2b^2 + (z_1-z_2)^2c^2$$
$$+ 2(x_1-x_2)(y_1-y_2)ab \cos\gamma$$
$$+ 2(y_1-y_2)(z_1-z_2)bc \cos\alpha$$
$$+ 2(z_1-z_2)(x_1-x_2)ca \cos\beta]^{1/2}$$

where x, y, and z are the fractional coordinates and a, b, and c are the corresponding unit cell dimensions.

(a) Use the data in 10.2 (a) to calculate the average Ni-O bond distance in α-$Ni(H_2O)_6SO_4$.

(b) Use the data in 10.2 (b) to calculate the S(1)-S(4) and S(2)-N(1) distances in S_4N_4.

10.4 *Unit cell volumes, crystal densities and numbers of molecules per unit cell.* The volume of a general parallelepiped with sides a, b, and c and angles α, between b and c; β, between a and c; and γ, between a and b is

$$V = abc(1 - \cos^2\alpha - \cos^2\beta - \cos^2\gamma + 2\cos\alpha\cos\beta\cos\gamma)^{1/2}$$

(a) Use the above formula along with the following data to calculate the number of molecules per unit cell in S_4N_4. Unit cell: monoclinic; a = 8.75Å,

b = 7.16Å, c = 8.65Å, β = 92.5°.

Density = 2.20 g/cm^3 (experimental)

(b) Bis-histidinenickel(II) monohydrate crystallizes in the orthorhombic system;

a = 15.18Å, b = 13.05Å, c = 7.72Å.

The density, ρ, is found experimentally to be 1.67 g/cm^3. How many molecules per unit cell are there?

10.5 *Choice of unit cell.* The basic criterion for choosing the conventionally acceptable unit cell is that the unit cell shall be the smallest parallelepiped that possesses all the symmetry of the lattice, such that it is repeated in three dimensions by translation alone. Additional, more or less arbitrary, conventions are established in each crystal class.

(a) One plane of a monoclinic lattice is shown below. Indicate the conventionally acceptable choice of unit cell axes and the origin.

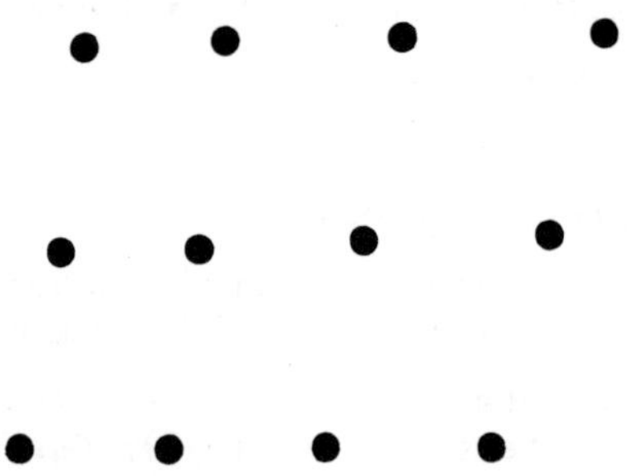

(b) In determining the crystal structure of $Zn(bipy)_3Br_2 \cdot 6H_2O$ the initial choice of parameters for the *hexagonal* unit cell was a = 7.77Å, c = 13.54Å. However, determination of the crystal density as 1.90 g/cm^3 led to the conclusion that this initial choice of unit cell parameters was wrong. Show why this is so. Use a sketch to show how this unit cell is related to an acceptable choice.

(c) For the illustrated lattice indicate both a hexagonal unit cell and an orthorhombic unit cell.

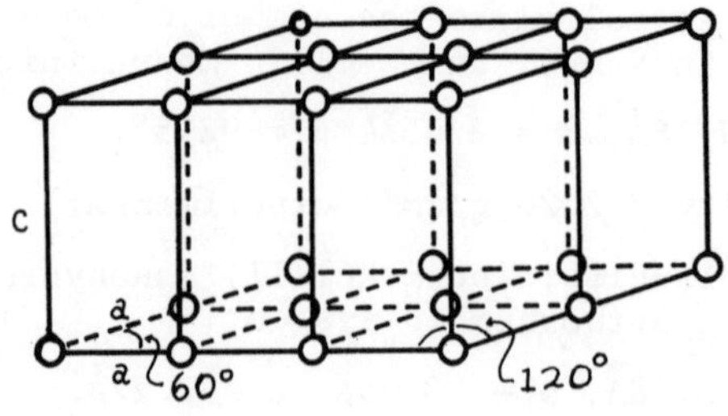

10.6 *Choice of unit cell.*

(a) Why are there no monoclinic space groups whose designations begin with B? Illustrate your answer.

(b) Why are there no tetragonal space groups whose designations begin with C? Illustrate your answer.

(c) Show that all the lattice points in the unit cell designated Fmmm are equivalent.

10.7 *Allowed crystallographic symmetries.* Although rotation axes of any order are allowed for molecular point groups, the translational periodicity and symmetry of a crystal lattice place strict limitations on the allowed values for crystallographic rotations. (An attempt to pack regular pentagons into a two-dimensional lattice with no gaps clearly indicates that 5-fold axes will not be allowed in either two- or three-dimensional lattices.) In fact, only 2-, 3-, 4-, and 6-fold rotation axes are allowed. Using simple trigonometry devise a proof of this statement.

10.8 *Improper axes of symmetry.* In Chapter 4 the rotation-reflection axes and operations, S_n, were treated. In discussion of crystal symmetry a different but equivalent set of axes and operations is used. These are designated $\bar{n}$ and are rotation *inversion* axes and operations. For some values of n, an S_n axis is identical in effect to a $\bar{n}$ axis but this is not usually so. Contrast the effects of the following:

S_4 vs $\bar{4}$, S_3 vs $\bar{3}$, S_6 vs $\bar{6}$, S_3 vs $\bar{6}$

10.9 *Hermann-Mauguin and Schoenflies notation.* Match the corresponding members of the two sets of point group symbols:

(a) m	(1) O_h
(b) mm2	(2) C_s
(c) m3m	(3) C_i
(d) $\bar{6}$m2	(4) C_{2v}
(e) $\bar{4}$	(5) D_{4h}
(f) 4/mmm	(6) D_3
(g) $\bar{1}$	(7) D_{3h}
(h) 32	(8) S_4

10.10 *Hermann-Mauguin notation.* For each of the crystallographic point groups in 10.9 interpret the Hermann-Mauguin symbol giving the orientation of all axes and planes.

10.11 *Notation for lines, points and planes in crystals.* (Read: Stout & Jensen, p. 21) Explain what is meant by each of the following:

(a) [010]

(b) 1/2, 1/4, 1/2

(c) (1 3 2)

(d) {1 1 1}

10.12 *Miller indices.* Determine the h, k and l values for planes having the following axis intercepts:

(a) 1, ∞, 1/2

(b) 1/3, 2/3, 1

(c) 3/4, 1/3, ∞

(d) -1/2, -1/3, ∞

(e) 1/4, 1/2, 1/2

10.13 *Bravais lattices.* By placing lattice points (that is, points whose surroundings differ only by translations) at the corners of the seven types of unit cells defining the seven crystal systems, we form the seven primitive lattices. There are seven other distinct three dimensional lattices involving only translations, which are derived from the primitive lattices by placing lattice points at the intersections of face and body diagonals. These fourteen are called the *Bravais* lattices. Not all types of centering give distinct new lattices; some are redundant and others symmetrically forbidden.

(a) How many lattice points are there in a C-centered orthorhombic unit cell?

(b) How many lattice points are there in the Bravais lattice I422?

(c) Give the symbol for a orthorhombic lattice with four lattice points per unit cell.

(d) Prove that all the lattice points in the lattice C2/m are all really equivalent points.

(e) Show that the two lattices C222 and P222 actually have the same point group symmetry.

10.14 *Glide planes and screw axes.* In addition to point symmetry, as exemplified by the 14 Bravais lattices, crystals also may possess various combinations of two other types of symmetry elements, both of which involve translation. The symbols and operations associated with each of these are as follows:

Glide plane; a, b, c; translation of a/2, b/2 or c/2 along a, b, or c followed by reflection in a plane containing a, b, or c. (Two other types of glide planes, n and d, involve translations along diagonals.)

Screw axis; n_p; rotation of $2\pi/n$ radians followed by translation of p/n in the direction of the axis.

Using the schematic diagram below illustrate the operation of the following:

(a) a two fold screw axis parallel to c

(b) a c-glide plane parallel to a

(c) a 4_1 axis parallel to c

(d) a 4_3 axis parallel to c

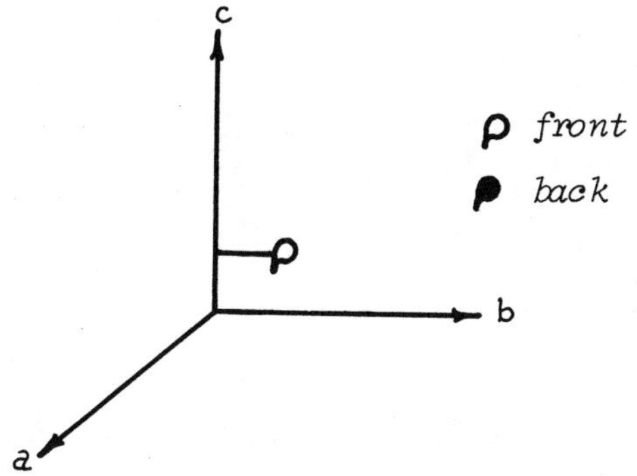

10.15 *Space groups.* The complete nonredundant set of all allowed combinations of the 32 crystallographic *point* groups with screw axes and glide planes comprises the 230 crystallographic *space* groups. The point group associated with a particular space group is found by replacing the indicated glide planes with mirror planes and the screw axes with simple axes and dropping the P,C,I, or F.

What are the point groups associated with the following space groups?

(a) I $2_1 2_1 2_1$

(b) Cc

10.16 *International Tables for X-ray Crystallography.* Volume I of *The International Tables for X-ray Crystallography* gives the properties of the 230 space groups. One of the most common of these is $P2_1/c$, No. 14, page 99 of Volume I. Using the key to the interpretation of the diagrams which is found on pages 47-50 of Volume I, answer the following:

(a) Where is the origin and how are the axes oriented in the diagrams?

(b) What does the symbol ⟶ 1/4 mean?

(c) How many equivalent general positions are there?

(d) Show how all the equivalent positions can be generated from any one by application of the symmetry elements.

(e) Show that no other points are generated by any combination of the symmetry operations than those given in the diagram.

10.17 *International Tables Symbolism.*

(a) Describe the symmetry elements indicated by the symbols in the diagram for the space group Pbca.

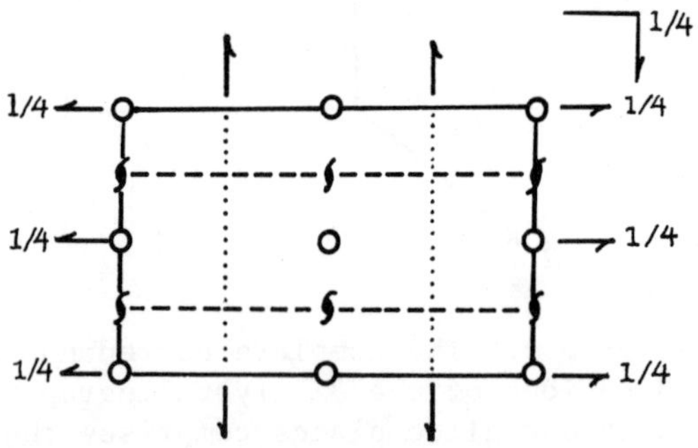

(b) Show how the eight equivalent positions are related by the symmetry operations.

(c) Show that the position (0,0,1/2) is a special position.

10.18 *Equivalent general and special positions.*

(a) The space group of $NPCl_2(NSOCl)_2$ is $P2_1/_n$; Z = 4. The four molecules are found to be in general positions. The coordinates of one of the phosphorus atoms are $x = -0.1701$, $y = 0.1539$, $z = 0.0514$. What are the coordinates of the other three phosphorous atoms? (*Acta Cryst.*, B25, 651 (1969)).

(b) The space group of bis-(diethyldithiophosphato)-nickel(II) is $P2_1/c$. The nickel atoms are found in special positions *a* (Wyckoff notation).

(1) How many nickel atoms are there per unit cell?

(2) What is the site symmetry of the nickel atoms?

(3) How many sulfur atom positions need to be specified in order to find where all the sulfur atoms are?

10.19 *Equivalent positions.* The space group of $Os(CO)_3\{P(C_6H_5)_3\}_2$ is $P\bar{3}c1$. The six osmium atoms per unit cell occupy two sets of positions, *a* and *d* (as designated by the *International Tables*).

(a) What is the minimum symmetry about each of the osmium atoms?

(b) How many crystallographically independent carbon atoms are there per molecule? How many independent carbon atoms are there per unit cell?

(c) What is the symmetry of the coordination sphere about each of the osmium atoms?

10.20 *Use of the International Tables.*

(a) Describe all the symmetry elements in the diagram on page 257 of Volume I of the *International Tables* for space group $P3_121$. Be sure to specify the handedness of screw axes.

(b) Generate the general positions of this space group.

(c) Show that $\bar{x}$, $\bar{x}$, 1/2 is a special position.

10.21 *Use of International Tables.* If $Zn(bipy)_3(IO_4)_2$ crystallizes in space group $P\bar{3}$ and there are two molecules per unit cell with the zinc atoms at 1/3, 2/3, 1/4 and 2/3, 1/3, -1/4, what is the minimum symmetry which the $Zn(bipy)_3^{2+}$ ion may have?

10.22 *Enantiomorphic Crystals.*

(a) Which of the following space groups characterize optically active crystals?

(1) $P2_1/c$

(2) $P4_12_12$

(3) $P3_12$

(4) C2

(b) What are the crystallographic characteristics of an "optically active" space group? Which are the crystal classes (point groups) to which all optically active crystals must belong?

(c) For which of those "optically active" space groups in (a) do enantiomorphic space groups with separate designations exist? Give the symbols of the enantiomorphic space groups.

10.23 *Pyro-electric crystals.*

(a) What are the characteristics of crystals which exhibit pyro-electricity?

(b) In the determination of a certain crystal structure the initial data left the choice of space groups undecided between C2 and C2/c. The crystals gave a positive pyro-electric effect. Which is the correct choice of space group?

(c) If in (b) the pyro-electric test had been negative, what would be the conclusion?

10.24 *Origin and properties of x-rays.*

(a) The wavelength of the $K\alpha$ radiation of an element is always longer than that of $K\beta$. Why is this so?

(b) Which element is normally used to filter out $K\beta$ radiation from a copper source? What is the general relationship between the atomic number of the source and the atomic number of the correct filter element?

(c) Which of the following elements in a crystal sample will cause serious absorption and fluorescence problems when determining the crystal structure with copper $K\alpha$ radiation: nickel, zinc, or iron?

10.25 *Von Laue's equations.*

(a) For the one-dimensional array of points illustrated below derive the equation relating the wave length of the x-rays λ, the spacing of the points a, the angle of incidence α_0 and the angle of emergence of the diffracted beam α (the angle where constructive interference of the radiation scattered by the points occurs).

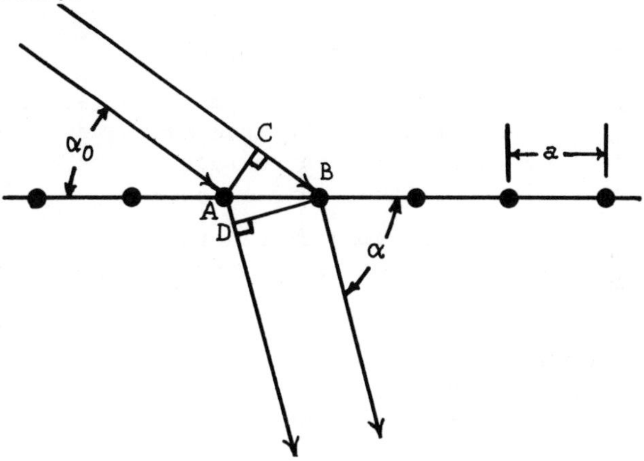

(b) Assume $\alpha_0 = 60°$, a = 4.0Å and λ = 0.50Å. Calculate α. Is more than one value possible?

(c) Assume $\alpha_0 = 30°$, a = 0.80Å and α must be *less than* 60°. What is the maximum value of λ allowed?

10.26 *Rotation photographs.* In determining cell dimensions and in aligning crystals for Weissenberg photographs, rotation photographs are taken. In this procedure the crystal is aligned in the camera so that it can be rotated about one of its axes. The x-ray beam is directed perpendicular to the rotation axis. The results of the experiment can be analyzed in terms of von Laue's equations. To predict the pattern of diffraction we consider the rows of atoms which lie in successive planes perpendicular to the rotation axis. All the diffracted beams will have to satisfy the diffraction conditions for a single such row of spots, which will have the spacing corresponding to the fundamental lattice translation along the rotation axis --*e.g.*, the a-dimension of the unit cell. Actually, of course, the analogous conditions related to b and c will also have to be satisfied and so it will only be at specific angles of rotation of the crystal that diffracted beams will emerge. Nevertheless, considering only the *a* condition will allow us to predict the pattern.

(a) Make a general sketch of this pattern.

(b) What will be the result if the angle of incidence is not 90°?

(c) Sketch the sort of pattern which the diffracted beams in (a) would make on a film held cylindrically about the rotation axis.

10.27 *Analysis of rotation photographs.*

(a) From the answer to problem 10.26 (a) and (c) derive the equation for the spacing between the lines of spots on the rotation photograph produced by rotating a crystal about its *a* axis. Assume $\alpha_0 = 90°$.

(b) From the following data determine the unit cell dimension along *a* of tris-(diethyldithiophosphato)-indium(III)
Rotation axis, *a*
X-ray wave-length $\lambda = 1.5405Å$
Diameter of film cylinder: 57.3 mm
Positions of rows of spots (as indicated on a cm scale laid over the film)

h = 3	8.20 cm
h = 2	8.90
h = 1	9.48
h = -1	10.52
h = -2	11.09
h = -3	11.83

10.28 *Bragg's Law.* Bragg's Law says that, when a set of parallel planes in a crystal are struck by an x-ray beam, a *diffracted* beam will emerge at the same angle as would a beam *reflected* by any one of the planes in the set. That is, the angle of incidence will equal the angle of "reflection".

(a) Use a geometrical construction to derive the relation between the angle of incidence and "reflection" θ, the wave length λ, and the interplanar spacing *d*. That is, derive Bragg's Law assuming the reflection analogy is correct. Note: the actual proof that the results of diffraction by a three dimensional array of points can be predicted by Bragg's reflection analog is not at first obvious. A concise demonstration of the proof is given by Stout and Jensen, Appendix B.

10.29 *Use of Bragg's Law.* For the following, assume λ = 1.5405Å.

(a) The space group of $Cu_2Cr(CN)_6$ is Fm3m and a = 10.24Å. Calculate θ for the following planes:

(1) (1 2 0)

(2) (1 3 2)

(b) The space group of BiSCl is Pnam and a = 7.70Å, b = 9.87Å, c = 4.00Å. Calculate θ for the following "reflections":

(1) (1 1 0)

(2) (3 1 1)

(c) The space group of $NaLiCO_3$ is P62m and a = 8.22Å, c = 3.27Å. Calculate θ for the following planes:

(1) (1 1 1)

(2) (2 1 0)

10.30 *Construction of crystal models.* One useful way to construct three dimensional crystal models is the following:

(a) Draw the a-b plane of the unit cell in a suitable scale (say, 1Å : 1 inch depending on the size of the cell) on a sheet of graph paper. For monoclinic cells use the a-c plane.

(b) Tape or glue this paper to the surface of a piece of urethane foam -- the tough resiliant type used in packing heavy objects is best.

(c) At the xy coordinates of each atom in the unit cell (xz for monoclinic) insert a suitable length -- say 18 inch -- stainless steel, 1/16 inch diameter, rod perpendicular to the plane of the paper (except for triclinic cells where the angles α and $\gamma \neq 90°$).

(d) Position cork or styrofoam balls or beads, color coded to distinguish different elements, on each of the rods at the correct values of z (y for monoclinic). (Cork balls drilled just under 1/16 inch are very easy to handle this way.)

In this way a model can fairly quickly be constructed using the data commonly given in the report of any new crystal structure. (That is, a, b, c, α, β, γ, the space group of the unit cell and x, y, and z for each independent atom.)

The positions of symmetrically equivalent molecules and atoms can be obtained from the essential data by the symmetry operations (or by using the equivalent positions listed in the *International Tables*.)

Use this method or a modification of it (depending on available materials) to construct a model of the unit cell of S_4N_4 from the data in problems 10.2 and 10.4.

SOLUTIONS

10.1 (a) orthorhombic, (b) rhombohedral, (c) triclinic (a = b accidentally, not by symmetry), (d) hexagonal, (e) monoclinic

10.2 The coordinates of the atoms in Ångstrom units are simply the products of the fractional coordinates x, y and z with the unit cell dimensions a, b, and c respectively.

(a)

	x(Å)	y(Å)	z(Å)
Ni	1.43	1.43	0.00
O(1)	1.164	-0.332	0.948
O(2)	3.19	1.662	1.032
O(3)	2.42	0.435	-1.56

(b)

	x(Å)	y(Å)	z(Å)
S(1)	0.0927	6.63	2.586
S(2)	1.312	5.15	0.815
S(3)	-1.300	5.70	0.564
S(4)	0.309	3.96	2.41

10.3 (a) For a tetragonal crystal the formula for ℓ reduces to:

$$\{[(x_1 - x_2)^2 + (y_1 - y_2)^2]\ a^2 + (z_1 - z_2)\ c^2\}^{1/2}$$

The results are

Ni-O(1)	2.02Å
Ni-O(2)	2.07Å
Ni-O(3)	2.10Å

(b) Only the terms containing cos α and cos γ drop out from the general expression for ℓ in a monoclinic crystal. The S(1)-S(4) distance is 2.702Å.

10.4 (a) For S_4N_4 $V = (8.75)(7.16)(8.65)(1 - \cos^2 92.5°)^{1/2}$

$= 529 \times 10^{-24}\ \text{cm}^3.$

The mass in grams per unit cell is given by V x ρ. Dividing this by the mass in grams of one S_4N_4 molecule gives Z, the number of molecules per unit cell

$$Z = \frac{V \times \rho}{M/N}$$

$$Z = \frac{(529) \times (2.20) \times 10^{-24}}{184.28 \times 1.664 \times 10^{-24}} = 3.79 \simeq 4$$

(b) Z = 4

10.5 (a) For the monoclinic system *b* is conventionally chosen as the unique axis. Therefore, the projection must be of the a - c plane since the angles are ≠ 90°. A right hand coordinate system is always used and for the monoclinic system the unit cell is chosen so that β is as near to 90° as possible and >90°. The conventional cell then is as illustrated.

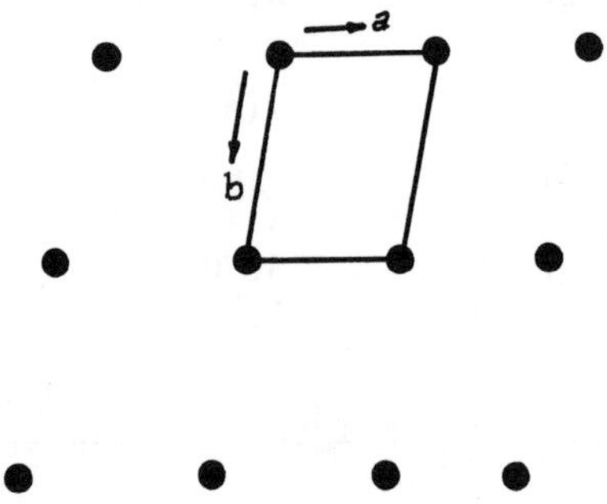

b perpendicular to *a* and *c*

(b) The initial choice can be shown to contain only 2/3 molecule (formula unit) per unit cell (see problem 10.4). The proper unit cell must contain an integral number of formula units.

(c)

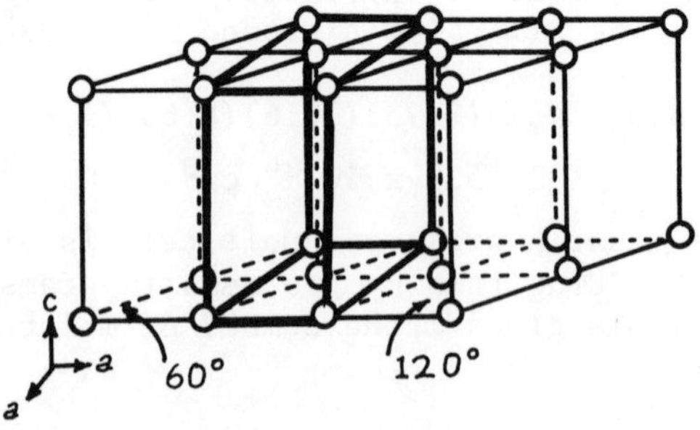

The orthorhombic cell is c-centered.

10.6 (a)

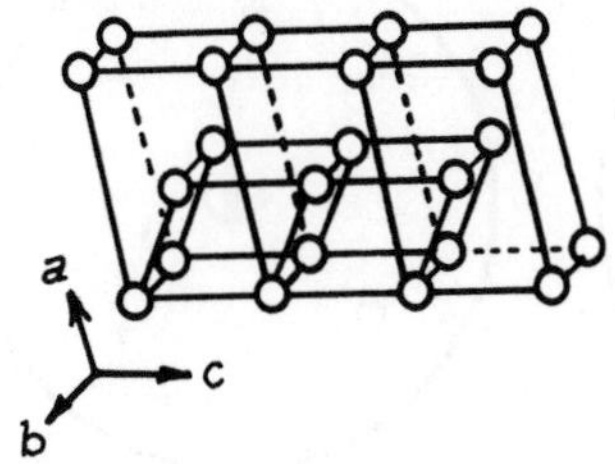

A primitive cell of 1/2 the volume can be chosen. This is preferable to the b-centered cell.

(b)

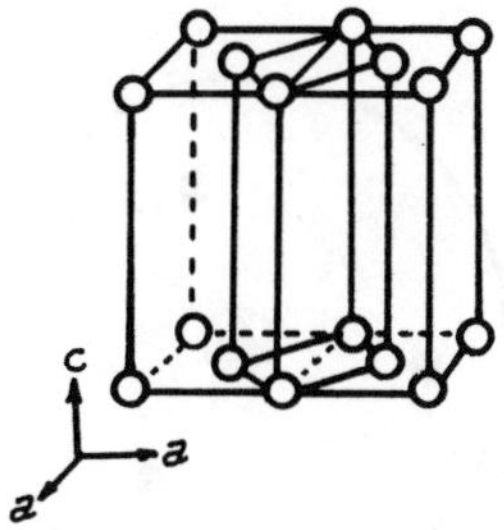

A primitive cell in a different orientation can always be chosen.

10.7 See Sands, page 53.

10.8 This is most easily done by means of stereographic projections. In these projections the axis is perpendicular to the plane of the paper. Points above the plane of the paper are indicated by + and below by O. All the equivalent points are generated by successive application of the operation S_n (or $\bar{n}$) on an arbitrary general point.

(a)

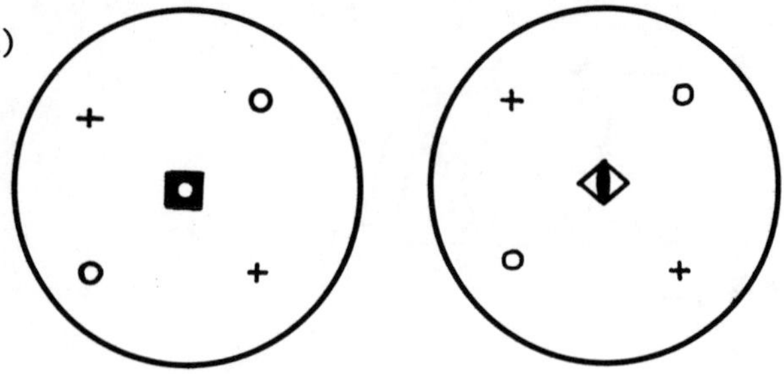

$\bar{4}$ S_4

$S_n = \bar{n}$ if $n = 4m$ where m is a positive integer.

(b)

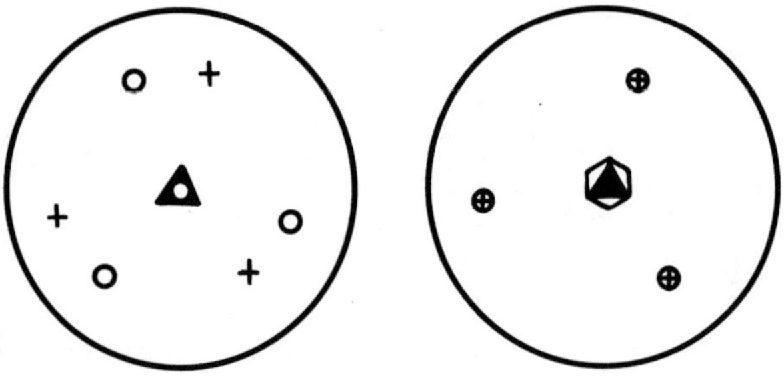

$\bar{3}$ S_3

10.9 (a) 2, (b) 4, (c) 1, (d) 7, (e) 8, (f) 5, (g) 3, (h) 6.

10.10 (a) A mirror plane only. Since this is a monoclinic group, the mirror plane must be perpendicular to y.

(b) A mirror plane perpendicular to x, a mirror plane perpendicular to y and a two-fold axis parallel to z.

(c) A 3 in the second position always denotes a cubic point group, *i.e.* one of the cubic classes. In this one there is a mirror perpendicular to the cube axis z, a set of four three-fold axes along the body diagonals of the cube and a set of six mirrors perpendicular to the face diagonals of the cube. Note: An additional exercise would be to show how these ten elements (one of the mirrors is redundant) generate the other elements which are involved in the forty-eight operations of the octahedral point group.

(d) A six-fold rotation inversion axis along *z*, three mirror planes containing *z* with a dihedral angle of 60° and three two-fold axes perpendicular to *z* bisecting the angles between the planes.

(f) A four-fold axis along *z* with a mirror plane perpendicular to it (4/m), mirror planes perpendicular to *x* and *y* and mirror planes perpendicular to (x + y) and (x - y).

10.12 (a) (102), (b) (632), (c) (490), (d) (230), (e) (422).

Note: These are general Miller indices. In their original form, as used by pre - x-ray crystallographers, Miller indices were always rationallized, that is, they contained no common factor. For example, (422) would have been considered equivalent to (211). Though useful in describing crystal faces, such a restriction is not practical in x-ray crystallography. Since the (422) set of planes has a different d-spacing from the (211) set, their x-ray diffraction properties are quite distinct and they must be considered as separate entities.

10.13 (a) Two. Each point at a corner is shared by eight unit cells and each point on a face is shared by two cells. Therefore, 8 x 1/8 + 2 x 1/2 = 2.

(b) Body centered cells (*innerzentrein*) always have two lattice points per unit cell.

(c) Fmmm or Fddd.

10.14 (a)

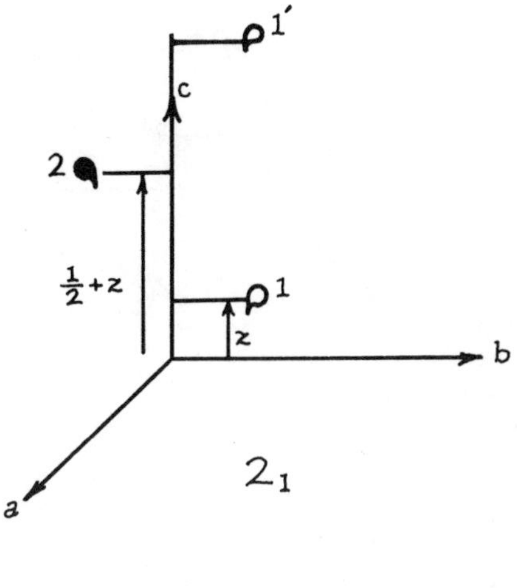

(b)

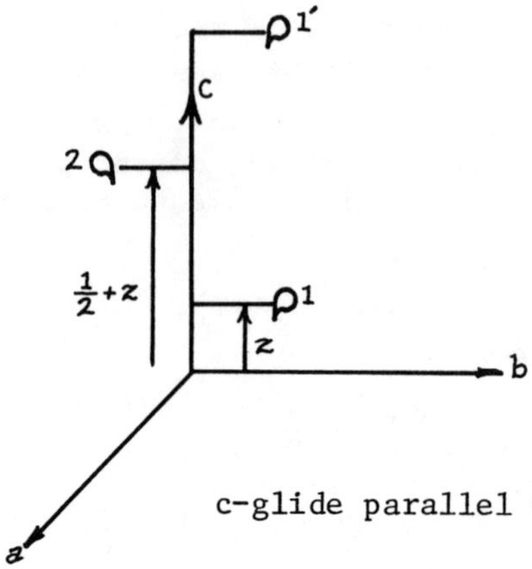

c-glide parallel to a

(c)

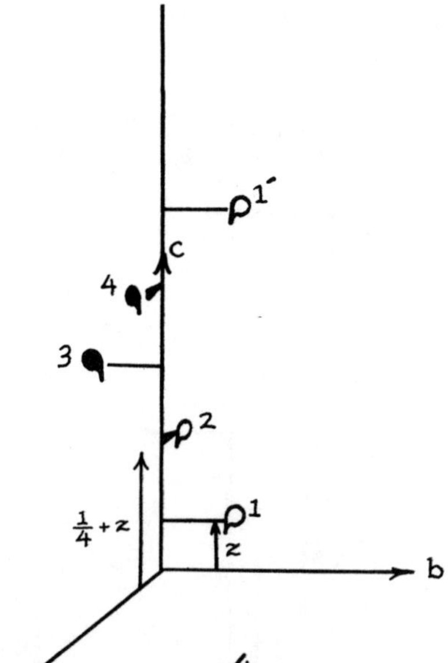

4_1

(d)

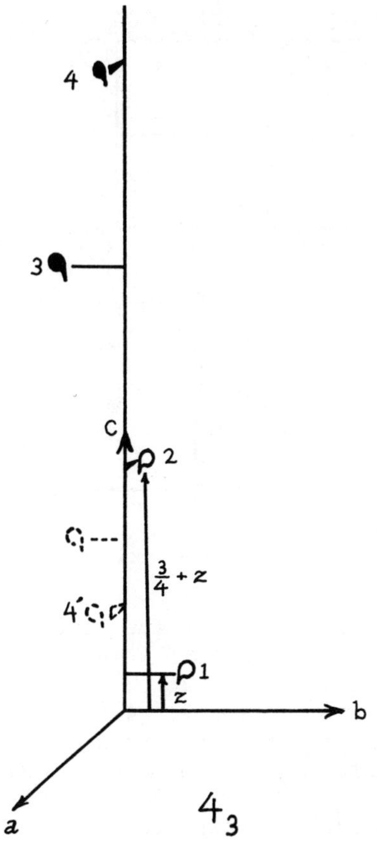

10.15 (a) 222, (b) m, (c) 3

10.16 (a) The origin is in the upper left corner always.

(b) ⟶1/4 means a two-fold screw axis along the direction indicated intersecting the *z*-axis at 1/4 *c*

(c) 4

(d)

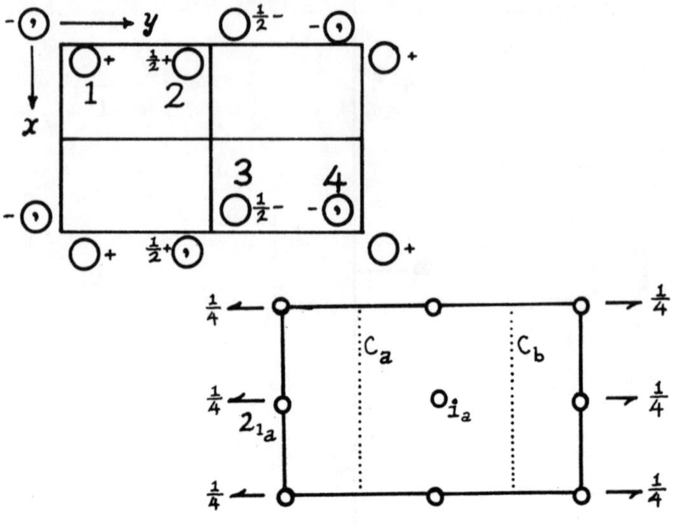

Origin at $\bar{1}$; unique axis b

From 1, 2 is generated by the glide plane c_a or by a combination of the two-fold screw axis 2_{1a} and the inversion center i_a. From 2, 3 is generated by the inversion center i_a. From 1, 4 is generated by the inversion center i_a or from 2 by a combination of the inversion center i_a and the glide plane c_b.

10.17 (a) The symmetry elements are

(1) ( ) 2-fold screw axes parallel to c at $x = 1/4$, $y = 0$; $x = 1/4$, $y = 1/2$; $x = 1/4$, $y = 1$; $x = 3/4$, $y = 0$; $x = 3/4$, $y = 1/2$; and $x = 3/4$, $y = 1$.

(2) (⟶ 1/4) 2-fold screw axes in the $z = 1/4$ plane parallel to y at $x = 0$, 1/2, and 1.

(3) ( ) 2-fold screw axes in the $z = 0$ plane parallel to x at $y = 1/4$ and 3/4.

(4) (o) inversion centers in the $z = 0$ plane at $x = 1/2$, $y = 1/2$ and at the corners and at the midpoints of each edge.

(5) (⋮) glide planes perpendicular to y at $y = 1/4$ and 3/4 with a translation of $c/2$ along z axis.

(6) (---) glide planes perpendicular to x at

$x = 1/4$ and $3/4$ with a translation of $b/2$ along the y axis.

(7) (⌐ 1/4) a glide plane perpendicular to the z axis at $z = 1/4$ with a translation of $a/2$ along the x axis.

(b)

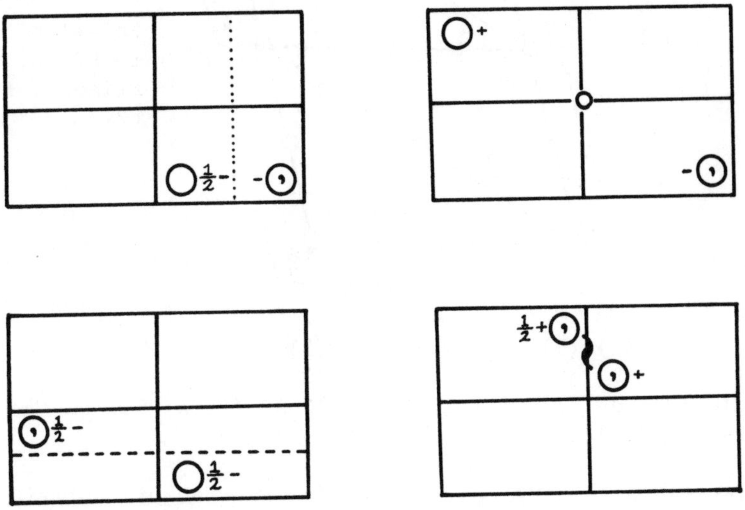

(c) Special positions are those which lie on a symmetry element and thus have fewer equivalent positions than do general positions (such as those illustrated in (b). The position (0,0,1/2) is a center of symmetry. (The center is in effect generated by the action of the 2-fold screw axis at $x = 1/4$, $y = 0$, on the center at $x = 1/2$, $y = 0$). The doubling of the equivalent positions due to the operation of inversion at (0,0,1/2) is lost if the initial position chosen is at the center.

10.18 (a) $P2_1/n$ differs from $P2_1/b$ only in the orientation of the c axis. In either case the general positions are at x, y, z; $\bar{x}$, $\bar{y}$, $\bar{z}$; $\bar{x}$, $1/2 - y$, $1/2 + z$; x, $1/2 + y$, $1/2 - z$. Therefore, the other three sulfurs are at:

x	y	z
0.1701	-0.1539	-0.0514
0.1701	0.3461	0.5514
-0.1701	0.6539	0.4486

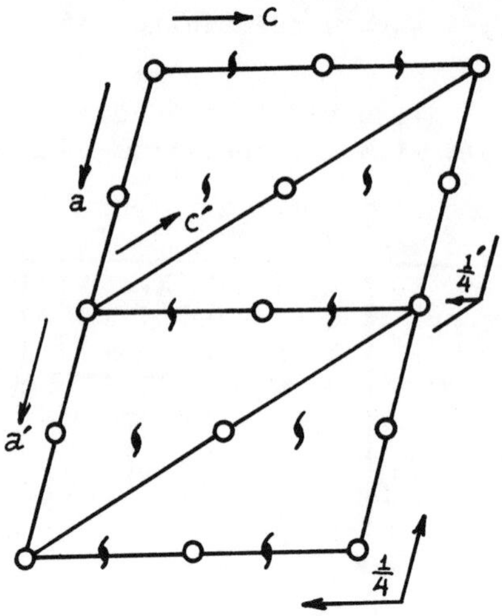

Relation of $P2_1/b$ to $P2_1/n$. Primes refer to $P2_1/n$.

(b) (1) 2, (2) i, (3) 2

10.19 See J. K. Stalick and J. A. Ibers, *Inorg. Chem.*, 8, 419 (1969).

10.21 C_3. (What is this in Hermann-Mauguin notation?)

10.22 (a) $P4_12_12$, tetragonal
$P3_12$, trigonal
C2, monoclinic

(b) Optically active crystals must belong to space groups which have no centers of symmetry or mirror planes. Of the 32 crystal classes only eleven fit this description. (See Buerger, pages 190-191)

(c) $P4_12_12$ and $P4_32_12$
$P3_12$ and $P3_22$

10.23 (a) Pyro-electric crystals must belong to space groups lacking a center of symmetry.

(b) C2 since a positive effect indicates that there is no center of symmetry.

(c) Negative results are inconclusive since the non centricity may be small and the charge effect unmeasurable, though finite.

10.24 (a) $K\alpha$ radiation results from the dropping of electrons from the L shell to the K shell, whereas $K\beta$ radiation is associated with a M→K transition.

(b) nickel

(c) iron

10.25 In order for diffracted intensity to be observed, rays scattered from the points must be parallel and in phase. Now, since radiation is scattered in all directions by each point, parallel rays will be found at any angle. However, only at certain angles will parallel rays be in phase. The path lengths of rays scattered by different points will differ. Only if they differ by an integral number of wave lengths will they be in phase. For example, for the points A and B in the figure, CB - AD must equal $n\lambda$. From basic trigonometry then the requirement is that $a(\cos\alpha_0 - \cos\alpha) = n\lambda$, or $\cos\alpha = \cos\alpha_0 - (n/a)\lambda$, where n is any positive or negative integer or zero.

(b) $\cos 60° = 0.50$

$$\cos\alpha = 0.50 - (n/4.0)(0.50)$$
$$= 0.50 - (1/8)n$$

n = 0 is always a solution, but gives a diffracted beam parallel to the incident beam, which isn't detectable. Other acceptable solutions include

n = 1	$\alpha = 67°58'$
n = -2	$\alpha = 41°25'$
n = 6	$\alpha = 104°29'$

10.26 (a)

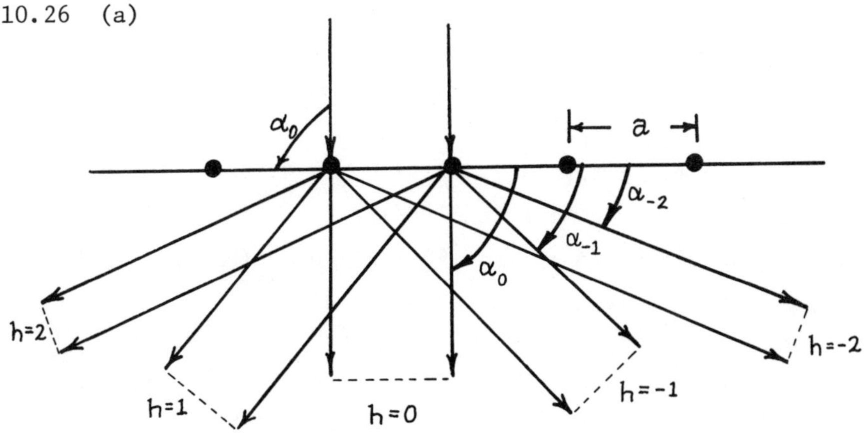

The locus of lines parallel to allowed diffraction directions is a set of coaxial cones whose half vertical angles are the allowed values of α for $h>0$ and $180 - \alpha$ for $h<0$. Since the real crystal is a three-dimensional array, the actual diffracted beams will appear only where these cones simultaneously intercept the allowed solutions of:

$$b(\cos \beta_0 - \cos \beta) = k\lambda$$
$$c(\cos \gamma_0 - \cos \gamma) = \ell\lambda.$$

Rotation of the crystal about the axis brings about many such coincidental events all of which will produce diffracted beams along the surfaces of the cones.

(b) The clue to visualizing the pattern produced when $\alpha_0 \neq 90$ is given by calculating the angles $\alpha_{\pm 1}$.

10.27 (a)

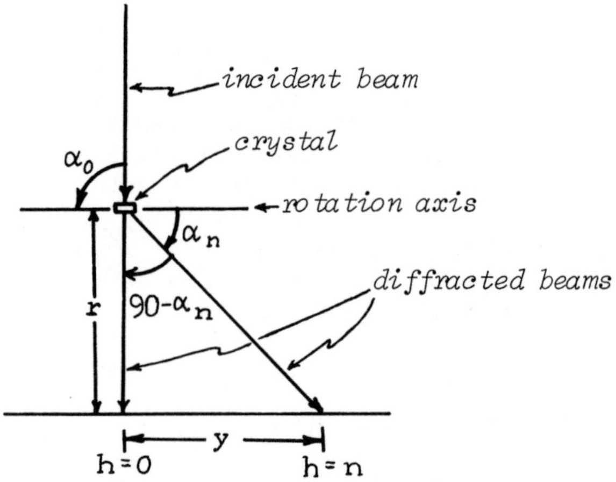

$a(\cos \alpha_0 - \cos \alpha) = -n\lambda \qquad \alpha_0 = 90°$

$a = -n\lambda/\cos \alpha$

$\alpha = \tan^{-1} r/y$

$\therefore a = -(n\lambda/\cos \tan^{-1} r/y)$

or since $\sin \alpha = \cos(90 - \alpha)$, $a = -(n\lambda/\sin \tan^{-1} y/r)$

where r is the radius of the cylinder of film and y is the distance from the $h = 0$ row of spots to the row $h = n$. (Note: a will always be positive. In the diagram above, for example, $n<0$ and $\cos \alpha>0$. When $n>0$, $\cos \alpha$ will be <0. Thus the minus sign will always cancel.)

(b) From the data $y_1 = 10.4\text{mm}/2 = 5.2\text{mm}$
$y_2 = 21.9\text{mm}/2 = 10.95\text{mm}$
$y_3 = 36.3\text{mm}/2 = 18.15\text{mm}$

$$a_1 = \frac{1.5405}{\sin \tan^{-1} 5.2/28.65} = \frac{1.5405}{\sin 10°17'} = \frac{1.5405}{0.1785} = 8.64\text{Å}$$

$$a_2 = \frac{2(1.5405)}{\sin \tan^{-1} 10.95/28.65} \; \frac{3.0810}{\sin 20°54'} = \frac{3.0810}{0.3567} = 8.64\text{Å}$$

$$a_3 = \frac{3(1.5405)}{\sin \tan^{-1} 18.15/28.65} \; \frac{4.6215}{\sin 32°20'} = \frac{4.6215}{0.5348} = 8.64\text{Å}$$

10.28 Just as in 10.22 the key concept is that, to be in phase, the path lengths of two rays must differ by an integral number of wave lengths.

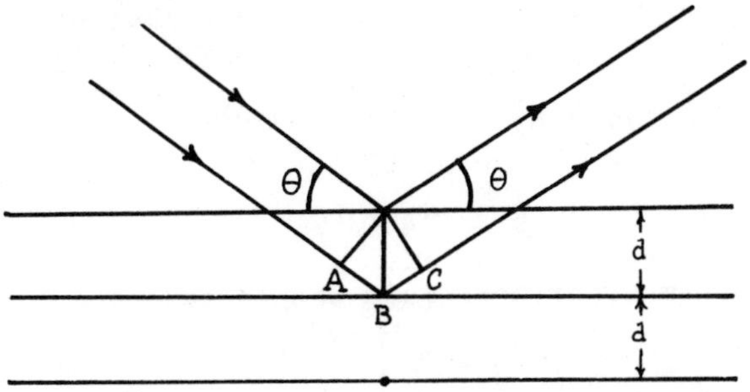

Thus AB + CD = $n\lambda$ is the condition and $\therefore$ $2d \sin \theta = n\lambda$ is the required relationship.

10.29 The general formula for the perpendicular distance between two planes d is

$$\frac{1}{d_{(hkl)}} = V[h^2b^2c^2 \sin^2\alpha + k^2a^2c^2 \sin^2\beta + \ell^2a^2b^2 \sin^2\gamma$$
$$+ 2h\ell ab^2c(\cos\alpha \cos\gamma - \cos\beta)$$
$$+ 2hkabc^2(\cos\alpha \cos\beta - \cos\gamma)$$
$$+ 2k\ell a^2cb(\cos\beta \cos\gamma - \cos\alpha)]^{1/2}$$

Where the volume of the unit cell

$$V = abc(1 - \cos^2\alpha - \cos^2\beta - \cos^2\gamma$$
$$+ 2\cos\alpha \cos\beta \cos\gamma)^{1/2}$$

(a) When $\alpha = \beta = \gamma = 90°$ the equation for d reduces to:

$$\frac{1}{d_{(hkl)}} = \left[\frac{h^2}{a^2} + \frac{k^2}{b^2} + \frac{\ell^2}{c^2}\right]^{1/2}$$

For cubic crystals this is simply

$$\frac{1}{d_{(hkl)}} = \left[\frac{h^2 + k^2 + \ell^2}{a^2}\right]^{1/2}$$

(1) $$\frac{1}{d_{(1\ 2\ 0)}} = \left[\frac{5}{105}\right]^{1/2} = 0.0476^{1/2} = 2.18 \times 10^{-1}$$

$$d = 4.58\text{Å}$$

Bragg's Law states that $\sin\theta = n\lambda/2d$; or if we allow the index n to be absorbed, $\sin\theta = \lambda/(2d/n) = \lambda/2d'$

$$\therefore \sin\theta_{(1\ 2\ 0)} = 1.5405/2(4.58)$$

$$= 0.168$$

$$\theta = 9°40'$$

(b) Pnam is orthorhombic and therefore $\alpha = \beta = \gamma = 90°$

$$\frac{1}{d} = \left[\frac{h^2}{a^2} + \frac{k^2}{b^2} + \frac{\ell^2}{c^2}\right]^{1/2}$$

$$\theta_{(1\ 1\ 0)} = 7°17'$$

(c) $P\bar{6}2m$ is hexagonal, $\gamma = 120°$

$$\theta_{(1\ 1\ 1)} = 17°27'$$

APPENDICES

Appendix A

Defined Values for Physical Constants

Meter (m)	1,650,763.73 wave lengths in vacuo of the unperturbed transition $^2P_{10} - {}^5D_5$ in 86_{Kr}.
Kilogram (kg)	mass of the international kilogram at Sèvres, France
Second (sec)	1/31,556,925.974,7 of the tropical year at 12^h ET, 0 January 1900
Degree Kelvin (°K)	defined in the thermodynamic scale by assigning 273.16° to the triple point of water (freezing point, 273.15°K = 0°C)
Unified atomic mass unit (u)	1/12 the mass of an atom of the ^{12}C nuclide
Mole (mol)	amount of substance containing the same number of atoms as 12g of pure ^{12}C
Standard acceleration of free fall (g_n)	9.806,60 m sec^{-2}, 980.665 cm sec^{-2}
Normal atmospheric pressure (atm)	101,325 N m^{-2}, 1,013,250 dyn cm^{-2}
Thermochemical calorie (cal_{th})	4.1840 J, 4.1840 x 10^7 erg
International Steam Table calorie (cal_{IT})	4.1868 J, 4.1868 x 10^7 erg
Liter (1)	0.001,000,028 m^3, 1,000.028 cm^3 (recommended by CIPM,1950)
Inch (in)	0.0254 m, 2.54 cm
Pound (avdp)	0.453,592,37 kg, 453.592,37 g

Appendix B

Nuclear Spins and Moments

Nucleus	Abundance %	I	g_N	$Q(10^{-24}cm^2)$
^{1}H	99.9844	1/2	5.585	0
^{2}D	0.0156	1	0.857	0.00274
^{7}Li	92.57	3/2	2.171	0.02
^{11}B	81.17	3/2	1.792	0.0355
^{13}C	1.108	1/2	1.405	0
^{14}N	99.635	1	0.403	0.02
^{15}N	0.365	1/2	-0.567	0
^{17}O	0.037	5/2	-0.757	-0.0265
^{19}F	100	1/2	5.257	0
^{23}Na	100	3/2	1.478	+1.0 or -.836
^{29}Si	4.70	1/2	-1.111	0
^{31}P	100	1/2	2.263	0
^{33}S	0.74	3/2	0.429	-0.064
^{35}Cl	75.4	3/2	0.548	-0.079
^{37}Cl	24.6	3/2	0.456	-0.062
^{39}K	93.08	3/2	0.261	0.113
^{47}Ti	7.32	5/2	-0.3153	-----
^{49}Ti	5.46	7/2	-0.3154	-----
^{51}V	99.8	7/2	1.471	+ 0.25
^{53}Cr	9.55	3/2	-0.3163	-----
^{55}Mn	100	5/2	1.387	+0.3
^{57}Fe	2.21	1/2	<.10	0
^{59}Co	100	7/2	1.328	+0.5
^{63}Cu	69.1	3/2	1.484	-0.16
^{65}Cu	30.9	3/2	1.590	-0.15

Appendix C

Conversion Factors for Energy Units

	electron volts	cal/mole	ergs/molecule	cm^{-1}
1 electron volt	1	2.3053×10^{4}	1.60203×10^{-12}	8.0675×10^{3}
1 cal/mole	4.3379×10^{-5}	1	6.9494×10^{-17}	0.34996
1 erg/molecule	6.2421×10^{11}	1.4390×10^{16}	1	5.0358×10^{15}
1 cm^{-1}	1.2395×10^{-4}	2.8575	1.9858×10^{-16}	1

Appendix D

Periodic Table of the Elements

I	II	Transition Elements										III	IV	V	VI	VII	0
H																	He
1																	2
Li	Be											B	C	N	O	F	Ne
3	4											5	6	7	8	9	10
Na	Mg											Al	Si	P	S	Cl	Ar
11	12											13	14	15	16	17	18
K	Ca	Sc	Ti	V	Cr	Mn	Fe	Co	Ni	Cu	Zn	Ga	Ge	As	Se	Br	Kr
19	20	21	22	23	24	25	26	27	28	29	30	31	32	33	34	35	36
Rb	Sr	Y	Zr	Nb	Mo	Tc	Ru	Rh	Pd	Ag	Cd	In	Sn	Sb	Te	I	Xe
37	38	39	40	41	42	43	44	45	46	47	48	49	50	51	52	53	54
Cs	Ba	*	Hf	Ta	W	Re	Os	Ir	Pt	Au	Hg	Tl	Pb	Bi	Po	At	Rn
55	56		72	73	74	75	76	77	78	79	80	81	82	83	84	85	86
Fr	Ra	**															
87	88																

* Lanthanides	La	Ce	Pr	Nd	Pm	Sm	Eu	Gd	Tb	Dy	Ho	Er	Tm	Yb	Lu
	57	58	59	60	61	62	63	64	65	66	67	68	69	70	71
** Actinides	Ac	Th	Pa	U	Np	Pu	Am	Cm	Bk	Cf	Es	Fm	Md	No	Lr
	89	90	91	92	93	94	95	96	97	98	99	100	101	102	103

Appendix E

Selected Fundamental Constants

Constant	Symbol	Value	Est.†† error limit	Centimeter-gram-second	
Speed of light in vacuum	c	2.997925	3	10^{10}	$cm\ sec^{-1}$
Elementary charge	e	1.60210	7	10^{-20}	$cm^{1/2}g^{1/2}$*
		4.80298	20	10^{-10}	$cm^{3/2}g^{1/2}sec^{-1}$†
Avogadro constant	N	6.02252	28	10^{23}	mol^{-1}
Electron rest mass	m_e	9.1091	4	10^{-28}	g
		5.48597	9	10^{-4}	u
Proton rest mass	m_p	1.67252	8	10^{-24}	g
		1.00727663	24	10^{0}	u
Neutron rest mass	m_n	1.67482	8	10^{-24}	g
		1.0086654	13	10^{0}	u
Faraday constant	F	9.64870	16	10^{3}	$cm^{1/2}g^{1/2}mol^{-1}$*
		2.89261	5	10^{14}	$cm^{3/2}g^{1/2}sec^{-1}mol^{-1}$†
Planck constant	h	6.6256	5	10^{-27}	erg sec
	$\hbar$	1.05450	7	10^{-27}	erg sec

Charge-to-mass ratio for electron	e/m_e	1.758796	19	10^{7}	$cm^{1/2}g^{-1/2}$*
		5.27274	6	10^{17}	$cm^{3/2}g^{-1/2}sec^{-1}$†
Rydberg constant	R_∞	1.0973731	3	10^{5}	cm^{-1}
Bohr radius	a_0	5.29167	7	10^{-9}	cm
Electron radius	r_e	2.81777	11	10^{-13}	cm
Thomson cross section	$8\ r_e{}^{2/3}$	6.6516	5	10^{-25}	cm^{2}
Gyromagnetic ratio of proton	γ	2.67519	2	10^{4}	rad $sec^{-1}G^{-1}$*
	$\gamma/2\pi$	4.25770	3	10^{3}	$sec^{-1}\ G^{-1}$*
Bohr magneton	μ_B	9.2732	6	10^{-21}	erg G^{-1}*
Nuclear magneton	μ_N	5.0505	4	10^{-24}	erg G^{-1}*
Proton moment	μ_p	1.41049	13	10^{-23}	erg G^{-1}*
Zeeman splitting constant	μ_B/hc	4.66858	4	10^{-5}	$cm^{-1}G^{-1}$*
Gas constant	R	8.3143	12	10^{7}	erg $°K^{-1}mol^{-1}$
Boltzmann constant	k	1.38054	18	10^{-16}	erg $°K^{-1}$

†† - Based on 3 std. dev. applied to last digits in preceding column.

* - Electromagnetic system.

† - Electrostatic system.

C - coulomb J - joule Hz - hertz W - watt N - newton T - tesla G - gauss

Appendix F

Atomic Masses of Elements Referred to C^{12} = 12.0000

Name	Symbol	Atomic Number	Atomic Weight	Name	Symbol	Atomic Number	Atomic Weight
Actinium	Ac	89	(227)	Neon	Ne	10	20.183
Aluminum	Al	13	26.9815	Neptunium	Np	93	(237)
Americium	Am	95	(243)	Nickel	Ni	28	58.71
Antimony	Sb	51	121.75	Niobium	Nb	41	92.906
Argon	Ar	18	39.948	Nitrogen	N	7	14.0067
Arsenic	As	33	74.9216	Nobelium	No	102	(253)
Astatine	At	85	(210)	Osmium	Os	76	190.2
Barium	Ba	56	137.34	Oxygen	O	8	15.9994
Berkelium	Bk	97	(249)	Palladium	Pd	46	106.4
Beryllium	Be	4	9.0122	Phosphorus	P	15	30.9738
Bismuth	Bi	83	208.980	Platinum	Pt	78	195.09
Boron	B	5	10.811	Plutonium	Pu	94	(242)
Bromine	Br	35	79.909	Polonium	Po	84	(210)
Cadmium	Cd	48	112.40	Potassium	K	19	39.102
Calcium	Ca	20	40.08	Praseodymium	Pr	59	140.907
Californium	Cf	98	(251)	Promethium	Pm	61	(145)
Carbon	C	6	12.01115	Protactinium	Pa	91	(231)
Cerium	Ce	58	140.12	Radium	Ra	88	(226)
Cesium	Cs	55	132.905	Radon	Rn	86	(222)
Chlorine	Cl	17	35.453	Rhenium	Re	75	186.2
Chromium	Cr	24	51.996	Rhodium	Rh	45	102.905
Cobalt	Co	27	58.9332	Rubidium	Rb	37	85.47
Copper	Cu	29	63.54	Ruthenium	Ru	44	101.07
Curium	Cm	96	(247)	Samarium	Sm	62	150.35

Dysprosium	Dy	66	162.50	Scandium	Sc	21	44.956
Einsteinium	Es	99	(254)	Selenium	Se	34	78.96
Erbium	Er	68	167.26	Silicon	Si	14	28.086
Europium	Eu	63	151.96	Silver	Ag	47	107.870
Fermium	Fm	100	(253)	Sodium	Na	11	22.9898
Fluorine	F	9	18.9984	Strontium	Sr	38	87.62
Francium	Fr	87	(223)	Sulfur	S	16	32.064
Gadolinium	Gd	64	157.25	Tantalum	Ta	73	180.948
Gallium	Ga	31	69.72	Technetium	Tc	43	(99)
Germanium	Ge	32	72.59	Tellurium	Te	52	127.60
Gold	Au	79	196.967	Terbium	Tb	65	158.924
Hafnium	Hf	72	178.49	Thallium	Tl	81	204.37
Helium	He	2	4.0026	Thorium	Th	90	232.038
Holmium	Ho	67	164.930	Thulium	Tm	69	168.934
Hydrogen	H	1	1.00797	Tin	Sn	50	118.69
Indium	In	49	114.82	Titanium	Ti	22	47.90
Iodine	I	53	126.9044	Tungsten	W	74	183.85
Iridium	Ir	77	192.2	Uranium	U	92	238.03
Iron	Fe	26	55.847	Vanadium	V	23	50.942
Krypton	Kr	36	83.80	Xenon	Xe	54	131.30
Lanthanum	La	57	138.91	Ytterbium	Yb	70	173.04
Lead	Pb	82	207.19	Yttrium	Y	39	88.905
Lithium	Li	3	6.939	Zinc	Zn	30	65.37
Lutetium	Lu	71	174.97	Zirconium	Zr	40	91.22
Magnesium	Mg	12	24.312				
Manganese	Mn	25	54.9380				
Mendelevium	Md	101	(256)				
Mercury	Hg	80	200.59				
Molybdenum	Mo	42	95.94				
Neodymium	Nd	60	144.24				

INDEX